T0213584

Undergraduate Lecture Notes in Physics

Series Editors

Neil Ashby, University of Colorado, Boulder, CO, USA

William Brantley, Department of Physics, Furman University, Greenville, SC, USA

Michael Fowler, Department of Physics, University of Virginia, Charlottesville, VA, USA

Morten Hjorth-Jensen, Department of Physics, University of Oslo, Oslo, Norway

Michael Inglis, Department of Physical Sciences, SUNY Suffolk County Community College, Selden, NY, USA

Barry Luokkala (iD), Department of Physics, Carnegie Mellon University, Pittsburgh, PA, USA

Undergraduate Lecture Notes in Physics (ULNP) publishes authoritative texts covering topics throughout pure and applied physics. Each title in the series is suitable as a basis for undergraduate instruction, typically containing practice problems, worked examples, chapter summaries, and suggestions for further reading.

ULNP titles must provide at least one of the following:

- An exceptionally clear and concise treatment of a standard undergraduate subject.
- A solid undergraduate-level introduction to a graduate, advanced, or non-standard subject.
- A novel perspective or an unusual approach to teaching a subject.

ULNP especially encourages new, original, and idiosyncratic approaches to physics teaching at the undergraduate level.

The purpose of ULNP is to provide intriguing, absorbing books that will continue to be the reader's preferred reference throughout their academic career.

More information about this series at http:/link.springer.com/series/8917

Samir Khene

Topics and Solved Exercises at the Boundary of Classical and Modern Physics

 Springer

Samir Khene
Department of Physics, Faculty of Sciences
Badji Mokhtar University
Annaba, Algeria

ISSN 2192-4791 ISSN 2192-4805 (electronic)
Undergraduate Lecture Notes in Physics
ISBN 978-3-030-87741-5 ISBN 978-3-030-87742-2 (eBook)
https://doi.org/10.1007/978-3-030-87742-2

This Springer imprint is published by the registered company Springer Nature Switzerland AG
The registered company address is: Gewerbestrasse 11, 6330 Cham, Switzerland

Preface

A number of physical phenomena which appeared towards the end of the nineteenth century could not be explained in the context of classical physics. These difficulties of interpretations were mainly related to the incompatibility of classical mechanics principles with the famous Michelson and Morley experiment. The resolution of these problems led to the development of the relativity theory.

Other difficulties of interpretation were related to thermodynamics. Indeed, the combining of thermodynamics with classical mechanics led to the Gibbs paradox of classical statistical mechanics, in which the entropy did not appear as a well-defined quantity. The black body radiation could not be explained without the introduction of the notion of quanta. As the experiments reached the atomic scale, classical mechanics failed to explain even the basic phenomena, such as the energy levels, atom size, or photoelectric effect. The combined efforts of several researchers to solve these problems led to the emergence of quantum mechanics.

Today, classical mechanics appears as an approximation of a much more general theory, namely quantum mechanics. Indeed, classical mechanics is a theory which is used to study the non-quantum mechanical motion of the low-energy particles in a weak gravitational field. This book summarizes some aspects of the atomic structure, matter-radiation interaction, black body, and macroscopic thermodynamics. It is more focused on the pedagogy than on the demonstrative rigor. It is thus recommended to the reader to refer to more developed works to deepen these notions. It has four chapters. Each chapter begins with a precise course, followed by exercises in order of increasing difficulty. These exercises are solved in a very detailed way; they allow students to assimilate the course and help them prepare for exams. This book is intended for first-year university students. It can also benefit bachelor's and doctoral students because of the deliberately simplified presentation of the treated concepts.

Atomic physics is introduced in Chap. 1. This discipline studies the atoms which form all the solid, liquid, and gaseous matter of the universe. It is the infinitesimal world where classical laws which govern our macroscopic daily life do not apply. Radioactivity, which has a direct impact on the human species, is also studied. It produces radiation and releases particles; it is invisible, odorless, and inaudible, and yet it radiates, contaminates, destroys, and kills. On another side, radioactivity benefits the medical field, industrial world, agro-food area, cultural field, and others; it is today essential for humanity's survival and development.

During interaction of electromagnetic radiation with matter, five phenomena can occur depending on the energy level of the incident photons: the Rayleigh scattering, photoelectric effect, Compton scattering, creation of electron-positrons pairs, and nuclear photo-production. In the second chapter, only the photoelectric effect and the Compton scattering are studied because, historically, it is these two phenomena that have defeated classical mechanics.

The third chapter presents the black body radiation theory. It is a purely fictitious ideal body from which the basic laws of thermal radiation are derived. In classical thermodynamics, its equivalent is the ideal gas, which is first studied before studying real gasses. The object which comes closest to the black body model is the inside of an oven. The study of black body stumbled upon the explanation of the ultraviolet catastrophe and led to the quantum revolution. This point is discussed in this chapter.

The fourth and final chapter studies macroscopic thermodynamics, which is the science of energy balance and prediction of the evolution of systems during modification of their external or internal parameters. It studies energy transformations, matter, and equilibrium states of systems. It presents a universal cháracter and applies to all physics and chemistry areas. The Gibbs paradox is one of the topics treated in this chapter.

In Appendix A, we show how the Michelson and Morley experiment led to the four-dimensional space notion.

In Appendix B, we give useful mathematical reminders in physics.

This book has been designed with a constant concern of pedagogy and the desire to make the treated topics simple and accessible to first-year university students. We will be very grateful to all those who would like to give us their comments and suggestions to improve the content of this book, and we apologize in advance for the inevitable errors which remain there.

Department of Physics   Prof. Dr. Samir Khene
Faculty of Sciences, Badji Mokhtar
University, Annaba, Algeria

Contents

Chapter 1
Atoms

Abstract Atomic physics is introduced in this chapter. This discipline studies the atoms which form all the solid, liquid, and gaseous matter of the universe. It is the infinitesimal world where classical laws which govern our macroscopic daily life do not apply. Radioactivity, which has a direct impact on the human species, is also studied. It produces radiations and releases particles; it is invisible, odorless, and inaudible, and yet it radiates, contaminates, destroys, and kills. On another side, the radioactivity benefits the medical field, industrial world, agro-food area, cultural field, and others; it is today essential for humanity's survival and development.

This chapter begins with a precise course, followed by exercises in order of increasing difficulty. These exercises are solved in a very detailed way; they allow students to assimilate the course and help them prepare for exams. It is intended for first-year university students. It can also benefit bachelor's and doctoral students, because of the deliberately simplified presentation of the treated concepts.

1.1 Introduction

All the solid, liquid, and gaseous matter of the universe is formed of atoms. Atomic physics studies those atoms. It is the world of the infinitely small, where the classical laws which govern our macroscopic daily life do not apply. The structure of atoms was totally unknown in 1895. It was the discovery of X-rays by Wilhelm Conrad Röntgen in 1895 and radioactivity by Henri Becquerel in 1896, and the works of Pierre and Marie Curie beginning in 1898 which opened the way for the understanding of the structure of atoms. Electrons were identified by Joseph John Thomson in 1897, then, in 1911, Ernest Rutherford highlighted the atomic nucleus existence. For many years, physicists thought that the nucleus was only formed of protons without being able to unravel the secret of its stability. At the same time as this and during the years 1924−1927, the development of quantum mechanics allowed scientists to accurately describe the behavior of electrons in atoms. Thus, in order, Dalton's, Thomson's, Rutherford's, Bohr's, Sommerfeld's, and Schrödinger's models were developed. The Schrödinger model included the major contributions of Heisenberg,

© The Author(s), under exclusive license to Springer Nature Switzerland AG 2021
S. Khene, *Topics and Solved Exercises at the Boundary of Classical and Modern Physics*, Undergraduate Lecture Notes in Physics,
https://doi.org/10.1007/978-3-030-87742-2_1

De Broglie, Pauli, Dirac, and Einstein. Only the nucleus remained an enigma. In 1931, Irène and Frédéric Joliot-Curie observed neutrons without understanding their nature. It was in 1932 that James Chadwick showed that the neutron was a neutral partner of the proton. The atomic structure was finally elucidated!

1.2   Atom Structure

An atom is composed of a nucleus around which electrons orbit. The nucleus of an atom has a radius of the order of 10^{-15} m. The atom can be considered as a sphere of the radius 10^{-10} m. The nucleus radius is thus 10^5 times smaller than that of the atom. Just like the solar system, the atom is essentially composed of emptiness!

1.2.1   Electron

All the electrons which orbit around the nucleus are identical. The electron is characterized by its charge $e = -1.602\ 176\ 634 \times 10^{-19}$ C and its mass m_e. Historically, it was Joseph John Thomson who determined in 1897 the charge-to-mass ratio of the electron. This charge is negative ($q = -e = -1.60218 \times 10^{-19}$ C). Electrochemists also use the faraday (F) as a unit of charge. The faraday is a charge of 96 500 C carried by one mole of electrons, that is an Avogadro's number, $N_A = 6.02 \times 10^{23}$, of electrons. The last two numbers we are using here are rounded off. Expressed in faradays, the Faraday constant equals to 1 faraday of charge per mole (96 500 C.mol^{-1}). The mass of the electron was historically calculated by Robert Andrews Millikan in 1911; it has a value of:

- In SI unit: $m_e = 9.109 \times 10^{-31}$ kg.
- In unified atomic mass unit: $m_e = 550 \times 10^{-6}$ u.
- In MeV/c^2: $m_e = 0.51$ MeV/c^2.

The u and MeV/c^2 units will be defined later. There is no exact determination of the electron radius because its value depends on the used method to measure it. An approximation gives a radius of the order of fentometer (1 fm $= 10^{-15}$ m). Depending on the atom, electrons can be from 1 to 118 to orbit around a nucleus. The electron is symbolized by e$^-$.

1.2.2   Nucleus

The nucleus is made up of neutrons and protons called *nucleons*. Almost all the mass of the atom is concentrated in the nucleus because the nucleus has a mass of several thousands of times larger than that of the electron and is about 10^5 times smaller than the atom. The nucleus is therefore very compact. Nuclei are positively charged.

1.2.2.1 Proton

The proton is a subatomic particle carrying a positive charge. Protons are present in the atomic nucleus, possibly linked with neutrons by a strong nuclear interaction. The proton is stable by itself outside the atomic nucleus. It is composed of three other particles: two up quarks and one down quark, making it a baryon. The proton mass is: $m_P = 1.67265 \times 10^{-27}$ kg $= 1.0072765$ u $= 938.272$ MeV/c^2. This mass is very close to that of the neutron (the other nucleon). If we limit ourselves to two significant figures, we can even consider that their masses are equal. The proton is a spherical particle of radius $r_p = 8.7 \times 10^{-16}$ m. In comparison, the radius of an atom is of the order of 10^{-10} m, namely, 10^4 times larger. The volume of a proton is $V = 2.8 \times 10^{-48}$ m^3. The proton carries a positive electric charge of which the approximate value is $q = +1.6 \times 10^{-19}$ C. It can be represented by the symbol p$^+$.

1.2.2.2 Neutron

The neutron is a neutral particle that makes up the nucleus of atoms with protons. For this reason, it is also called, just like the proton, *a nucleon*. It was discovered in 1932 by James Chadwick. It is not an elementary particle, no more than the proton moreover, because it is also composed of three quarks: one up quark up and two down quarks of which electrical charges are respectively +2/3 and −1/3 making it a baryon of no electrical charge. Its quarks are linked by a strong interaction transmitted by gluons. The neutrons of the atomic nucleus are generally stable, but free neutrons are unstable because they disintegrate in a little less than 15 minutes. Free neutrons are produced by nuclear fission and fusion reactions. The mass of the neutron is equal to $m_N = 1.67493 \times 10^{-27}$ kg $= 1.0086655$ u $= 939.565$ MeV/c^2. The neutron is 1.0014 times as massive as the proton. The proton can be represented by the symbol n^0.

1.2.3 Notation of Nuclides

The nuclide is a type of atom or atomic nucleus characterized by the number of protons and neutrons which it contains. It is noted by:

$$_Z^A X \tag{1.1}$$

where:

- X is the chemical symbol of the nuclide (of the atom).
- Z is the charge number (atomic number). It represents the number of protons.
- A is the mass number. It represents the number of nucleons which make up the nucleus.

Knowing that N denotes the number of neutrons, we can write $A = Z + N$. The mass of the atom is the sum of the mass of its various constituents, namely:

$$m_{atom} = m_{nucleus} + m_{electrons} = (Zm_P + Nm_N) + Zm_e. \qquad (1.2)$$

Since the atom is electrically neutral, Z also represents the number of electrons. A nuclide is therefore a chemical species that differs from other known chemical species either by its mass number A or by its atomic number Z:

$$^{4}_{2}He \quad ^{35}_{15}Cl \quad ^{37}_{17}Cl \quad ^{16}_{8}O$$

The proton, the neutron, and the electron are noted in the same way:

$$^{1}_{1}p \quad ^{1}_{0}n \quad ^{0}_{-1}e$$

We call:

- *Isotopes*: two atoms which have the same charge number Z but a different mass number A:
$$^{12}_{6}C \quad ^{16}_{6}C \quad ^{14}_{6}C$$

- *Isobars*: two atoms which have the same mass number A but a different charge number Z:
$$^{17}_{7}N \quad ^{17}_{8}O \quad ^{17}_{9}F$$

- *Isotones*: two atoms which have the same number of neutrons N:
$$^{51}_{23}V \quad ^{52}_{24}Cr \quad ^{54}_{26}Ca$$

In this example, the number of neutrons is 26. In nuclear physics, the chemical symbol represents one atom, whereas in chemistry this symbol represents 1 mole of atoms that is to say 6.02×10^{23} atoms.

1.2.4 Mass Units

The mass unit in the international system is the kilogram (kg). The electron, for example, has a mass $m_e = 9.11 \times 10^{-31}$ kg. As we see, this unit is far too big for this kind of infinitesimal particle. That is why we use two other units better adapted to physical phenomena observed on this scale.

1.2.4.1 Unified Atomic Mass Unit

The unified atomic mass unit (u) is defined as being one-twelfth (1/12) of the mass of one atom of carbon 12. As 1 mole of this carbon weighs 12 g, it becomes:

$$1\,u = \frac{12 \times 10^{-3}\,\text{kg}}{(6.02214 \times 10^{23})\,(12)} = 1.66054 \times 10^{-27}\,\text{kg}. \tag{1.3}$$

1.2.4.2 MeV/c² Unit

In special relativity, Einstein has shown that at any mass m_0 corresponds to energy $E = m_0 c^2$ called *rest mass energy* where E is in joules, m_0 in kilograms, and c, the speed of light in m/s. From this equation, we deduce that:

$$m_0 = \frac{E}{c^2}. \tag{1.4}$$

Equation (1.4) shows that a mass can be expressed by energy divided by c^2. Although the joule is the SI unit of energy, in nuclear physics, the used unit is the electronvolt (eV) because it is better adapted to nuclear phenomena with:

$$1\,\text{eV} = 1.6 \times 10^{-19}\,\text{J} \Rightarrow 1\,\text{J} = 0.625 \times 10^{19}\,\text{eV} = 0.625 \times 10^{13}\,\text{MeV}. \tag{1.5}$$

On the basis of (1.4) and (1.5) and to determine the mass of a particle in MeV/c², it is enough to calculate this mass in joules then to convert it in MeV ($1\,\text{MeV} = 10^6\,\text{eV}$).

1.3 Atomic Models

1.3.1 Dalton's Model

Ancient Greek philosophers such as Empedocles of Agrigento (VI century BC) considered that the nature of things was explained by a clever mixture of four elements: fire, air, water, and earth. In the fourth century BC, the philosopher Democritus thought that matter was formed of indivisible grains of matter which he called *atomos* (atoms in Greek). For him, these atoms were all full but not all similar: they were round or crooked, smooth or rough; they assembled to form objects. Democritus had no experimental proof of what he said. His intellectual approach, purely philosophical, was mere speculation. The Greek philosopher Aristotle (384–322 BC) rejected this theory and took up the idea of four elements. It is on this false conception that the works of alchemists will rest for more than 20 centuries. In 1803, the British scientist John Dalton (1766–1844) revived the idea of Democritus by declaring that matter was composed of atoms. This is the first scientist to try to explain the composition of matter. His atomic model is known today as the billiard ball model. In this model, all the atoms of the same element are identical and the atoms are different from one element to another. In 1808, he succeeded in determining certain relative atomic masses of chemical elements opening the way to the creation of the periodic table that we know today. He compared different elements with hydrogen. Some values are not correct, but this information and its new atomic model constitute a good starting point for modern chemistry (Fig. 1.1).

1.3.2   Thomson's Model

On the basis of his experiments on the rays which propagate in the cathode tubes of Crookes, the British physicist Sir Joseph John Thomson discovered the electron. A theorist and excellent experimentalist, he proposed in 1897 a theory on the atomic structure known as *plum pudding model*, in which electrons are considered as negative grapes embedded in bread of positive matter (Fig. 1.2).

In this model:

(i) The atom is a sphere of uniform density.
(ii) The atom is composed of negative particles and positive particles.
(iii) The number of negative particles is equal to the number of positive particles justifying the electrical neutrality of the atom.

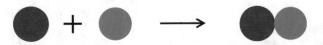

Fig. 1.1 The Dalton model: all the atoms of the same element are identical and the atoms are different from one element to another

Fig. 1.2 The Thomson model: electrons are considered as negative grapes embedded in the bread of positive matter

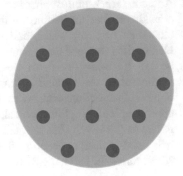

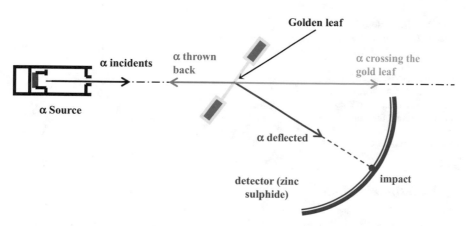

Fig. 1.3 Rutherford bombarded a thin sheet of gold with α-particles (helium atom that has lost two electrons). He observed that the majority of particles passed through the gold leaf without deflection and damage to the gold leaf. He also observed that some particles were slightly deflected and others were thrown back

1.3.3 Rutherford's Model

To verify the Thomson model, the British physicist and chemist Ernest Rutherford realized in 1911 an experiment which allowed him to develop a new model of the atom (Fig. 1.3).

He bombarded a thin sheet of gold with α-particles (helium atom that has lost two electrons). He observed that the majority of particles passed through the gold leaf without deflection and damage to the gold leaf. He also observed that some particles were slightly deflected and others were thrown back. These observations were incompatible with the Thomson model. To explain them, Rutherford proposed a new model in which the atom was not full. It consisted of a positively charged nucleus that contained almost all the mass of the atom and electrons that revolved around it such as planets around the Sun. Between the nucleus and the electrons, there was a vacuum, even a great vacuum. Indeed, the size of the atom is of the order of 10^{-10} m and that of the nucleus of the order of 10^{-15} m. The ratio between the size of the atom and that of the nucleus is thus 10^5. By comparing the atomic nucleus to an apple of 5 cm of diameter, the atom would have a diameter of 5 km! The atom is thus essentially made of vacuum. All its mass is concentrated in a tiny, positively charged nucleus (Fig. 1.4).

Rutherford's discovery fails to explain why an electron that revolves around its nucleus does not radiate energy. Indeed, because of the kinetic energy loss due to its rotational movement, the electron should, in principle, fall nucleus in spiral causing the destruction of the atom. But, except radioactive elements, all atoms are stable. It does not explain either the discontinuous spectra of atoms.

Fig. 1.4 Rutherford's
model: the atom is not full. It
consists of a positively
charged nucleus that
contains almost all the mass
of the atom and electrons
that orbit around it such as
planets around the Sun

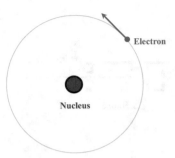

1.3.4 Bohr's Model

The Bohr model is a complement of the Rutherford model. To explain the fact that
the electron in its movement around the nucleus does not emit light, the Danish
physicist Niels Henrik David Bohr adds two constraints:

(i) The electron radiates no energy when it is in a stable orbit called *stationary
    orbit*. These orbits are quantized. These are the only allowed orbits in which
    electrons can orbit around the nucleus.
(ii) The electron emits or absorbs energy only during an orbit change. Thus, the
    possible orbits of the electron are described as quantized concentric circles.

In this model, the electron describes a uniform circular motion around a nucleus
formed of a proton (the case of hydrogen atom). It is subjected to the Coulomb force.
By applying the fundamental equation of dynamics to the electron, we obtain:

$$\sum \vec{F} = m_e \, \vec{a}. \tag{1.6}$$

By projecting (1.6) on an axis directed towards the center of the circle of radius r,
we obtain:

$$\frac{1}{4\pi\varepsilon_0} \frac{q_e^2}{r^2} = r m_e \omega^2 = m_e \frac{v^2}{r}. \tag{1.7}$$

Since the mechanical energy E of the electron is the sum of its kinetic energy E_C
and the potential energy E_P (Fig. 1.5):

$$E = E_C + E_P = \frac{1}{2} m_e v^2 - \frac{q_e^2}{4\pi\varepsilon_0 r}, \tag{1.8}$$

it becomes

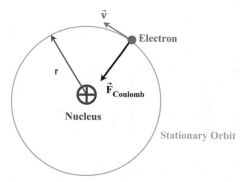

Fig. 1.5 Bohr's model: it is a complement of the Rutherford model. To explain the fact that the electron in its movement around the nucleus does not emit light, Bohr postulates that the electron emits or absorbs some energy only when it changes its orbit. In this model, the electron describes a uniform circular motion around a nucleus formed of a proton (the case of hydrogen atom). It is subjected to the Coulomb force

$$E = \frac{1}{2} \times \frac{q_e^2}{4\pi\varepsilon_0 r} - \frac{q_e^2}{4\pi\varepsilon_0 r} = -\frac{q_e^2}{8\pi\varepsilon_0 r} \tag{1.9}$$

where $\vec{F}$ is the Coulomb force. We obtain after rearrangements:

$$m_e v^2 = \frac{1}{4\pi\varepsilon_0} \frac{q_e^2}{r}. \tag{1.10}$$

To avoid making this demonstration unwieldy, let us put:

$$\frac{q_e^2}{4\pi\varepsilon_0} = e^2. \tag{1.11}$$

The expression of the kinetic energy is then written as:

$$E_C = \frac{1}{2} \times \frac{q_e^2}{4\pi\varepsilon_0 r} = \frac{e^2}{2r}. \tag{1.12}$$

The potential energy then takes the following form:

$$E_P = -\frac{e^2}{r} \tag{1.13}$$

which gives for the total energy of the electron E:

$$E = \frac{e^2}{2r} - \frac{e^2}{r} = -\frac{e^2}{2r}. \tag{1.14}$$

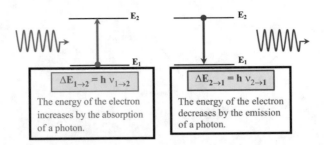

Fig. 1.6 To explain the discontinuous spectrum of the atom of hydrogen and to understand why the electron does not fall on the nucleus, Bohr formulates two hypotheses: (i) The electron orbits into a permitted orbit. (ii) There is energy absorption for $\Delta E_{1\rightarrow 2}$ and emission for $\Delta E_{2\rightarrow 1}$

Equation (1.14) shows that the electron energy is a continuous function of the radius r and that r varies continuously. So, this result does not allow to explain the discontinuous spectrum of the atom of hydrogen and to understand why the electron does not fall on the nucleus in spiral. Bohr then formulates two hypotheses (Fig. 1.6):

(i) The electron revolves into a permitted orbit.
(ii) There is energy absorption for $\Delta E_{1\rightarrow 2}$ and emission for $\Delta E_{2\rightarrow 1}$.

According to the Planck quantum theory, the energy of the quantum emitted during the transition of the electron between the layers m and n such that $m > n$ is:

$$\Delta E_{m\rightarrow n} = E_m - E_n = h\nu_{m\rightarrow n} = h\frac{c}{\lambda_{m\rightarrow n}}. \qquad (1.15)$$

As the frequency of the emitted radiation $\nu_{m\rightarrow n}$ can take only certain discrete values, the energy emitted is thus quantized. It is the same for the norm of the kinetic moment of the electron:

$$\sigma = m_e v r = n\hbar = n\frac{h}{2\pi} \qquad (1.16)$$

where $\hbar$ is Planck's constant reduced by a factor 2π and n is a positive integer or zero. This number, called *main quantum number*, suffices to characterize any orbit (radius, energy, moment, etc.). To facilitate calculations, consider the case where $n = 1$. The previous expression is then written as:

$$m_e v_1 r_1 = \hbar. \qquad (1.17)$$

By squaring (1.17), we obtain:

$$m_e^2 \, v_1^2 \, r_1^2 = \hbar^2 \qquad (1.18)$$

what we can also write:

$$m_e \left(\frac{m_e \, v_1^2}{r_1} \right) r_1^3 = \hbar^2. \tag{1.19}$$

Given Eqs. (1.9) and (1.12) of the kinetic energy, we get:

$$\frac{m_e \, v_1^2}{r_1} = \frac{e^2}{r_1^2} \tag{1.20}$$

and (1.20) transforms into:

$$m_e \left(\frac{e^2}{r_1^2} \right) r_1^3 = \hbar^2 \tag{1.21}$$

or again in:

$$m_e e^2 r_1 = \hbar^2. \tag{1.22}$$

We finally get:

$$r_1 = \frac{\hbar^2}{m_e e^2}. \tag{1.23}$$

This radius is called *Bohr's radius*. It has the value $r_1 = 0.53$ Å. The radius associated with the electronic shell n has the following expression:

$$r_n = n^2 r_1. \tag{1.24}$$

By starting from (1.14) and (1.23), the energy of the electron for $n = 1$ has for expression:

$$E_1 = -\frac{e^2}{2r_1} = -\frac{e^2}{\dfrac{2\hbar^2}{m_e e^2}} = -\frac{m_e e^4}{2\hbar^2}. \tag{1.25}$$

This energy is called *Rydberg's energy*. It has the value $E_1 = -13.6$ eV. It is a negative energy, meaning that the electron is more stable in the atom than outside. This energy will have to be supplied to extract it from the atom and make it a free electron. The energy associated with the n layer has the following expression:

$$E_n = \frac{E_1}{n^2}. \tag{1.26}$$

The interpretation of the atomic spectrum of hydrogen is based on the fact that in one orbit, the electron does not radiate. It absorbs or releases energy only by passing from one orbit to another by absorbing or emitting a photon of energy proportional to the light wave frequency:

$$E = h\nu. \tag{1.27}$$

By starting from (1.11), (1.25), (1.26), and (1.27), we obtain the energy emitted during the jump from the orbit m to the orbit n:

$$E_m - E_n = \frac{E_1}{m^2} - \frac{E_1}{n^2} = E_1 \left(\frac{1}{m^2} - \frac{1}{n^2} \right) = -\frac{m_e e^4}{2\hbar^2} \left(\frac{1}{m^2} - \frac{1}{n^2} \right) = \frac{m_e e^4}{2\hbar^2} \left(\frac{1}{n^2} - \frac{1}{m^2} \right) =$$

$$\frac{m_e \left(\frac{q_e^2}{4\pi\varepsilon_0} \right)^2}{2\hbar^2} \left(\frac{1}{n^2} - \frac{1}{m^2} \right) = \frac{m_e q_e^4}{32\pi^2\varepsilon_0^2\hbar^2} \left(\frac{1}{n^2} - \frac{1}{m^2} \right) = \frac{m_e q_e^4}{32\pi^2\varepsilon_0^2 \left(\frac{h}{2} \right)^2} \left(\frac{1}{n^2} - \frac{1}{m^2} \right) = \frac{m_e q_e^4}{8\varepsilon_0^2 h^2} \left(\frac{1}{n^2} - \frac{1}{m^2} \right). \tag{1.28}$$

Knowing that:

$$\lambda_{m \to n} = \frac{c}{\nu_{m \to n}} = \frac{hc}{h\nu_{m \to n}} = \frac{hc}{E_m - E_n} \tag{1.29}$$

and by taking the inverse of the wavelength, we finally arrive at:

$$\frac{1}{\lambda_{m \to n}} = \frac{E_m - E_n}{hc} = \frac{m_e q_e^4}{8\varepsilon_0^2 h^3 c} \left(\frac{1}{n^2} - \frac{1}{m^2} \right) = R_H \left(\frac{1}{n^2} - \frac{1}{m^2} \right). \tag{1.30}$$

It is the Ritz formula where R_H is the Rydberg constant. It is about $R_H = 1.097 \times 10^7 \text{ m}^{-1}$. This value is very close to the experimental value obtained from the spectrum of the atom of hydrogen from the great success of Bohr's model. The spectrum of hydrogen is the set of wavelengths which the atom of hydrogen is capable of emitting or absorbing. This light spectrum is composed of discrete wavelengths of which values are given by the Ritz formula. The main series of spectral lines of the atom of hydrogen are the series of Lyman, Balmer, Paschen, Brackett, Pfund, and Humphreys. Thus, the atoms with high energy levels emit a discrete light spectrum, made of photons of frequencies given during transitions of electrons to lower energy levels. All these frequencies constitute the spectral emission lines of the atom. In the same way, certain wavelengths are absorbed by the atom passing its electrons towards higher energy levels; these are the spectral absorption lines of the atom. These two sets of lines are characteristic of the atom. Bohr's model explains these spectra. It also justifies the experimental formula of Balmer. The Franck and Hertz experiment confirms the validity of this model of the energy levels of the atom of hydrogen which can be extended to hydrogen-like ions (Z protons, one electron). However, Bohr's model does not explain the spectrum of other atoms and molecules. In addition, the discovery of the fine structure of hydrogen lines puts in the wrong Balmer's formula. Arnold Sommerfeld proposes a modification of Bohr's model by introducing elliptical orbits to take into account these new data. It is the advent of quantum theory with the Schrödinger equation that ultimately allows a complete treatment of the atom of hydrogen.

1.3.5   The Sommerfeld-Wilson Model

Bohr's model does not take into account two elements:

(i) The real orbit of the electron is elliptical and noncircular.
(ii) The actual form of the nucleus of the atom is not punctual. In addition, this nucleus is in perpetual motion relative to a center of gravity.

On the basis of the following two modifications, that is, (i) the quantization of elliptical orbits and (ii) the introduction of relativistic mechanics, the German theoretical physicist Arnold Johannes Wilhelm Sommerfeld leads to a more complex expression of the electron energy:

$$E_{n,l} = E_n \left[1 + \frac{\alpha^2 Z^2}{n^2} \left(\frac{n}{l+1} - \frac{3}{4} \right) \right] \tag{1.31}$$

where E_n is the Bohr energy and α the constant of fine structure and with this time, the presence of two quantum numbers n and $l \in (0, 1, 2, \ldots, n-1)$. With this model, we go from the three levels of Bohr (K,L,M) to a greater number of levels following the introduction of the two quantum numbers n and l:

$$\begin{cases} K \\ L \\ M \end{cases} \rightarrow \begin{cases} K \\ L_0 & L_1 \\ M_0 & M_1 & M_2 \end{cases} . \tag{1.32}$$

These levels will be defined later. However, the experiment has revealed the existence of:

(i) A fine structure of three lines instead of two for the level L.
(ii) A fine structure of five lines instead of three for the level M.

To expand his theory, Sommerfeld then suggested the introduction of an additional quantum number j which he called *internal quantum number*, without being able to provide further details. Despite all those advances, his theory remained incomplete (Fig. 1.7).

Fig. 1.7 The Sommerfeld-
Wilson model: from a
spectral point of view, this
theory brings nothing more
than the Bohr model except
for the elliptical shape of
orbits

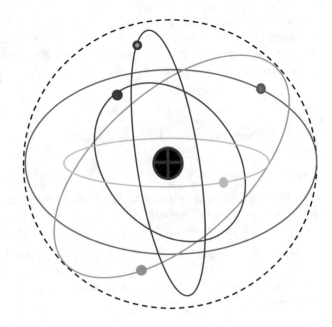

1.3.6   De Broglie's Model

In 1924, the French mathematician and physicist Louis Victor de Broglie developed
an interpretation of the quantization of the Bohr atomic model because the wave-
particle duality of radiation stood out as a necessary step in the explaining and
interpreting historical experiments linked to the matter-radiation interaction (photoelectric effect, Compton scattering, black body, emission spectrum, etc.). De Broglie
thus exploited the close link between matter and radiation to produce the hypothesis
according to which material corpuscles might have an undulatory aspect. This
hypothesis was one of the main catalysts that allowed the genesis of wave mechanics. This hypothesis states that to every material particle of energy E and momentum
$\vec{P}$ is associated a plane wave of which characteristics are the angular frequency ω:

$$\omega = \frac{E}{\hbar} \tag{1.33}$$

and the wave vector $\vec{k}$:

$$\vec{k} = \frac{\vec{P}}{\hbar}. \tag{1.34}$$

The de Broglie wavelength associated with the particle is defined by:

$$\lambda = \frac{h}{P} = \frac{h}{mv} \tag{1.35}$$

where $P = mv$ is the particle momentum. With regard to the magnitude of the Planck constant which influences the value of the wavelength, the wave character associated with a particle cannot be manifested for macroscopic objects. On the other hand, at the atomic or subatomic scale, the wave character of material particles (electrons, protons, neutrons, etc.) can be proved only if their wavelengths are comparable to the length scales which characterize their movement. For example, an electron moving in the cathode tube of an oscilloscope will have a wavelength very small compared to the size of the tube. Classical mechanics is sufficient to treat the movement of this electron. On the other hand, the movement of an electron bound to an atom will be characterized by a wavelength of the order of Angstrom, thus comparable to the atomic dimension. In this case, the wave character of the electron cannot be ignored and requires other theoretical tools than those of classical mechanics to describe its motion. The Louis de Broglie hypothesis on the undulatory nature of material particles must find an experimental justification. This is why works were carried out later to demonstrate irrefutably the wave character of material corpuscles. The realized experiment principle was based on diffraction and interference phenomena using a particle beam (electrons, atoms, neutrons, etc.). The Davisson and Germer experiment in 1927 was one of the most brilliant demonstrations of de Broglie's hypothesis on the existence of matter waves. Other experiments using a beam of atoms were also carried out, allowing conclusion of the wave nature of particles like the helium beam diffraction made by Estermann and Stern in 1932.

1.3.7 Schrödinger's Model

The Bohr model failed to explain all the emission lines of atoms with more than one electron and was incompatible with the Heisenberg uncertainty which stated that for a microscopic particle, it was not possible to accurately determine its position and its speed. This forced scientists to definitely abandon the idea of the orbit of an electron mentioned in the Bohr model. Classical mechanics having reached its limits, it is quantum mechanics that takes over by defining the electron not by its orbit, but by its energy and its probability of presence at a point in space. The Schrödinger equation, conceived by the Austrian physicist Erwin Rudolf Josef Alexander Schrödinger in 1925, was a fundamental equation in nonrelativistic quantum physics. It described the evolution over time of a nonrelativistic massive particle and thus fulfilled the same role as the dynamics fundamental equation in classical mechanics. In quantum mechanics, the state at the instant t of a system is described by an element written in bra-ket notation of Paul Dirac $|\Psi(t)\rangle$ of the Hilbert complex space. This element represents the probabilities of results of all possible measurements of a system. The temporal evolution of $|\Psi(t)\rangle$ is described by:

$$\widehat{H} \mid \Psi(t)\rangle = i\,\hbar\,\frac{\partial}{\partial t} \mid \Psi(t)\rangle = \frac{\vec{P}^2}{2m} \mid \Psi(t)\rangle + V\left(\vec{r},t\right) \mid \Psi(t)\rangle \qquad (1.36)$$

where:

- i is the imaginary unit.
- $\hbar$ is the reduced Planck constant with $\hbar = h/2\pi = 1.05457 \times 10^{-34}$ J.s.
- $\partial/\partial t$ indicates a partial derivative with respect to time t.
- $\mid \Psi(t)\rangle$ is the wave function of the quantum system.
- $\widehat{H}$ is the Hamiltonian operator depending on time. This observable corresponds to the total energy of the system (kinetic energy + potential energy).
- $\vec{r}$ is the observable position.
- $\vec{P}$ is the observable impulse.
- $V\left(\vec{r},t\right)$ is the observable potential energy.

This equation is a postulate. It was supposed correct after Davisson and Germer experimentally confirmed the Louis de Broglie hypothesis (wave-particle duality). Its resolution allowed obtaining the energy values accessible to the electron and the mathematical functions that govern its behavior. This equation has solutions only for certain energy values called *eigenenergies* justifying the energy quantization principle. The associated functions, called *atomic orbitals* or *eigenfunctions*, allow access to the probability of the electron presence at a point in space but not at its exact position. For one eigenenergy value, it is possible to have several eigenfunctions that verify this equation called *degenerate eigenfunctions*. Moreover, this equation can only be solved rigorously for a one-electron building.

1.4   Energy Levels of the Atom of Hydrogen

According to (1.26), the energy of the electron associated with the quantum number n has the following expression:

$$E_n = \frac{E_1}{n^2} = \frac{-13.6 \text{ eV}}{n^2}. \qquad (1.37)$$

- For $n > 1$: the atom is in an excited and unstable state.
- For $n = 1$: the atom is in a fundamental state and stable.
- For $n = \infty$: the atom loses its electron. It becomes a positively charged ion.

1.4.1 Series of Lines of Lyman

According to (1.30), we have:

$$\frac{1}{\lambda_{m \to n}} = \frac{E_m - E_n}{hc} = R_H \left(\frac{1}{n^2} - \frac{1}{m^2} \right). \tag{1.38}$$

Lyman's lines are obtained when electronic transitions reach the fundamental level $n = 1$ with $m = 2, 3, 4, 5$, etc. The emission of this series of spectral lines is done in the ultraviolet. In this case, (1.38) is reduced to:

$$\frac{1}{\lambda_{m \to 1}} = \frac{E_m - E_1}{hc} = R_H \left(1 - \frac{1}{m^2} \right). \tag{1.39}$$

The limit wavelength is obtained when $m \to \infty$, i.e.:

$$\frac{1}{\lambda_{\infty \to 1}} = R_H \Rightarrow \lambda_{\infty \to 1} = \frac{1}{R_H} = 91.1 \text{ nm}. \tag{1.40}$$

1.4.2 Series of Lines of Balmer

Balmer's lines are obtained when electronic transitions reach the level $n = 2$ with $m = 3, 4, 5$, etc. The emission of this series of spectral lines is done in the visible. In this case, (1.38) is reduced to:

$$\frac{1}{\lambda_{m \to 4}} = \frac{E_m - E_2}{hc} = R_H \left(\frac{1}{4} - \frac{1}{m^2} \right). \tag{1.41}$$

The limit wavelength is obtained when $m \to \infty$, i.e.:

$$\frac{1}{\lambda_{\infty \to 4}} = R_H \left(\frac{1}{4} \right) \Rightarrow \lambda_{\infty \to 4} = \frac{4}{R_H} = 364.4 \text{ nm}. \tag{1.42}$$

1.4.3 Series of Lines of Paschen

Paschen's lines are obtained when electronic transitions reach the level $n = 3$ with $m = 4, 5$, etc. The emission of this series of spectral lines is done in the infrared. In this case, (1.38) is reduced to:

$$\frac{1}{\lambda_{m \to 9}} = \frac{E_m - E_2}{hc} = R_H \left(\frac{1}{9} - \frac{1}{m^2} \right). \tag{1.43}$$

The limit wavelength is obtained when $m \to \infty$, i.e.:

$$\frac{1}{\lambda_{\infty \to 9}} = R_H \left(\frac{1}{9} \right) \Rightarrow \lambda_{\infty \to 9} = \frac{9}{R_H} = 819.9 \text{ nm.} \tag{1.44}$$

1.4.4 Energy Diagram

On the basis of (1.37), we deduce the following values:

n	1	2	3	4	5	…………	∞
E_n (eV)	−13.6	−3.4	−1.5	− 0.85	− 0.54	………..	0

The emission spectrum of a substance is its identity card (Fig. 1.8). The analysis of the emitted light by a star allows determining its composition and its temperature without setting foot there.

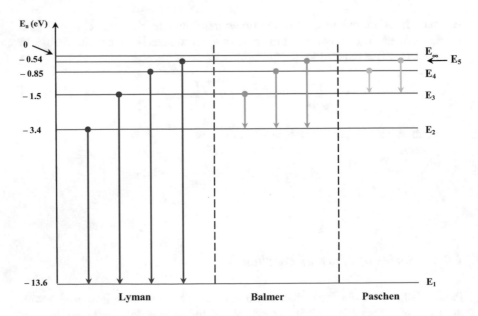

Fig. 1.8 Hydrogen energy diagram of first five levels

1.4.5   Quantum Numbers

In classical mechanics, a particle is entirely characterized by six parameters: three position parameters (x,y,z) and three velocity parameters (v_x,v_y,v_z). In quantum mechanics, a microscopic particle such as the electron is defined not by the six parameters but by four quantum parameters: n, l, m, and s, *of which* the physical meaning is not always obvious:

- *Number n*: It is the principal quantum number. It characterizes the electron shell:

  - $n = 1$ for the K-shell.
  - $n = 2$ for the L-shell.
  - $n = 3$ for the M-shell and so on and so forth. This number gives the energy level of the electron.

- *Number l*: It is the angular or azimuthal or second quantum number. It is an integer that varies from 0 to $n - 1$. It defines the subshell s, p, d, f, etc. So:

  - $l = 0$ corresponds to the s-subshell,
  - $l = 1$ corresponds to the p-subshell,
  - $l = 2$ corresponds to the d-subshell,
  - $l = 3$ corresponds to the f-subshell,

and the excited states g, h, I, etc. for $l = 4$, 5, 6, etc. This number tells us about the geometry of the atomic orbital, that is to say the spatial region in which the electron moves. For example, the shell $n = 1$ gives two second numbers $l = 0$ and $l = 1$, hence two subshells $1s$ and $2p$ (Fig. 1.9).

- *Number m*: It is the magnetic quantum number. It takes values between $-l$ and $+l$ (including the values $-l$ and $+l$). It determines the orientation of orbitals in space. For example, for $l = 1$, we have $m = -1, 0, +1$ so three orientations corresponding to three axes of a three-dimensional system, that is to say three orbitals p of the same energy (p_x, p_y, p_z) or three quantum boxes. A more naive way of defining this number is to say that it corresponds to the kinetic moment, which gives the

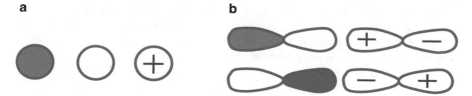

Fig. 1.9 Atomic orbitals. (**a**) Representation of $1s$: the orbital radius is most likely the electron-nucleus distance. The wave function is positive. (**b**) Representation of $2p$: the wave function is of opposite sign on both sides of the nodal plane (the plane of probability of the presence of the electron null)

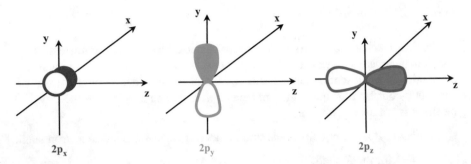

Fig. 1.10 Orientation of 2*p* orbitals in space. The orbital size increases as *n* increases

direction of rotation of the electron around the nucleus. Since the electron is an electric charge, its rotational motion creates a magnetic field (Fig. 1.10).

- *Number s*: It is the quantum number of spin or simply spin. This number can take two values, $m_S = +1/2$ and $m_S = -1/2$. If $m_S = +1/2$, it is customary to represent the electron by an arrow pointing upwards ↑ (spin up). If $m_S = -1/2$, it is represented by a down arrow ↓ (spin down). The spin is an intrinsic form of angular momentum. It is the only quantum observable that has no equivalent in classical physics, unlike the position, momentum, or energy of a particle. Its measurement gives discontinuous values and is subject to Heisenberg's uncertainty principle. Naively, we can consider the two values of the spin give the two directions of rotation of the electron on itself.

1.4.6   Electronic Configuration of an Atom

The electronic configuration of an atom is the numerical distribution of electrons in its energy shells. Energy shells are the different regions around the atom where electrons are statistically most likely to be there. The electronic configuration allows quickly and simply informing on the number of shells of energies which an atom possesses as well as on the number of electrons that populate each shell of this atom. There are two electronic configuration nomenclatures: the simplest one is the configuration of type *K*, *L*, *M*, etc., which allows understanding how electrons are distributed over different electronic shells. The second method is the fruit of quantum mechanics; it is more complex, but more fruitful: the electronic configuration is described as *s*, *p*, *f*, etc.

1.4.6.1   Electronic Configuration of Type K, L, M, etc.

In this model, the rules of nomenclature are as follows: let us consider a chemical element of the periodic table of Mendeleïev of atomic number Z. The corresponding neutral atom carries Z electrons which will be distributed on different electronic shells at different distances from the nucleus. The more the shell is taken away from the nucleus, the more it will contain electrons. Thus, the first shell contains a maximum of 2 electrons, the second 8, the third 18, the fourth 32, etc., in accordance to the rule which stipulates that on the nth shell we must have $2n^2$ electrons. Each shell is characterized by a capital letter: the first shell is the K shell, the second L, the third M, the fourth N, and so on and so forth in alphabetical order. So, to determine the electronic configuration of a chemical element, we consider the total number of electrons which it contains, and we write in exponent the number of electrons that are distributed on each shell. Let us apply this simple rule to the following cases:

- Hydrogen (H) has for atomic number $1 \Rightarrow Z = 1$: its configuration is K^1.
- Lithium (Li) has for atomic number $3 \Rightarrow Z = 3$: its configuration is $K^2 L^1$.
- Carbon (C) has for atomic number $6 \Rightarrow Z = 6$: its configuration is $K^2 L^4$.
- Magnesium (Mg) has for atomic number $12 \Rightarrow Z = 12$: its configuration is $K^2 L^8 M^2$.
- Tin (Sn), $Z = 50$: its configuration is $K^2 L^8 M^{18} N^{22}$.
- Uranium (U), $Z = 92$: its configuration is $K^2 L^8 M^{18} N^{32} O^{32}$.

This type of configuration has a limited interest, that is, to introduce the notion of electronic configuration in order to be able to realize Lewis structures. It is a method that greatly simplifies the representation of molecules and its great power is to contain in a few details most of the information needed to understand the phenomena involved in chemical reactions. The most interesting method currently used to describe the electronic configuration of the atom is the atomic orbital (O.A) method.

1.4.6.2   Atomic Orbitals Method

This method is directly derived from quantum mechanics. To determine the four quantum numbers of an electron in an atom, it is necessary to respect four rules: rule of stability, Pauli's exclusion principle, the Klechkowski rule, and the Hund rule.

Stability Rule

When the atom is in the ground state, electrons occupy the lowest energy levels.

Pauli's Exclusion Principle

In 1925, the Austrian physicist Wolfgang Ernst Pauli states a principle according to
which two electrons of the same atom cannot have the same quantum characteristics.
There is a difference of at least one quantum number between two electrons. In other
words, we can put in each electronic shell at most $2n^2$ electrons where n is the
number of the shell. This principle will afterwards be generalized to any fermion or
particle of half-integer spin such as neutrinos and quarks, as well as compound
particles such as protons, neutrons, etc.

Levels of Energy

Each of the shells and subshells actually represents a level of energy. These levels
increase from the center to the outside of the atom. In the graph below, we give a
representation of different shells and subshells and the energy level associated with
them. The number of atomic orbitals for each value of n is n^2. So (Fig. 1.11):

- In the $n = 1$ shell, we have 1 atomic orbital because $n^2 = 1^2 = 1$.
- In the $n = 2$ shell, we have 4 atomic orbitals because $n^2 = 2^2 = 4$.
- In the $n = 3$ shell, we have 9 atomic orbitals because $n^2 = 3^2 = 9$.

Klechkowski's Rule

The Klechkowski rule, named after the Russian chemist Vsevolod Klechkowski, is
an empirical method that allows one to predict with relatively good precision the

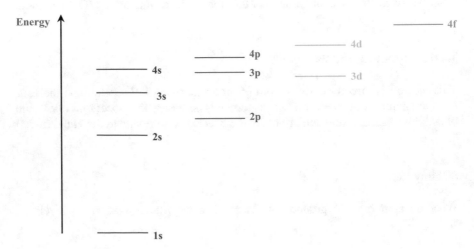

Fig. 1.11 Representation of different shells and subshells and the energy level associated with
them. We notice that the $3d$ energy level is close to the $4s$ level

order of filling of electrons in subshells of electrically neutral atoms in the ground state. If n and l are respectively the main quantum number and the azimuthal quantum number, the Klechkowski rule states that electronic subshells filling of electrically neutral atoms in the ground state arranged by increasing atomic number obey the following rules:

- In ascending order of values of $n + l$ defining electronic subshells.
- In ascending order of values of n when several subshells have equal $n + l$ values.

The quantum number n is an integer satisfying $n \geq 1$, while l is an integer satisfying $0 \leq l \leq n - 1$. Since the values $l = 0$, 1, 2, 3, etc. define respectively the subshells s, p, d, f, etc., the filling order of electron subshells deduced from Klechkowski's rule can be summarized in Table 1.1 and Fig. 1.12.

The order of filling of electronic subshells, of atoms electrically neutral, in the fundamental state, arranged by increasing atomic number, is thus: $1s \rightarrow 2s \rightarrow 2p \rightarrow 3s \rightarrow 3p \rightarrow 4s$, etc. The Klechkowski diagram allows to find this sequence by means of a simple construction: we place the subshells s in the first column, the subshells p, d, f, etc. are added in succession on the same line, and the reading is done diagonally, each diagonal representing a given value of $n + l$. Thus, the filling of energy levels is:

The Klechkowski rule tells us only about the order of filling of subshells and gives no indication of the number of electrons that each subshell can contain. This rule applies only to atoms electrically neutral at their ground state, the electronic configuration of ions and excited atoms can indeed significantly deviate.

Table 1.1 Order of filling of electron subshells

Quantum numbers			
Sum	Principal	Azimuthal	Subshells
$n + l = 1$	$n = 1$	$l = 0$	$1s$
$n + l = 2$	$n = 2$	$l = 0$	$2s$
$n + l = 3$	$n = 2$	$l = 1$	$2p$
	$n = 3$	$l = 0$	$3s$
$n + l = 4$	$n = 3$	$l = 1$	$3p$
	$n = 4$	$l = 0$	$4s$

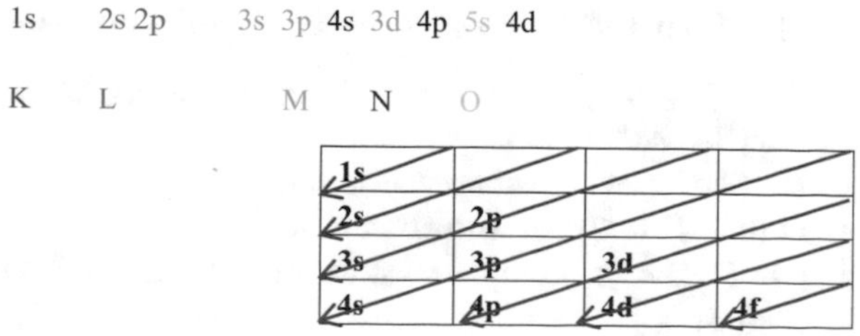

Fig. 1.12 Most common representation of the Klechkowski diagram: here, each diagonal red arrow corresponds to a value of $n + l$

Hund's Rule

About one in five chemical elements have an electronic configuration in the ground state different from that deduced from the Klechkowski rule. This stems from the fact that the latter results from an approximation of the energy of electronic subshells taking into account only the quantum numbers n (principal) and l (azimuth), whereas the energy of electrons also involves their spin s. More precisely, Hund's rule indicates that the lowest energy spectroscopic term is the one with the highest spin multiplicity $2s + 1$, which means that in the same subshell, the most stable configuration is obtained when the number of electrons with identical spins is maximal. Hund's rule is negligible in front of the Klechkowski rule for the elements of the s-block and the p-block of the Mendeleïev periodic table, so that the Klechkowski rule is always observed; it can, however, be decisive for certain elements of the d-block and f-block, that is to say for transition metals, lanthanides and actinides, because the energy levels of electronic subshells of the valence shell of these elements are quite similar, so that it may be energetically more favorable to redistribute electrons by observing the Hund rule than to follow the Klechkowski rule.

Symbolic Notation

Once we have gathered all the aforementioned rules which allow detailing of the electronic structure, let us see now how to write it by considering the example of carbon 6. This one contains 6 electrons. Its electronic structure is thus: $1s^2 \, 2s^2 \, 2p^2$ where the numbers represent the shell, the letter, the subshell and the exponent, the number of electrons contained in the subshell. The sum of exponents obviously gives the total number of electrons of the atom. To be able to start the filling of a new shell, all of the lower energy levels have beforehand to be performed because the atom is in its fundamental state. We can also use the abbreviated notation by considering, for example, calcium (Ca) which contains 20 electrons. Its electronic structure can be written as: $[Ar] \, 4s^2$. This means that the calcium atom has the same electron configuration as the rare gas argon $+ \, 4s^2$. This abbreviated notation works on the following principle: we give between brackets the rare gas symbol (He, Ar, Kr, Xe, Rn) and then we indicate what is in addition to this configuration of the noble gas. The goal of this approach is to relieve the writing. For elements with an atomic number from 1 to 10, the abbreviated notation does not exist, that is:

- Cu (29 e^-): structure: $1s^2 2s^2 2p^6 3s^2 3p^6 3d^{10} 4s^1$; abbreviated notation: $[Ar] \, 3d^{10} 4s^1$.
- He (2 e^-): $1s^2$ (no abbreviated notation).
- O (8 e^-): $1s^2 2s^2 2p^4$ (no abbreviated notation).
- Cl (17 e^-): complete structure: $1s^2 2s^2 2p^6 3s^2 3p^5$; abbreviated notation: $[Ne] \, 3s^2 3p^5$.

We can also represent orbitals and layers by quantum boxes. Each box represents an orbital (thus 2 e^- antiparallel spins) and the set of contiguous boxes will symbolize the shell; thus:

- ☐ Box of the $1s$ shell with its two electrons. Each electron is represented by an arrow of which direction symbolizes the electron spin. $1s$-orbitals can receive a maximum of 2 electrons.

- ☐ Box of the p-subshell; p-orbitals can receive a maximum of 6 electrons.

- ☐ Box of the d-subshell; d-orbitals can receive a maximum of 10 electrons.

In view of what has just been said, the electronic structure of sulfur (16 e^-), for example, is: $1s^2 2s^2 2p^6 3s^2 3p^4$

1.4.7   De-Excitation of the Atom

An atom is in an excited state if one of electrons of a lower electronic shell jumps to an upper electronic shell as a result of an external energy supply. The excited state being unstable, the electron returns very quickly to its initial electronic shell (after 10^{-8} to 10^{-15} s) by redirecting the energy it received in the form of a monochromatic radiation that is at the origin of the spectral emission of each chemical species. Atomic de-excitation is a variation of energy between higher and lower electronic levels as we emphasized in Sect. 1.3.4, which deals with the Bohr model and other related paragraphs:

$$E = E_{\text{sup}} - E_{\text{inf}} = h = \frac{hc}{\lambda}. \tag{1.45}$$

The variation ΔE causes the emission of a quantum of energy, of zero mass, moving at the speed of light $c = 3 \times 10^8$ m/s where h is the universal constant of Planck ($h = 6.266 \times 10^{-34}$ J.s), ν the frequency of the monochromatic light in hertz, and λ the wavelength of the radiation in meters. The energy ΔE is expressed in joules in the international system and preferentially in electronvolts (eV) in nuclear physics.

If the external energy supply is sufficiently large, an electron of the K-shell can be ejected outwards causing the atom ionization. To do this, an incident photon, for example, must have the energy of an X-photon or that of a γ-photon. Once the electron of the K-shell was torn off, this atom is in an excited state. A de-excitation process is automatically triggered: an outermost L-shell electron fills a K-shell vacancy left by the ejected electron, which causes either the emission of an X-photon (also called *X-ray* or *X-ray fluorescence*) or the direct transmission of

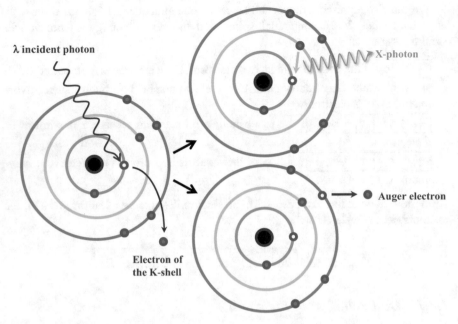

Fig. 1.13 At first, the γ-photon tears an electron from the K-shell. The atom that has lost one of its internal electrons is in an excited state. An outermost L-shell electron fills a K-shell vacancy left by the ejected electron, which causes either the emission of an X-photon during this transition (the X-ray fluorescence) or the direct transmission of this energy to another electron of any shell, called Auger electron, which is then ejected. The Auger emission is in competition with the X-emission

this energy to another electron of any shell, called *Auger electron*, named after the French physicist Pierre Victor Auger (1899–1993), which is then ejected with a relatively low kinetic energy (Fig. 1.13). The Auger emission is in competition with the X-emission. Emitted within a dense matter, the X-ray is generally absorbed after a short course.

1.5   Energy Levels of the Nucleus

Nuclei are a priori very different from atoms because the nucleus is very compact, whereas the atomic space essentially consists of vacuum. Nevertheless, nuclei and atoms have common features. Indeed, when the nucleus is in an excited state, it has a supplement of energy. It returns to the state of minimum energy called *fundamental state* by getting rid of this excess energy by emission of a γ-photon of characteristic energy. These photons are of the same nature as X-photons emitted by the electronic cloud of atoms, with the difference being that their energy is much greater, usually of the order of MeV (Fig. 1.14).

Fig. 1.14 Typical diagram of de-excitation of a nucleus: the horizontal lines in red represent permitted excitation energies. When the nucleus is in an excited state, it returns to its fundamental state by emitting one or more γ-photons with energies characteristic of the nuclide

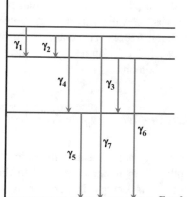

Excitation energy (MeV)

Fundamental level

Like the atom, the nucleus is composed of shells. Binding energies of nucleons of the nucleus can only take discrete values corresponding to as many shells. Nuclei are also governed by the laws of quantum mechanics. The nucleus can only be found in a limited number of states. These states are characterized first of all by their energy. In addition to this layered structure, the nucleus may have collective movements that correspond to new states. Unlike the atom where we consider the state of an individual electron through its energy level and its rotation, for a nucleus we can only consider all of its nucleons. In addition, the nucleons constituting the nucleus can vibrate. The energies of these vibrations have well-defined values. Finally, the nucleus is not necessarily spherical; it can deform and undergo a collective rotational movement. The energies of these rotational states are also quantized.

1.5.1 Binding Energy

The nucleus of an atom has Z protons and N neutrons. The binding energy of nucleons is given by:

$$E_B = \left(Z\, m_P c^2 + N\, m_N c^2\right) - m_{\text{nucleus}} c^2 \tag{1.46}$$

where:

- $m_{\text{nucleus}} c^2$ is the mass energy of the nucleus.
- $Z\, m_P c^2$ is the mass energy of protons contained in the nucleus with $m_P c^2 = 938.28$ MeV, mass energy of one proton.
- $N\, m_N c^2$ is the mass energy of neutrons contained in the nucleus with $m_N c^2 = 939.59$ MeV, mass energy of one neutron.

- E_B is the binding energy of nucleons called *mass defect*. This energy can be considered as a potential well in which protons and neutrons are kept locked in the nucleus as the water which gives some energy to the external environment to the point that it freezes. This energy is positive since it corresponds to the energy that the external environment must provide to the nucleus (study system) to separate its nucleons.

1.5.2 Binding Energy per Nucleon and Aston's Curve

The binding energy per nucleon is the ratio of the binding energy E_B of the nucleus and the number A of nucleons present in that nucleus. It is usually expressed in MeV/nucleon. This energy is not a universal constant because it depends on nuclides.

The Aston curve of the name of the English physicist Francis William Aston gives the binding energy per nucleon E_B/A as a function of the number A of nucleons that make up the nucleus. But, it is more telling to plot the curve $- E_B/A = f(A)$ because it allows to compare the stability of different atomic nuclei by proceeding as in an energy diagram where the most stable nuclei occupy the lower part of the curve. This curve reveals that the binding energy per nucleon is minimal for $60 < A < 90$. This corresponds to the most stable nuclei with $E_B/A \approx 9$ MeV. This curve reaches its minimum for a number of nucleons equal to 70. On both sides of this point, nuclei can undergo modifications that bring them closer to this point (Fig. 1.15). The nucleus energy son obtained after modification is greater than the

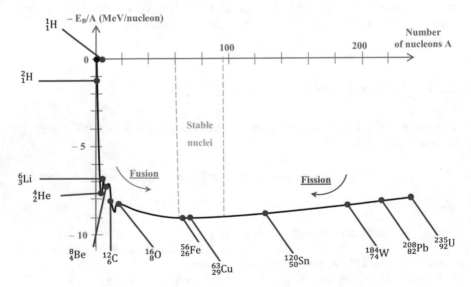

Fig. 1.15 Aston's curve: it gives the binding energy per nucleon E_B/A as a function of the number A of nucleons that make up the nucleus

energy of the nucleus father which generated it. This change is manifested by a decrease in the mass of the system and a release of a very great energy towards the external environment. There are two ways to obtain nuclei of high binding energy per nucleon:

- The nucleus is broken into two nuclei, causing a nuclear reaction of fission.
- Two nuclei are assembled into a single nucleus, causing a nuclear reaction of fusion.

Light nuclei such as $_1^2H$, $_1^3H$ and $_2^4H_e$ change by fusion, whereas heavy nuclei such as $_{92}^{235}U$, $_{92}^{235}U$ and $_{94}^{239}Pu$ change by fission.

1.5.3   Nuclear Fission

1.5.3.1   Introduction

In 1932, James Chadwick discovered the neutron. Later, physicists realized that the nucleus could absorb neutrons. The goal was to create new nuclei by neutron bombardment which in the case of uranium 235 produced two smaller nuclei: uranium 235 had fissioned. This phenomenon was discovered in 1938 by Otto Hahn, Lise Meitner and Fritz Strassmann.

1.5.3.2   Description

In the nucleus of an atom, three interactions coexist: electromagnetic, strong, and weak. The weak interaction is responsible for the nuclear disintegration; it does not intervene here. As protons are positively charged, the electromagnetic interaction causes their repulsion and the nucleus bursting, while the strong interaction leads on the contrary to the nucleon attraction. For light nuclei, a balance between these three forces is established, making them particularly stable. In the case of a heavy nucleus which contains many nucleons, the attraction between nucleons by strong interaction is exerted only between nucleons quite close to each other. The nuclear building is fragile. At the slightest disturbance, and even spontaneously, it can be divided into several pieces: this is what is called *nuclear fission*. A nucleus, uranium 235, for example, which undergoes the fission, is said to be *fissile*. If it is collided by a neutron, it forms uranium 236, which then quickly breaks, by emitting two nuclei, neutrons and γ-radiation. There are several fission possibilities of uranium 235. One of the most common is:

$$_{92}^{235}U + _0^1n \rightarrow _{92}^{236}U \rightarrow _{38}^{94}Sr + _{54}^{140}Xe + 2\,_0^1n + \gamma. \qquad (1.47)$$

The formed nuclei, as well as the emitted neutrons (two or three on average), carry some kinetic energy. Taking also into account the energy of the γ-radiation, the fission is the energy producing for heavy nuclei, of the order of 200 MeV for the fission of uranium 235. From a macroscopic point of view, this energy is manifested by a release of heat. The produced nuclei, called *nuclei sons*, are less unstable than the initial uranium nucleus (nucleus father), but are not very stable. They are mostly highly radioactive because they emit β^- and γ-rays.

1.5.3.3   Chain Reaction

If the emitted neutrons by fission strike new nuclei of uranium 235, for example, they can fission them in their turn. This is the principle of chain reaction, observed by Frédéric Joliot-Curie in 1939. This reaction only occurs if neutrons collide with the nuclei with an adequate kinetic energy, of the order of 0.02 MeV. They are called *thermal neutrons* or *slow neutrons*. If the neutron is too slow, it will bounce on the nucleus. On the other hand, if it is too fast, it will pass through the nucleus without interacting with it.

1.5.3.4   Fission Applications

The fission has two major applications:

- *Nuclear weapons (A-bombs)*: The chain reaction is not controlled. A fission reaction induces two or three others, which themselves induce others, etc. The number of fissions increases exponentially, which generates a nuclear explosion of very high intensity. Nuclear weapons have been used twice in war, both times by the United States of America against Japan near the end of the Second World War. On August 6, 1945, the US Army Air Forces detonated a uranium gun-type fission bomb nicknamed *Little Boy* over the Japanese city of Hiroshima; 3 days later, on August 9, the US Army Air Forces detonated a plutonium implosion-type fission bomb nicknamed *Fat Man* over the Japanese city of Nagasaki. These bombings caused injuries that resulted in the deaths of approximately 200 000 civilians and military personnel. These two bombs were based on this mode of operation.
- *Nuclear power plants*: By controlling the chain reaction, nuclear power plants generate heat in reactors used to run turbines and produce electricity. In pressurized water plants, this water serves both to transport the heat produced (heat transfer fluid) and to slow the fast neutrons formed after each fission to make them suitable for other fissions.

1.5.4 Thermonuclear Fusion

1.5.4.1 Description

The fusion is the opposite of the fission. It consists of forming one nucleus from two others. This reaction concerns light nuclei. To achieve this, the electrical repulsion between the protons of each nucleus to be fused must be overcome. To do this, one must place oneself in conditions of very high temperatures and pressures. Here are three examples of fusion with isotopes of hydrogen and the respective energies they emit:

$$
\begin{aligned}
&{}_1^2\text{H} + {}_1^2\text{H} \rightarrow {}_2^3\text{He} + {}_0^1\text{n} \ (3.3 \text{ MeV}) \\
&{}_1^2\text{H} + {}_2^3\text{H} \rightarrow {}_2^4\text{He} + {}_1^1\text{p} \ (18.3 \text{ MeV}) \\
&4\,{}_1^1\text{H} \rightarrow {}_2^4\text{He} + 2\,{}_1^1\text{e} \ (27 \text{ MeV}).
\end{aligned}
\tag{1.48}
$$

The isotope ${}_1^2\text{H}$ is named *deuterium* and the isotope ${}_1^3\text{H}$ is the *tritium*, and it is radioactive. The nuclear fusion is the source of the energy of stars, including the Sun. At its heart, the latter fuses hydrogen to produce some helium, at a temperature of about 15 million of kelvins with a density of 150 tons/m^3. For the hottest stars and/or at certain periods of their evolution, it is even possible to observe the fusion of heavy elements such as neon, carbon, and oxygen.

1.5.4.2 Current and Future Applications

Current and future applications can be summarized as follows:

- *H-bombs*: The H-bomb was developed during the Cold War. By the explosion of a fission sub-A-bomb, the temperature conditions required for the fusion are reached, and a fusion H-bomb then explodes. H-bombs are currently the most powerful weapons which humanity built. They constitute an example of uncontrolled fusion.
- *Controlled fusion*: As for the fission, the main goal is to control the fusion to produce some electricity. If current or future projects can produce the fusion by powerful lasers, another studied track concerns tokamaks. These torus-shaped structures maintain some ionized material (plasma) in levitation by magnetic fields. It is then brought to high temperatures, of the order of 100–200 million of kelvins, which allow the fusion.

1.5.5 De-Excitation of the Nucleus

When a nucleus receives an energy supply from the outside, it goes into an excited state. As electronic shells of the atom, the energy states of the nucleus are quantized. The nucleus de-excites into two ways, by γ-emission and by internal conversion.

1.5.5.1   De-Excitation by γ-Emission

γ-Rays are of the same nature as X-rays, with the difference being that the energy they carry is much higher, from a few tens of thousands of electronvolts to several million electronvolts. This emission generally follows a phenomenon of decay β^-, β^+, α, or neutron capture by a nucleus. During these decays, the topology of nucleons in the nucleus is not ideal and the nucleus finds itself in an excited state, that is to say with a supplement of energy with regard to its fundamental state. Nucleons rearrange themselves in the nucleus by getting rid of this excess of energy in one or several stages, emitting each time a γ-photon. The γ-transition is almost always immediate. It can exceptionally occur with a certain delay, just like for the excited state of technetium, which lasts several hours, making it useful in hospitals as a source of γ-radiation for therapeutic purposes. Like the atom, the nucleus also has well-defined energy states. The jump from one energy state to another is done by emitting a γ-photon of a characteristic energy during the transition. The measurement of the energy of γ-photons emitted constitutes a means of identification of the nature of the emitting nucleus. Because of the energy they carry, γ-rays are very penetrating. To stop some of them, several tens of centimeters of lead or several meters of concrete are needed. As an example, the de-excitation of nickel 60 which passes to its fundamental state by emitting a γ-photon:

$$\,^{60}_{28}\mathrm{N}_i^* \rightarrow \,^{60}_{28}\mathrm{N}_i + \gamma. \tag{1.49}$$

1.5.5.2   De-Excitation by Internal Conversion

The internal conversion is a process also related to the emission of a γ-photon because it happens that the excess of energy which is in the nucleus is transmitted directly to an electron of electronic shells, generally of the K-shell which is ejected of the atom. This electron is called *electron of conversion*. The space left vacant in electronic shells is subsequently filled by an electron from upper shells and so on and so forth. There is therefore a rearrangement of electronic shells with Auger electron emission or X-rays characteristic of the nuclide. Indeed, the energy transferred to the electron is that of the γ-photon minus the binding energy of the electron on its electronic shell. The energy of the γ-photon and the binding energy of the electron on its shell have well-defined values. As a result, the energy of the ejected electron takes a series of well-defined values.

1.6   Radioactivity

1.6.1   Definition

Radioactivity is a physical phenomenon by which unstable atomic nuclei called *radionuclides* or *radioisotopes* are transformed into other atoms by means of a process of nuclear decay by simultaneously emitting particles of matter such as electrons, helium nuclei, neutrons, etc. and energy mainly in the form of γ-photons. It was discovered in 1896 by the French physicist Henri Becquerel in uranium and very quickly confirmed by Marie Curie in radium. If the decay is spontaneous, it is said *natural radioactivity*. If it is caused by a nuclear reaction, it is called *artificial* or *induced radioactivity*.

1.6.2 Nuclear Reactions

A nuclear reaction is described by an equation which concerns only nuclei and which must verify the laws of conservation of the charge number Z and the mass number A. It is written in the following general form:

$$\ _{Z}^{A}X \rightarrow \ _{Z_1}^{A_1}Y^* + \ _{Z_2}^{A_2}W \tag{1.50}$$

where X is the radioactive nucleus father, Y^* the excited nucleus son, and W the emitted particle. Since the nucleus son is in an excited state, it returns to the fundamental level by getting rid of the energy excess in the form of a γ-photon according to the following equation:

$$\ _{Z_1}^{A_1}Y^* \rightarrow \ _{Z_1}^{A_1}Y + \gamma. \tag{1.51}$$

1.6.3 Main Radioactive Emissions

1.6.3.1 α-Radioactivity

α-Radioactivity corresponds to the helium nucleus emission composed of two protons and two neutrons. Emitted α-particles are not very penetrating and can be stopped by a sheet of paper or by superficial layers of the skin, but these particles may tear away electrons from the material which they cross, making them particularly ionizing. This type of radiation is used for heavy nuclei. The α-decay is written as follows:

$$\ _{Z}^{A}X \rightarrow \ _{Z-2}^{A-4}Y + \ _{2}^{4}He. \tag{1.52}$$

This transformation is not isobaric since nuclides father and son do not have the same mass number A. Polonium 210, for example, is α-radioactive. Its nucleus is transformed as follows:

$$^{210}_{84}\text{Po} \rightarrow {}^{206}_{82}\text{N}_{\text{i}} + {}^{4}_{2}\text{He}. \tag{1.53}$$

1.6.3.2 β^--Radioactivity

β^--Radioactivity corresponds to the emission of electrons. It concerns nuclei containing an excess of neutrons with respect to the number of protons, both forming the nucleus $(N > Z)$. β^--Rays are moderately penetrating: they can cross the superficial layers of the skin, but some millimeters of aluminum are enough to stop them. β^--Decay is written as follows:

$$^{A}_{Z}\text{X} \rightarrow {}^{A}_{Z+1}\text{Y} + {}^{0}_{-1}\text{e}^-. \tag{1.54}$$

This transformation is isobaric since father and son nuclides have the same mass number A. Carbon 14, for example, is β^--radioactive. Its nucleus is transformed as follows:

$$^{14}_{6}\text{C} \rightarrow {}^{14}_{7}\text{Y} + {}^{0}_{-1}\text{e}^-. \tag{1.55}$$

1.6.3.3 β^+-Radioactivity

β^+-Radioactivity corresponds to the emission of positrons. These particles have the same mass as electrons, but are positively charged. It concerns artificial nuclei containing an excess of protons with respect to the neutron number, both forming the nucleus $(Z > N)$. The penetration of positrons is similar to that of electrons. But at the end of its path, a positron annihilates itself with an electron met on its passage by forming two photons γ of 511 Kev each, emitted at $180°$ from each other which returns the problem back to the case of the γ-radiation. The β^+-decay is written as follows:

$$^{A}_{Z}\text{X} \rightarrow {}^{A}_{Z-1}\text{Y} + {}^{0}_{+1}\text{e}^+. \tag{1.56}$$

This transformation is isobaric since father and son nuclides have the same mass number A. Oxygen 15, for example, is β^+-radioactive. Its nucleus is transformed as follows:

$$^{15}_{8}\text{O} \rightarrow {}^{15}_{7}\text{N} + {}^{0}_{+1}\text{e}^+. \tag{1.57}$$

1.6.3.4 Electronic Capture

Radioactivity by electron capture concerns unstable nuclides containing an excess of protons relative to the number of neutrons $(Z > N)$. These nuclei disintegrate by β^+-decay, or by electronic capture, or by both of these two processes with a different probability for each of them. During the capture of an electron, there is the emission of a neutrino. The simplest way to conceive the neutrino is to imagine it as an electron of a very low mass, without electrical charge and without known structure. That is why the neutrino stemming from the radioactivity is called *neutrino-electron*. This name is used to distinguish it from two other types of neutrinos, *neutrino-mu* and *neutrino-tau*. The decay by electronic capture is written as follows:

$$ {}^A_Z X + {}^0_{-1} e^- \rightarrow {}^A_{Z-1} Y + {}^0_0 \nu. \tag{1.58} $$

This transformation is isobaric since father and son nuclides have the same mass number A. Cobalt 58, for example, is radioactive by electronic capture. Its nucleus is transformed as follows:

$$ {}^{58}_{27} Co + {}^0_{-1} e^- \rightarrow {}^{58}_{26} Fe + {}^0_0 \nu. \tag{1.59} $$

1.6.4 Universal Law of Radioactive Decay

The disintegration of radioactive nuclei is random at the microscopic level, but at the macroscopic level, the average number of nuclides at an instant t follows a well-defined law called *radioactive decay*. For this purpose, let us consider the number of radioactive nuclides $n(t)$ of a given species present at the instant t. Between the instants t and dt, a number of nuclides of this species have disintegrated and changed in nature. Let dn be the variation of the number of nuclides between these two instants. This variation depends on the number $n(t)$, the nature of nuclide, and the length of time dt:

$$ dn = -n(t)\,\lambda\,dt \Rightarrow \frac{dn}{n(t)} = -\lambda\,dt \Rightarrow n(t) = n_0\,e^{-\lambda t} \tag{1.60} $$

where λ is a constant which depends on the nuclide nature; it is expressed in s^{-1}. It determines the probability of disintegration of a nucleus per unit of time (not to be confused with the wavelength that is expressed with the same symbol). n_0 is the nuclide number present in the sample at the initial moment $t = 0$. The minus sign indicates a decrease of $n(t)$. Equation (1.60) shows that the decay of radioactive nuclei follows a decreasing exponential law.

1.6.5 Half-Life

The half-life $t_{1/2}$ of a radioactive nucleus is the time necessary for the decay of half of nuclei present in a sample of this species. From (1.60) we deduce:

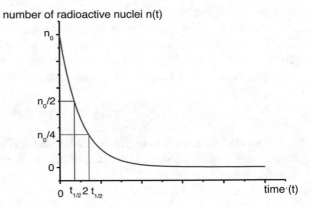

Fig. 1.16 Curve of the radioactive decay where the number of the radioactive nucleus is halved at the end of each half-life

$$\frac{n(t)}{n_0} = e^{-\lambda t_{1/2}} = \frac{1}{2} \Rightarrow t_{1/2} = \frac{\ln 2}{\lambda} = \frac{0.693}{\lambda}. \tag{1.61}$$

The half-life $t_{1/2}$ is very different from one nucleus to another. The half-life has only a statistical value. It indicates that a radioactive nucleus has a half chance to disappear after a half-life. It is greater than 10^{30} years for vanadium and less than 2.96×10^{-7} s for polonium 212. In Fig. 1.16, we give the curve of the radioactive decay where the number of radioactive nuclei is divided by two at the end of each radioactive half-life.

1.6.6   Activity of a Radioactive Source

The activity A of a radioactive source at a given instant is equal to the decay rate of radioactive nuclei constituting it at this moment (not to be confused with the mass number which expresses itself with the same symbol A):

$$A = -\frac{dn}{dt} = \lambda n. \tag{1.62}$$

The activity unit is the becquerel (Bq) (1 Bq = 1 decay per second). This number depends on the decay constant λ and the average number $n(t)$ of nuclei present in the sample at time t. Just like $n(t)$, the activity decreases exponentially in time.

1.6.7   Mean-Life

The mean-life $\bar{t}$ of n_0 atoms present at $t = 0$ corresponds to the time at the end of which this number has decreased by a factor $1/e$:

$$\overline{n(t)} = \frac{n_0}{e} \tag{1.63}$$

where $e = 2.718$. It is a transcendental number which forms the base of natural logarithms. It verifies $ln\ e = 1$.

1.6.8   Measurement of Radioactivity

The knowledge of the decay number, the absorbed radioactive energy, and its biological effects allows completely identifying the radioactivity of a source. For this purpose, three units are used: becquerel, gray, and sievert.

1.6.8.1   Becquerel

The becquerel (Bq) is the unit of measurement of the radioactivity of a source. It characterizes the number of spontaneous decays per second. The more this number is important, the more this source is strongly radioactive. In the human body, radioactivity is naturally of 120 Bq/kg.

1.6.8.2   Gray

The gray (Gy) measures the radioactive dose absorbed by an irradiated body during an accidental or a therapeutic exposure. It corresponds to the amount of energy received per unit mass (1 Gy = 1 J/kg of the irradiated material).

1.6.8.3   Sievert

The sievert (Sv) measures the biological effects of ionizing radiation on the living matter. At equal doses, these effects vary depending on the nature of the ionizing radiation. At equal energy, the impact of α-radiation is twice the radiation β or γ. To translate these differences, a dimensionless factor Q (weighting factor) is introduced to relate the absorbed dose D in Gr to the equivalent dose H in Sv:

$$H = D \times Q. \tag{1.64}$$

The weighting factor Q takes into account the nature of the radiation and its effects on the tissues of the living matter: $Q = 1$ for γ-rays and 20 for α-particles. The admissible dose threshold for a person is 1 mSv/year beyond the natural radioactivity estimated on average at 2.4 mSv/year by excluding doses received in medicine, on average 1.3 mSv/year.

1.6.8.4   Measuring Devices

Radioactivity being imperceptible by man, the latter has invented instruments that measure it in a given place or person. All these devices measure the ionization, i.e., the number of electrons tearing from atoms, and the excitation, that is to say energy transmissions to atoms that made them switch into an excited state. These measuring instruments are grouped into four categories:

- *Dosimeters*: they measure the external exposure of a person to different radiations (photographic dosimeter, thermo-luminescent dosimeter).
- *Geiger-Muller counters*: they detect β- and γ-radiations. These meters are widely used in the nuclear industry.
- *Gaseous meters*: they detect α-radiations.
- *Neutron detectors*: they detect neutron radiations.

1.6.9   Dangers of Radioactivity

The human body can be exposed to a radioactive radiation in two ways:

- *By external irradiation*: The radiation is emitted by an external source to the human body. This radiation can affect the whole body or certain organs and tissues.
- *By internal contamination*: The radiation is emitted by an internal source to the body during ingestion or inhalation of radioactive substances.

1.6.10   Radioprotection Rules

Protective measures aim to limit risks related to radioactivity by taking the precaution of staying away from radioactive sources; by using protective screens made of concrete, lead, etc.; and by limiting the time of exposure to radioactive radiations.

1.6.11   Radioactivity Harms

Radioactivity produces radiations and releases particles; it is invisible, odorless, and inaudible, and yet it radiates, contaminates, destroys, and kills. α-Radiations are stopped by a simple sheet of paper, but β- and γ-radiations require strengthened protections. In a nuclear power plant in distress, two types of particles are particularly followed: iodine 131, which attaches primarily on the thyroid gland and can cause cancers. Its life is fortunately short, a week. The second is cesium 137 which has a lifetime of 30 years. This isotope is in particular used for the following up of the radioactive cloud. This is one of the fission products of uranium. Its study allowed highlighting peaks of radioactivity in the atmosphere in 1960 further to atmospheric tests of A-bombs and in 1986 after the disaster of Chernobyl and now Fukushima. The Chernobyl nuclear accident was classified at level 7, the highest on the international scale of nuclear events that took place on April 26, 1986, in Lenin power plant, located at the time in Ukraine in the USSR. The accident was caused by an uncontrolled increase of the power of the No. 4 reactor leading to the melting of its core. This has led to an explosion and the release of large quantities of radioactive elements in the atmosphere, causing a very large contamination to the environment, and many deaths and diseases had arisen immediately or in the long term due to irradiation or contamination. The consequences of Chernobyl disaster, while controversial, are important in terms of health, ecology, economics, and politics. More than 200 000 people had been permanently evacuated. The report of the International Atomic Energy Agency (IAEA) established in 2005 listed nearly 30 deaths by acute radiation syndrome directly attributable to the accident and estimated that 5% of deaths of military liquidators would be related to the disaster. In the local populations, 4000 thyroid cancers were diagnosed between 1986 and 2002, the vast majority of which is attributed to the disaster. In addition to that, the management of radioactive waste from nuclear power plants is a major problem today. Indeed, nuclear materials are transported in different physical forms (gases, powders, liquids, metallic assemblies) and use trains as well as trucks, cargo ships, and even airplanes, and this transport of materials so dangerous is not without risk. The security measures necessary for the functioning of a nuclear power plant are extremely expensive and a power station requires the presence of a river in order to cool the reactor. The resulting water is warmed, causing fauna destruction.

The effects of a nuclear bomb are devastating for humans, flora, and fauna. From the first millionth of one second, thermal energy released in the atmosphere transforms the surrounding air into a fireball about one kilometer in diameter and several million degrees. On the ground, the temperature reaches several thousand degrees. In a radius of one kilometer, everything is instantly reduced to ashes. Up to 4 km from the epicenter, buildings and humans spontaneously catch fire; people within 8 km are burned to the third degree. After the heat, the shock wave generated by the phenomenal pressure due to the expansion of hot gases progresses like a solid air wall at a vertiginous speed of a thousand kilometers per hour. It reduces to dust all that is on its trajectory within a radius of 2 km. Of the 90 000 buildings in the city of Hiroshima, 62 000 were completely destroyed. These damages estimated further to the atomic bombing of the Japanese city of Hiroshima, but can dramatically increase with the type and the bomb power. The third effect is that of residual radioactive radiation. They cause myeloma (bone marrow cancers); leukemia (blood cancers); tumors of the breast, the skin, lungs, the bladder, and the thyroid; etc. It is all the more so terrifying that these effects only appear days, months, or even years after the explosion. Irradiation also causes hereditary genetic abnormalities.

The greater the dose is strong, the greater the consequences for the body are important with the risk of cancer. From 500 mSv, we can observe rapid effects on the body. A single dose of 1000 mSv causes redness, nausea, vomiting, dizziness, and even bleeding, but is not life-threatening. Above 5000 mSv, the exposure can lead to death, severely affecting the lymphocytes, red blood cells, and cells of the digestive system. If the dose exceeds 10 000 mSv, the radiation affects the central nervous system causing death. In the case of prolonged or repeated exposure, the risk of cancer increases even for low levels of radiation. Radiation damages DNA molecules and increases the frequency of mutations, favoring the development of bone, blood, and thyroid cancers and infantile malformations.

1.6.12 Radioactivity Benefits

In the medical field, X-ray diagnosis allows to explore the human body and to detect certain diseases. The principle is to inject into the body some radioelements and observe their course and behavior. In this way, we are able to accurately establish the morphology of examined organs and to detect anatomical and functional abnormalities. The nuclear medicine uses many techniques related to radioactive properties. The examination by scintigraphy allows to visualize on a computer the information collected by highly sophisticated detectors such as γ-cameras. The positron emission camera examination reveals complex processes of certain organs by following the way in which radionuclides that have been injected into the patient are metabolized. Radiotherapy uses the energy of ionizing radiation to destroy cancer cells. This radiation can be applied from the outside by transcutaneous radiotherapy (cobalt-60 bomb) or from the inside by endo-brachytherapy (by means of needles implanted near the tumor). In 2016, we estimated at 35 million the number of persons who used the nuclear medicine, either for a diagnosis or for a therapy.

In today's industrial world, 17% of the world electricity is produced by nuclear power plants. The industrial radiography, widely used in metallurgy or in aeronautics, serves to X-ray metal parts and to verify welds. Radioisotope gauges measure the intensity of a radiation at the source and on arrival. They allow to gauge the level of a liquid (verification of the filling of a tank, a silo) or to control the thickness, the density, or the homogeneity of a material. Ionization detectors allow highlighting the presence of different gases in ambient air. Their use is multiple (fire detectors, firedamp dosing in mines). Industrial irradiation allows developing stronger and lighter materials. Its applications are numerous in medicine and in industry (lighter prostheses, more resistant electrical cables). Industrial tracers serve to detect the leaks of liquid or gas in the buried or inaccessible pipelines by means of the property that radioelements have to be detected in an extremely precise way. They also allow to study underground routes of water or possible pollutants and to follow the movements of sands and vases during the study of ports and estuaries.

In the agro-food area, the ionization of agro-food products (potatoes, onions, strawberries, etc.) by γ-rays, electron beams, or X-rays allows obtaining better conditions of conservation: the stop of the germination and destruction of parasites and microorganisms. This technique, commonly used in many countries, does not make the product radioactive and does not affect its nutritional quality. The improvement of cultures is possible by radio-genesis: the exposure of plants (wheat, barley, rice, etc.) to γ-rays causes the mutation of some of their genes. Mutant strains most resistant to diseases or bad weather or those that adapt to unfavorable soil are then selected. Irradiating treatments have been developed to sterilize certain types of male insects such as tsetse flies to protect humans and crops: the pest population decreases progressively by radio-sterilization, thus without the use of insecticides.

In the cultural field, carbon dating in archeology is a practical application of the radioactive decay rule. This property allows calculating the moment when the radioelement has been incorporated into the sample that we wish to date. Carbon 14, thorium 232, and potassium 40 can be used to date fossils, bones, or minerals up to 1 billion years old. The cultural heritage conservation uses the γ-ray irradiation treatments that eliminate insects, fungi, and bacteria that are responsible for the often irreversible degradation of invaluable paintings. Ramses II's mummy, for example, benefited from such a treatment in 1976. The treatments by impregnation of a resin, hardened under the effect of radioelements, also allow consolidating the materials constituting the cultural work to be restored.

1.7 Solved Exercises

1.7.1 Structure of the Atom

Exercise 1
1. *What is the charge of the electron, the proton, and the neutron?*
2. *In which case does the atom become electrically charged?*
3. *Where is the major part of the mass of an atom?*
4. *How are the protons confined in the nucleus of the atom?*
5. *Let us consider the following atoms and ions:*

$$^{14}_{7}N \qquad Al^{3+} \; \textit{from} \; ^{27}_{13}Al \qquad O^{2-} \; \textit{from} \; ^{16}_{8}O.$$

Give the composition of the nucleus and the number of electrons for each chemical species.

Solution
1. The electron carries a negative charge $q_e = -e = -1.6 \times 10^{-19}$ C. The proton carries a positive charge $q_p = +1.6 \times 10^{-19}$ C. The neutron is neutral.
2. The atom becomes electrically charged if it wins or loses one or several electrons.
3. Almost all the mass of the atom is concentrated in its nucleus.

4. In the nucleus of an atom, three interactions coexist: electromagnetic, strong, and weak. The weak interaction is responsible for nuclear disintegration. As protons are positively charged, the electromagnetic interaction causes their repulsion and the nucleus bursting, while the strong interaction leads on the contrary to the nucleon attraction. For light nuclei, a balance between these three forces is established, making them particularly stable. In the case of a heavy nucleus which contains many nucleons, the attraction between nucleons by strong interaction is exerted only between nucleons quite close to each other. The nuclear building is fragile. At the slightest disturbance, and even spontaneously, it can be divided into several pieces: this is what is called *nuclear fission*.

5. For the nitrogen atom $^{14}_{7}N$, we have $A = 14$ and $Z = 7$. Since Z represents the number of protons in the nucleus, this atom contains seven protons. As the number A represents the number of nucleons (protons + neutrons), the number of neutrons in the nucleus is therefore $N = A - Z = 14 - 7 = 7$ neutrons. Since the atom is neutral, the number of protons is necessarily equal to the number of electrons, so the nitrogen atom contains seven electrons.

- For the Al^{3+} ion stemming from $^{27}_{13}Al$, we have $Z = 13$, so this ion contains 13 protons and $A - Z = 27 - 13 = 14$ neutrons. This ion is a cation that has $a + 3e$ charge. It therefore lacks three electrons with respect to the neutral atom $^{27}_{13}Al$. It therefore contains $Z - 3 = 10$ electrons.
- For the O^{2-} ion stemming from $^{16}_{8}O$, we have $Z = 8$, so this ion contains 8 protons and $A - Z = 16 - 8 = 8$ neutrons. This ion is an anion that has $a - 2e$ charge. It therefore has an excess of two electrons with respect to the neutral atom $^{16}_{8}O$. It therefore contains $Z + 2 = 10$ electrons.

Exercise 2

The nucleus of 1 sulfur atom contains 32 nucleons and 16 neutrons.
1. *Determine the value of the mass number A.*
2. *Deduce the value of the atomic number Z.*

Solution

1. *Value of the mass number A:*

The mass number is the number of nucleons contained in the nucleus of an atom. As a result, $A = 32$.

2. *Value of the atomic number Z:*

The atomic number of an atom is its charge number Z, that is to say the number of protons contained in its nucleus, i.e., $Z = A - N = 32 - 16 = 16$ protons. This is the atomic number of sulfur mentioned in the periodic table of Mendeleïev.

Exercise 3

The nucleus of an atom contains six protons and six neutrons.
1. *How many electrons this atom has?*
2. *What is the number of nucleons A in this atom?*
3. *What is its atomic number Z?*
4. *Identify this atom.*

Solution

1. *Number of electrons*:

Since there are $Z = 6$ protons in this atom and since it is electrically neutral, it necessarily contains 6 electrons.

2. *Number of nucleons*:

The number of nucleons is the sum of protons and neutrons confined in the nucleus: $A = Z + N = 6$ protons + 6 neutrons = 12 nucleons.

3. *Atomic number*:

The atomic number is the number of protons, $Z = 6$.

4. *Identification of the atom*:

Referring to the table of periodic elements of Mendeleïev, we find that it is carbon $^{12}_{6}C$.

Exercise 4

What is the number of neutrons, protons, and electrons present in each of the following atoms or ions and what is the corresponding chemical element?

$$^{55}_{25}Mn \qquad ^{40}_{18}Ar \qquad ^{207}_{82}Pb^{2+} \qquad ^{31}_{15}P^{3-}.$$

Solution

The resolution of this exercise is based on a simple arithmetic. It is based on the meaning of the two numbers A (mass number) and Z (mass number) associated with the symbol of the element $^{A}_{Z}X$ and on the relationship $A = Z + N$ which links the mass number A to the number of protons Z and to the number of neutrons N contained in the nucleus. As for the number of electrons that orbit around the nucleus, in an atom it is equal to that of protons Z, but in an ion, it is different from it, one or several electrons less if it is positively charged and one or several electrons in addition if it is negatively charged.

- *For manganese $^{55}_{25}Mn$*:

This atom contains $Z = 25$ protons and $N = A - Z = 55 - 25 = 30$ neutrons. Since the atom is neutral, there are as many electrons as there are protons, i.e., $Z = 25$ electrons.

- *For silver* $^{40}_{18}$Ar:

This atom contains $Z = 18$ protons and $N = A - Z = 40 - 18 = 22$ neutrons. Since the atom is neutral, there are as many electrons as there are protons, i.e., $Z = 18$ electrons.

- *For lead ion* $^{207}_{82}$Pb^{2+}:

This ion contains $Z = 82$ protons and $N = A - Z = 207 - 82 = 125$ neutrons. As the ion carries two positive charges, it therefore lacks two electrons with respect to the neutral atom; the number of electrons which orbit around the nucleus is thus $Z = 82 - 2 = 80$ electrons.

- *For phosphorus ion* $^{31}_{15}$P^{3-}:

This ion contains $Z = 15$ protons and $N = A - Z = 31 - 15 = 16$ neutrons. As the ion carries three negative charges, it therefore has an excess of three electrons with respect to the neutral atom; the number of electrons which orbit around the nucleus is thus $Z = 15 + 3 = 18$ electrons.

Exercise 5

Let us consider the aluminum atom $^{27}_{13}$Al.
1. *What is the mass of the nucleus of this atom?*
2. *What is the mass of electrons of this atom?*
3. *What is the mass of the atom? Conclude. We give:*
   - *Proton mass* $m_P = 1.67 \times 10^{-27}$ kg.
   - *Neutron mass* $m_N = 1.67 \times 10^{-27}$ kg.
   - *Mass of the electron* $m_e = 9.1 \times 10^{-31}$ kg.

Solution

1. *Mass of the nucleus of this atom*:

This atom contains $Z = 13$ protons and $A - Z = 27 - 13 = 14$ neutrons. The mass of the nucleus is the sum of masses of its different constituents:

$$m_{\text{nucleus}} = Z m_P + Z m_N = (13)(1.67 \times 10^{-27} \text{ kg}) + (14)(1.67 \times 10^{-27} \text{ kg})$$
$$= 4.509 \times 10^{-26} \text{ kg}. \tag{1}$$

2. *Mass of electrons*:

Since this atom is electrically neutral, it necessarily contains the same number of electrons as there are protons, i.e., 13 electrons. The mass of these electrons is therefore:

$$m_{\text{electrons}} = Zm_e = (13)\left(9.1 \times 10^{-31}\ \text{kg}\right) = 1.183 \times 10^{-29}\ \text{kg.} \qquad (2)$$

3. *Mass of the atom:*

The mass of the atom is the sum of the mass of the nucleus and that of electrons:

$$m_{\text{atom}} = m_{\text{nucleus}} + m_{\text{electrons}} = 4.409 \times 10^{-26}\ \text{kg} + 0.001183 \times 10^{-26}\ \text{kg}$$
$$= 4.510 \times 10^{-26}\ \text{kg.} \qquad (3)$$

By comparison between the mass of the atom (3) and that of the nucleus (1), we realize that these two masses are almost equal, meaning that almost all the mass of the atom is concentrated in the nucleus.

Exercise 6
The nucleus of the zinc atom (Zn) contains 64 nucleons and the electrical charge of the nucleus is $Q = +4.80 \times 10^{-18}$ C. What is the notation of this nuclide? We give the elementary charge $q = +1.6 \times 10^{-19}$ C.

Solution
The nucleus contains 64 nucleons. Its mass number is therefore $A = 64$. The electrical charge of the nucleus is that of the Z protons which compose it, each carrying an elementary charge q from which we deduce the number Z of protons:

$$Z = \frac{Q}{q} = \frac{4.80 \times 10^{-18}\ \text{C}}{1.6 \times 10^{-19}\ \text{C}} = 30\ \text{protons.} \qquad (1)$$

The notation of the nuclide is therefore:

$$^{64}_{30}\text{Zn.} \qquad (2)$$

Exercise 7
If we say about the bismuth ion Bi^{3+} that it has 127 neutrons, 83 protons, 81 electrons, and 210 nucleons, which of these data are certainly accurate, possibly exact, and certainly false knowing that the atomic mass of the natural element is 209 and its atomic number is 83.

Solution
Bismuth has for atomic number 83 so the number of protons is necessarily 83; otherwise we would have another chemical element. The ion Bi^{3+} has 3 electrons less than the neutral atom so that the number of 81 electrons given in the wording is necessarily false because the exact number of electrons that orbit around the nucleus is $Z = 83 - 3 = 80$ electrons. The number of neutrons ($N = 127$ neutrons) and the total number of nucleons ($A = 210$ nucleons) indicated in the wording can be possibly accurate if the isotope of bismuth 210 exists especially as we are not far from the atomic mass of the natural element which is 209.

Exercise 8

The nucleus of the nickel atom (Ni) contains 30 neutrons and the electrical charge of the nucleus is $Q = +4.48 \times 10^{-18}$ C. What is the notation of this nuclide? We give the charge $q = +1.6 \times 10^{-19}$ C.

Solution

The electrical charge of the nucleus is that of the Z protons that compose it, each carrying a charge q from which we deduce the number Z of protons:

$$Z = \frac{Q}{q} = \frac{4.48 \times 10^{-18} \text{ C}}{1.6 \times 10^{-19} \text{ C}} = 28 \text{ protons.} \tag{1}$$

Knowing that the number of neutrons is $N = 30$, the mass number is therefore $A = Z + N = 30 + 28 = 58$. The notation of the nuclide is therefore:

$$^{58}_{28}\text{Ni.} \tag{2}$$

Exercise 9

An X ion has 10 electrons and 12 protons.
1. Is it an anion or a cation?
2. Is the formula of this ion X^{2-} or X^{2+}?

Solution

1. *Nature of the ion*:

The ion X has two fewer electrons than the protons. This ion is therefore positively charged. It is a cation.

2. *Formula of the ion*:

$$X^{2+}. \tag{1}$$

Exercise 10

The natural element of iron consists of four isotopes:
- ^{54}Fe *Fe (6.04%), atomic mass $= 53.953$*
- ^{56}Fe *Fe (91.57%), atomic mass $= 55.948$*
- ^{57}Fe *Fe (2.11%), atomic mass $= 56.960$*
- ^{58}Fe *Fe (0.28%), atomic mass $= 57.959$*

What atomic mass can be predicted for natural iron? Knowing that the experimental atomic mass of natural iron is 55.85, what can you conclude?

Solution

To find the atomic mass of the natural iron, it is sufficient to calculate the weighted average of masses of the four isotopes:

$$m_{ave} = \frac{(53.953)(6.04)}{100} + \frac{(55.948)(91.57)}{100} + \frac{(56.960)(2.11)}{100}$$
$$+ \frac{(57.959)(0.28)}{100}$$
$$= 55.854. \tag{1}$$

This value is very close to the value of the atomic mass given by the experiment.

Exercise 11

The mass of one carbon atom is $m = 2 \times 10^{-23}$ g. What is the number of carbon atoms contained in one carbon pencil lead of the mass $M = 0.50$ g?

Solution

The number of required carbon atoms is:

$$n = \frac{M}{m} = \frac{0.5 \text{ g}}{2 \times 10^{-23} \text{ g}} = 0.25 \times 10^{23} \text{ atoms.} \tag{1}$$

Exercise 12

An ion has 16 protons and 18 electrons.
1. Is it an anion or a cation?
2. Express the charge of this ion as a function of the electron charge in coulombs (C).

Solution

1. Nature of the ion:

This ion has two electrons more than protons. It is therefore negatively charged. It is an anion.

3. Charge of the ion:

$$Q(\text{ion}) = -2\,e = -2\left(1.6 \times 10^{-19} \text{ C}\right) = -3.2 \times 10^{-19} \text{ C.} \tag{1}$$

Exercise 13

The natural lithium is a mixture of two isotopes ^{6}Li and ^{7}Li, of which atomic masses are 6.017 and 7.018, respectively. Its atomic mass is 6.943. What is the isotopic composition in % of each isotope?

Solution

Let x and y be the respective isotopic percentages of ^{6}Li and ^{7}Li. Starting from the exercise hypotheses, we obtain a system of two equations with two unknowns:

$$\{x + y = 1 \tag{1}$$

$$\{ 6.017\,x + 7.018\,y = 6.943. \tag{2}$$

From (1), we deduce:

$$x = 1 - y. \tag{3}$$

By injecting (3) into (2), we obtain:

$$6.017\,(1 - y) + 7.018\,y = 6.943 \;\Rightarrow\; y = 0.925 = 92.5. \tag{4}$$

By injecting (4) into (3), we find the value of x:

$$x = 1 - y = 1 - 0.925 = 0.075 = 7.5. \tag{5}$$

Exercise 14
Nuclei are characterized by the following pairs of values (Z,A):
(7,14); (14,28); (27,59); (13,27); (14,29); (7,15).
1. *Is a chemical element characterized by the value of the charge number Z or by the value of the mass number A?*
2. *Deduce the represented number of chemical elements.*
3. *Identify isotopes.*

Solution
1. *Characterization of a chemical element*:

A chemical element is characterized by the value of its charge number Z. This number is also called *atomic number*.

2. *Number of chemical elements*:

	Isotope 1	Isotope 2	Element name
Element 1	7.14	7.15	Nitrogen (N)
Element 2	13.27		Aluminum (Al)
Element 3	14.28	14.29	Silicon (Si)
Element 4	27.59		Cobalt (Co)

3. *Isotope identification*:
   Isotopes have the same number of protons and different numbers of neutrons:
   - Two isotopes of nitrogen: $^{14}_{7}\text{N}$ and $^{15}_{7}\text{N}$.
   - Two isotopes of silicon: $^{28}_{14}\text{Si}$ and $^{29}_{14}\text{Si}$.

Exercise 15
A button of copper jeans of mass M = 4.27 g contains n = 4 × 10^{22} atoms. What is the mass of one copper atom?

Solution
The mass of one atom of copper is:

$$m = \frac{M}{n} = \frac{4.27 \text{ g}}{4 \times 10^{22} \text{ atoms}} = 1.07 \times 10^{-22} \text{ g}. \tag{1}$$

Exercise 16

One neutral atom contains 16 neutrons and has a total charge of $Q = +2.56 \times 10^{-18}$ C.
1. What is the atomic number of the nucleus?
2. What is its mass number A?
3. What is the number of electrons that orbit around the nucleus? We give $e = +1.6 \times 10^{-19}$ C.

Solution

1. *Atomic number of the nucleus*:

The total positive charge Q is contained in the nucleus in the form of protons of which number is:

$$Z = \frac{Q}{e} = \frac{2.56 \times 10^{-18} \text{ C}}{1.6 \times 10^{-19} \text{ C}} = 16. \tag{1}$$

The charge number Z, also called *atomic number*, characterizes the chemical element which, in our case, is sulfur (S).

2. *Mass number A*:

The mass number is the sum of the number of protons and that of neutrons contained in the nucleus, i.e.: $A = Z + N = 16 + 16 = 32$.

3. *Number of electrons that revolve around the nucleus*:

As the atom is neutral, there are as many electrons as there are protons, that is to say $Z = 16$ electrons.

Exercise 17

1. Give the notation of the hydrogen nuclide and that of the uranium nuclide knowing that the first consists of only 1 proton and the second consists of 92 protons and 143 neutrons.
2. Calculate the mass m_H of the hydrogen atom and the mass m_U of the uranium atom.
3. Compare the masses of two atoms. Conclude. We give the mass of one proton $m_P = 1.673 \times 10^{-27}$ kg and that of one neutron $m_N = 1.675 \times 10^{-27}$ kg, and we neglect the mass of electrons.

Solution

1. • *Notation of the hydrogen nuclide*:

The nucleus of hydrogen contains only one proton and no neutrons so $Z = 1$ and $A = Z + N = 1 + 0 = 1$. Its notation is:

$$\mathstrut_1^1 \text{H}. \tag{1}$$

- *Uranium nuclide notation*:

    The nucleus of the uranium atom contains 92 protons and 143 neutrons $\Rightarrow$ $Z = 92$ and $A = Z + N = 92 + 143 = 235$. Its notation is:

$$\mathstrut_{92}^{235} \text{U}. \tag{2}$$

2. • *Mass m_H of the hydrogen atom*:

$$m_H = m_{\text{nucleus}} = Z m_P = (1)\left(1.673 \times 10^{-27} \text{ kg}\right) = 1.673 \times 10^{-27} \text{ kg}. \tag{3}$$

- *Mass m_U of the uranium atom*:

$$m_U = Z m_P + Z m_N = (92)\left(1.673 \times 10^{-27} \text{ kg}\right) + (143)\left(1.675 \times 10^{-27} \text{ kg}\right)$$
$$= 393.441 \times 10^{-27} \text{ kg}. \tag{4}$$

3. *Comparison of the masses of two atoms*:

    By comparing the masses of the uranium atom with the mass of the hydrogen atom, we obtain:

$$\frac{m_U}{m_H} = \frac{393.441 \times 10^{-27} \text{ kg}}{1.673 \times 10^{-27} \text{ kg}} = 235.17. \tag{5}$$

    The mass of the uranium atom is thus about 235 times greater than that of the hydrogen atom.

Exercise 18

A gold nugget weighs $M = 543$ g.
1. *Evaluate the mass m_{Au} of one gold atom of charge number $Z = 79$ and of mass number $A = 197$.*
2. *How many golden atoms this nugget contains? We give $m = m_P = m_N = 1.67 \times 10^{-27}$ kg, and we neglect the mass of electrons.*

Solution

1. *Mass m_{Au} of one atom of gold*:

    By neglecting the mass of electrons, we consider that almost all the mass of the atom is concentrated in the nucleus, of which the number of nucleons is hypothetically $A = 197$. In addition, the mass of the proton is taken equal to that of the neutron. On the basis of these data, the mass of the gold atom is:

$$m_{Au} = A \times m = (197)\left(1.67 \times 10^{-27} \text{ kg}\right) = 3.29 \times 10^{-25} \text{ kg}. \tag{1}$$

2. *Number of gold atoms in the nugget*:

It is given by:

$$n = \frac{M}{m_{Au}} = \frac{543 \times 10^{-3} \text{ kg}}{3.29 \times 10^{-25} \text{ kg}} = 1.65 \times 10^{24} \text{ atoms}. \tag{2}$$

Exercise 19

Calculate the mass of the bromine atom $^{79}_{35}Br$ and the mass of the corresponding ion Br^- in two ways:
1. *By neglecting the mass of electrons.*
2. *By taking into account the mass of electrons.*
   We give:
   • *Proton mass $m_P = 1.673 \times 10^{-27}$ kg.*
   • *Proton mass $m_N = 1.675 \times 10^{-27}$ kg.*
   • *Mass of the electron $m_e = 9.1 \times 10^{-31}$ kg.*

Solution
1. *Masses of the atom and the ion of bromine by neglecting the mass of electrons*:

Since the mass of electrons is negligible by hypothesis, the mass of the bromine atom is that of its nucleus containing $Z = 35$ protons and $A - Z = 79 - 35 = 44$ neutrons:

$$M_{Br} = Zm_P + Nm_N = (35)\left(1.673 \times 10^{-27} \text{ kg}\right) + (44)\left(1.675 \times 10^{-27} \text{ kg}\right)$$
$$= 132.26 \times 10^{-27} \text{ kg}. \tag{1}$$

The bromine ion Br^- is the bromine atom that has won one electron. As the mass of electrons is negligible by hypothesis, we can consider as a first approximation that the bromine ion and the bromine atom masses are equal.

2. *Masses of the bromine atom and the ion of bromine by taking into account the mass of electrons*:

The mass of the bromine atom is therefore that of its nucleus containing $Z = 35$ protons and $A - Z = 79 - 35 = 44$ neutrons + the mass of $Z = 35$ electrons:

$$M_{Br} = Zm_P + Nm_N + Zm_e = (35)\left(1.673 \times 10^{-27} \text{ kg}\right) + (44)\left(1.675 \times 10^{-27} \text{ kg}\right) +$$
$$(35)\left(9.1 \times 10^{-31} \text{ kg}\right) = 132.29 \times 10^{-27} \text{ kg}. \tag{2}$$

The bromine ion Br^- is the bromine atom that has gained one electron. Its mass is:

$$M_{Br^-} = M_{Br} + m_e = 132.29 \times 10^{-27} \text{ kg} + 9.1 \times 10^{-31} \text{ kg}$$
$$= 132.2909 \times 10^{-27} \text{ kg}. \tag{3}$$

The difference between the masses of the Br atom and the Br^- ion once calculated by neglecting the mass of electrons and a second time by taking into account the mass of these electrons is very small. This fully justifies the calculation of masses of these two quantities by neglecting the mass of electrons because almost all the mass of the atom is concentrated in its nucleus.

Exercise 20
The nucleus of the hydrogen atom is likened to a tennis ball of radius r = 3.3 cm. By preserving the proportions, what would be the corresponding radius R attributed to the atom of this element? We give:
- *Proton radius $r_P = 1.2 \times 10^{-15}$ m.*
- *Radius of the hydrogen atom $r_H = 53 \times 10^{-12}$ m.*

Solution
Since the hydrogen atom contains only one proton and no neutrons, the proton radius is that of the nucleus. By making a rule of three, we deduce the radius that the hydrogen atom would have on our scale in proportional terms:

$$R = \frac{r_H \cdot r}{r_P} = \frac{(53 \times 10^{-12} \text{ m}) (3.3 \text{ m})}{1.2 \times 10^{-15} \text{ m}} = 145.8 \times 10^3 \text{ m} = 145.8 \text{ km}. \tag{1}$$

We become aware through this result that the hydrogen atom is essentially made of vacuum!

Exercise 21
The mass of one iron bolt is M = 2.6 g. Calculate the number of iron atoms it contains knowing that the number of nucleons is A = 56. We consider that the masses of the proton and the neutron are equal $m = m_P = m_N = 1.67 \times 10^{-27}$ kg, and we neglect the mass of electrons.

Solution
Since the mass of electrons is negligible, the mass of the iron atom is concentrated in the nucleus:

$$m_{Fe} = A \times m = (56)(1.67 \times 10^{-27} \text{ kg}) = 9.35 \times 10^{-26} \text{ kg}. \tag{1}$$

The number of iron atoms contained in the bolt is:

$$n = \frac{M}{m} = \frac{2.6 \times 10^{-3} \text{ kg}}{9.35 \times 10^{-26} \text{ kg}} = 2.8 \times 10^{22} \text{ atoms.} \tag{2}$$

Exercise 22

The nucleus of one copper atom is represented by $^{63}_{29}Cu$.
1. *What is this atom made up of?*
2. *Calculate the mass of this nucleus.*
3. *Calculate the mass of the atom by neglecting the mass of electrons. What mistake do you make in making this approximation? We give:*
   - *Proton mass $m_P = 1.673 \times 10^{-27}$ kg.*
   - *Proton mass $m_N = 1.675 \times 10^{-27}$ kg.*
   - *Mass of the electron $m_e = 9.1 \times 10^{-31}$ kg.*

Solution

1. *Composition of the nucleus:*

The copper atom contains $Z = 29$ protons and $N = A - Z = 63 - 29 = 34$ neutrons. Since this atom is electrically neutral, there are as many protons as there are electrons. So the number of electrons is $Z = 29$ electrons.

2. *Mass of this nucleus:*

The mass of the nucleus is the mass of the nucleons that compose it:

$$m_{Cu} = Zm_P + Nm_N = (29)(1.673 \times 10^{-27} \text{ kg}) + (34)(1.675 \times 10^{-27} \text{ kg})$$
$$= 1.05 \times 10^{-25} \text{ kg.} \tag{1}$$

3. *Mass of the atom by neglecting the mass of electrons:*

By neglecting the mass of electrons, we consider that almost all the mass of the atom is concentrated in its nucleus. The mass of the iron atom is:

$$m_{\text{atom}} \approx m_{\text{nucleus}} = 1.05 \times 10^{-25} \text{ kg.} \tag{2}$$

In the previous calculation, we neglected the mass of electrons, the number of which is 29, that is to say a total mass of:

$$m_e = Zm_e = (29)(9.1 \times 10^{-31} \text{ kg}) = 2.6 \times 10^{-29} \text{ kg.} \tag{3}$$

The relative error committed by neglecting the mass of these electrons is estimated by reporting it to the mass of the atom:

$$\frac{m_e}{m_{Cu}} \times 100 = \frac{2.6 \times 10^{-29} \text{ kg}}{1.05 \times 10^{-25} \text{ kg}} \times 100 = 0.025\%. \tag{4}$$

The made error is acceptable!

Exercise 23

Let us consider one ion of charge $Q_{ion} = +4.8 \times 10^{-19}$ C. Its nucleus contains 28 neutrons and carries a total charge of $Q_{nucleus} = +3.84 \times 10^{-18}$ C.
1. *What is the atomic number of the nucleus? Identify the chemical element.*
2. *What is its number of nucleons?*
3. *How many electrons orbit around the nucleus and what is the symbol of the ion? We give the electron charge $e = +1.6 \times 10^{-19}$ C.*

Solution

1. *Atomic number of the nucleus*:

The atomic number is the charge number Z of the nucleus corresponding to the number of protons contained in this nucleus:

$$Z = \frac{Q_{nucleus}}{e} = \frac{3.84 \times 10^{-18} \text{ C}}{1.6 \times 10^{-19} \text{ C}} = 24. \tag{1}$$

The chemical element is chromium (Cr).

2. *Number of nucleons*:

The number of nucleons is the sum of protons and neutrons contained in the nucleus: $A = Z + N = 24 + 28 = 52$.

3. *Number of the electrons that orbit around the nucleus*:

As the ion is positively charged, it comes from an atom that has lost electrons, of which the number is given by:

$$n = \frac{Q_{ion}}{e} = \frac{4.8 \times 10^{-19} \text{ C}}{1.6 \times 10^{-19} \text{ C}} = 3 \text{ lost electrons.} \tag{2}$$

The number of electrons that orbit around the nucleus is therefore $Z - 3 = 24 - 3 = 21$ electrons. The ion is symbolized by Cr^{3+}.

Exercise 24

Calculate the mass of one electron, one proton, one neutron, and one α-particle (helium atom) in unified atomic mass unit (u), joules (J), and MeV/c^2. What is the unit of 1 u in MeV/c^2. We give:
- *Proton mass $m_P = 1.673 \times 10^{-27}$ kg.*
- *Neutron mass $m_N = 1.675 \times 10^{-27}$ kg.*
- *Mass of the electron $m_e = 9.109 \times 10^{-31}$ kg.*

Solution

The unit of mass in the international system is the kilogram (kg). This unit is far too big for infinitesimal particles. That is why we use two other units better adapted to the physical phenomena observed on this scale.

• *Unified atomic mass unit (u):*

The unified atomic mass unit u is defined as being one-twelfth (1/12) of the mass of one carbon atom 12. However, 1 mole of this carbon weighs 12 g, i.e.:

$$1 \text{ u} = \frac{12 \times 10^{-3} \text{ kg}}{(6.0221 \times 10^{23})(12)} = 1.661 \text{ } 10^{-27} \text{ kg}. \tag{1}$$

• *Unit MeV/c²:*

In restricted relativity, Einstein has shown that at any mass m_0 corresponds an energy $E = m_0 c^2$ called *rest mass energy* where E is in joules, m_0 in kilograms, and c, the speed of light, in m/s. From this equation, we deduce:

$$m_0 = \frac{E}{c^2}. \tag{2}$$

Equation (2) shows that a mass can be expressed by an energy divided by c^2. Although the joule is the SI unit of energy, in nuclear physics, the used unit is the electronvolt (eV) because it is better adapted to nuclear phenomena with:

$$1 \text{ eV} = 1.6 \text{ } 10^{-19} \text{ J} \Rightarrow 1 \text{ J} = 0.625 \text{ } 10^{19} \text{ eV} = 0.625 \text{ } 10^{13} \text{ MeV}. \tag{3}$$

On the basis of (2) and (3) and to determine the mass of one particle in MeV/c², it suffices to calculate this mass in joules then to convert it into MeV with $1 \text{ MeV} = 10^6 \text{ eV}$.

• *Proton mass $m_P = 1.673 \times 10^{-27}$ kg:*

   – *In unified atomic mass unit (u):*

$$m_P = \frac{(1.673 \times 10^{-27} \text{ kg}) (1 \text{ u})}{1.661 \times 10^{-27} \text{ kg}} = 1.00722 \text{ u}. \tag{4}$$

   – *In joules (J):*

$$E = m_P c^2 = (1.673 \times 10^{-27} \text{ kg}) (3 \times 10^8 \text{ m.s}^{-1})^2 = 15.057 \times 10^{-11} \text{ J}. \tag{5}$$

- *In (MeV/c^2):*

$$m_P = \frac{(0.625 \times 10^{13} \text{ MeV}) (15.057 \times 10^{-11} \text{ J})}{1 \text{ J}} = 941.06 \text{ MeV}. \qquad (6)$$

By using (2), we find:

$$m_P = \frac{E}{c^2} = 941.06 \frac{\text{MeV}}{c^2}. \qquad (7)$$

- The neutron mass in kilograms is $m_N = 1.675 \times 10^{-27}$ kg; we obtain:
  - *In unified atomic mass unit (u):*

$$m_N = \frac{(1.675 \times 10^{-27} \text{ kg}) (1 \text{ u})}{1.661 \times 10^{-27} \text{ kg}} = 1.00843 \text{ u}. \qquad (8)$$

- *In joules (J):*

$$E = m_N c^2 = (1.675 \times 10^{-27} \text{ kg}) (3 \times 10^8 \text{ m.s}^{-1})^2 = 15.075 \times 10^{-11} \text{ J}. \qquad (9)$$

- *In (MeV/c^2):*

$$m_N = \frac{(0.625 \times 10^{13} \text{ MeV}) (15.075 \times 10^{-11} \text{ J})}{1 \text{ J}} = 942.19 \text{ MeV}. \qquad (10)$$

By using (2), we get:

$$m_N = \frac{E}{c^2} = 942.19 \frac{\text{MeV}}{c^2}. \qquad (11)$$

- *The mass of one electron in kilograms is $m_e = 9.109 \times 10^{-31}$ kg:*
  - *In unified atomic mass unit (u):*

$$m_e = \frac{(9.109 \times 10^{-31} \text{ kg}) (1 \text{ u})}{1.661 \times 10^{-27} \text{ kg}} = 5.4841 \times 10^{-4} \text{ u}. \qquad (12)$$

- *In joules (J):*

$$E = m_e c^2 = (9.109 \times 10^{-31} \text{kg}) (3 \times 10^8 \text{ m.s}^{-1})^2 = 81.891 \times 10^{-15} \text{ J}. \qquad (13)$$

– *In (MeV/c²):*

$$m_e = \frac{\left(0.625 \times 10^{13} \text{ MeV}\right) \left(81.891 \times 10^{-15} \text{ J}\right)}{1 \text{ J}} = 0.522 \text{ MeV}. \qquad (14)$$

By using (2), we obtain:

$$m_e = \frac{E}{c^2} = 0.522 \frac{\text{MeV}}{c^2}. \qquad (15)$$

• *Mass of one α-particle:*

The α-particle is the helium atom that has lost its two electrons. Its nucleus is formed with two protons and two neutrons, of which the mass is:

$$m_a = Zm_P + Nm_N = (2)\left(1.673 \times 10^{-27} \text{ kg}\right) + (2)\left(1.675 \times 10^{-27} \text{ kg}\right)$$
$$= 6.696 \times 10^{-27} \text{ kg}. \qquad (16)$$

– *In unified atomic mass unit (u):*

$$m_a = \frac{\left(6.696 \times 10^{-27} \text{ kg}\right) (1 \text{ u})}{1.661 \times 10^{-27} \text{ kg}} = 4.0313 \times 10^{-4} \text{ u}. \qquad (17)$$

– *In joules (J):*

$$E = m_a c^2 = \left(6.696 \times 10^{-27} \text{ kg}\right) \left(3 \times 10^8 \text{ m.s}^{-1}\right)^2 = 60.264 \times 10^{-11} \text{ J}. \qquad (18)$$

– *In (MeV/c²):*

$$m_a = \frac{\left(0.625 \times 10^{13} \text{ MeV}\right) \left(60.264 \times 10^{-11} \text{ J}\right)}{1 \text{ J}} = 3766.5 \text{ MeV}. \qquad (19)$$

By using (2), we obtain:

$$m_a = \frac{E}{c^2} = 3766.5 \frac{\text{MeV}}{c^2}. \qquad (20)$$

- *One u in MeV/c²:*

  - *In joules (J):*

$$E = m_a c^2 = \left(1.661 \times 10^{-27} \text{ kg}\right) \left(3 \times 10^8 \text{ m.s}^{-1}\right)^2 = 14.949 \times 10^{-11} \text{ J.} \quad (21)$$

  - *In (MeV/c²):*

$$m_\alpha = \frac{\left(0.625 \times 10^{13} \text{ MeV}\right) \left(14.949 \times 10^{-11} \text{ J}\right)}{1 \text{ J}} = 934.31 \text{ MeV.} \quad (22)$$

By using (2), we find:

$$m_\alpha = \frac{E}{c^2} = 934.31 \frac{\text{MeV}}{c^2}. \quad (23)$$

1.7.2   Atomic Models

Exercise 25

In the Bohr model, the atom is a two-point system consisting of a nucleus, one proton of mass m_P and of charge +e, and one electron M, of mass m_e and of charge −e. Since the proton is much heavier than the electron, we consider it as fixed in the supposed Galilean study referential $R\left(O, \overrightarrow{e_x}, \overrightarrow{e_y}, \overrightarrow{e_z}\right)$ where the origin O coincides with the nucleus of the atom.

I. *First postulate of Bohr: The electron moves only in certain circular orbits called stationary states. The electron has a circular motion of radius r and speed v around the nucleus. The gravitational field is negligible at the atomic scale and the electron is subjected only to the force of electrostatic interaction:*

$$\overrightarrow{F} = -\frac{e^2}{4\pi\varepsilon_0 r^2} \overrightarrow{e_r}.$$

1. *Show that the circular motion of the electron around the nucleus is uniform and express v^2 as a function of r, e, m_e, and ε_0.*
2. *Express the kinetic energy E_C, the potential energy E_P, and the mechanical energy E of the electron.*

II. *Second and third postulates of Bohr: The accelerated electron by the proton cannot radiate continuously, but must wait to jump from a permitted orbit n to another orbit of lower energy m to emit a radiation in the form of a photon of energy $h\nu_{n\rightarrow m} = E_n - E_m$ (with $n > m$) where E_n and E_m are the energies of the two states n and m, h is Planck's constant, and $\nu_{n\rightarrow m}$ is the frequency of radiation corresponding to the transition $n \rightarrow m$. To quantify the energy of the* electron, Bohr adds a third postulate or quantization condition: the only allowed circular trajectories are those for which the orbital kinetic momentum is an integer multiple of the reduced Planck's constant $\hbar$:

$$\sigma(M) = n\hbar = n\,\frac{h}{2\pi}.$$

3. Determine the speed v of the electron as a function of r, m_e, h, and the main quantum number n ($n \geq 1$).
4. The radii of circular orbits in which the electron may travel increase as n^2 such as $n = n^2 r_1$, where r_1 is the Bohr radius. Calculate in picometer (pm) the radius r_1.
5. Deduce the total energy of the quantized electron in the form:

$$E_n = -\frac{E_1}{n^2}.$$

6. By assuming the electron in its fundamental state ($n = 1$), calculate its speed v_1 and the ionization energy of the atom in eV. Is the electron relativistic?
7. Determine the literal expression of the R_H constant of Rydberg relative to the hydrogen atom and calculate its value knowing that:

$$\frac{1}{\lambda_{n \to m}} = \frac{\nu_{n \to m}}{c} = R_H \left(\frac{1}{m^2} - \frac{1}{n^2} \right)$$

with $n > m$, and c, the speed of light in vacuum. We give:
- Planck's constant $h = 6.62 \times 10^{-34}$ J.s.
- Speed of light $c = 3 \times 10^8$ m/s.
- Mass of the electron $m_e = 9.1 \times 10^{-11}$ kg and charge of the electron

  $e = 1.6 \times 10^{-19}$ C.
- Permittivity of vacuum $\varepsilon_0 = 8.854 \times 10^{-12}$ F/m.
- 1 eV $= 1.6 \times 10^{-19}$ J and 1 pm $= 10^{-12}$ m.

Solution
1. Nature of the electron movement and $v^2 = f(r,e,m_e,\varepsilon_0)$:

The considered system is the electron which orbits around the nucleus of a circular motion in a supposed Galilean terrestrial reference R_g. In any circular motion, the radius r of the trajectory is constant. In the Frenet coordinate system $\left(\vec{n}, \vec{\tau} \right)$, the vector-speed of the circular motion is tangent to the trajectory:

$$\vec{v} = v\,\vec{\tau}. \tag{1}$$

The vector-acceleration has two components:

$$\vec{a} = \frac{d\vec{v}}{dt} = a_\tau \vec{\tau} + a_n \vec{n} = \frac{dv}{dt}\vec{\tau} + \frac{v^2}{r}\vec{n} = r\ddot{\theta}\,\vec{\tau} + r\dot{\theta}^2\,\vec{n} \qquad (2)$$

where $\dot{\theta}$ and $\ddot{\theta}$ are the angular speed and the angular acceleration, respectively. The vector-acceleration is always directed towards the interior of the trajectory. The value of the tangential component a_τ can be positive, negative, or zero. The normal component a_n can be positive or zero. During its rotation around the nucleus, the electron is subjected to the Coulomb central force and to its weight.

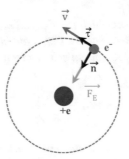

Since the weight of the electron is negligible compared to the Coulomb force, let us apply the fundamental equation of dynamics:

$$\sum \vec{F} = \frac{e^2}{4\pi\varepsilon_0 r^2}\vec{n} = m_e\,\vec{a} = m_e\left(\frac{dv}{dt}\vec{\tau} + \frac{v^2}{r}\vec{n}\right) = m_e\left(r\ddot{\theta}\,\vec{\tau} + r\dot{\theta}^2\,\vec{n}\right). \qquad (3)$$

- By the projection of (3) on the axis $\vec{\tau}$ of the Frenet-Serret frame, we obtain:

$$\frac{dv}{dt} = 0 \Rightarrow v = r\dot{\theta} = \text{Cte.} \qquad (4)$$

Equation (4) shows that the modulus of the linear speed $|\vec{v}|$ is constant as well as the angular speed $\dot{\theta}$ since the radius of the trajectory is constant $\Rightarrow$ the electron orbits around the nucleus in a uniform circular movement. The vector-acceleration is, in this case, directed towards the nucleus:

$$\vec{a} = \frac{v^2}{r}\vec{n} = r\dot{\theta}^2\,\vec{n}. \qquad (5)$$

- By the projection of (3) on the axis $\vec{n}$ of the Frenet-Serret frame, we find $v = f(r, e, m_e, \varepsilon_0)$:

$$\frac{e^2}{4\pi\varepsilon_0 r^2} = \frac{m_e v^2}{r} \implies v = \frac{e}{\sqrt{4\pi\varepsilon_0 m_e r}}. \tag{6}$$

2. *Expression of the kinetic energy E_C, the potential energy E_P, and the mechanical energy E of the electron:*

- *Kinetic energy E_C:*

  The kinetic energy of the electron in the frame (R_g) is:

$$E_C = \frac{1}{2} m_e v^2. \tag{7}$$

By starting from (6) and on the basis of (7), we obtain:

$$m_e v^2 = \frac{e^2}{4\pi\varepsilon_0 r} \implies E_C = \frac{e^2}{8\pi\varepsilon_0 r}. \tag{8}$$

- *Potential energy E_P:*

To determine the potential electrostatic energy, it is necessary to start from the elementary work provided by the Coulomb force $\vec{F}$:

$$dW\left(\vec{F}\right) = \vec{F}.d\overrightarrow{OM} = -\vec{F}.d\left(r\vec{n}\right) = \left(\frac{e^2}{4\pi\varepsilon_0 r^2}\vec{n}\right) \cdot \left(dr\,\vec{n} + rd\,\vec{\tau}\right)$$

$$= -\frac{e^2}{4\pi\varepsilon_0 r^2} dr = -dE_P.. \tag{9}$$

By integration of (9), we get:

$$E_P = -\frac{e^2}{4\pi\varepsilon_0 r} + \text{Cte}. \tag{10}$$

By taking $E_P(r \to \infty) = 0$, it becomes:

$$E_P = -\frac{e^2}{4\pi\varepsilon_0 r}. \tag{11}$$

- *Mechanical energy E:*

  The mechanical energy is given by:

$$E = E_C + E_P = \frac{e^2}{8\pi\varepsilon_0 r} - \frac{e^2}{4\pi\varepsilon_0 r} = -\frac{e^2}{8\pi\varepsilon_0 r}. \tag{12}$$

3. *Speed v of the electron as a function of r, m_e, h, and the main quantum number n*
   (n ≥ 1):

   The kinetic moment of the electron evaluated at the point O is:

   $$\vec{\sigma}(M) = \overrightarrow{OM} \times m_e \vec{v} = r\,\vec{n} \times m_e\,v\,\vec{\tau} = m_e r\,v\,\vec{k}. \tag{13}$$

   Since this kinetic moment is according to the Bohr postulate quantized, we get:

   $$\sigma(M) = m_e r\,v = n\frac{h}{2} \quad \Rightarrow \quad v = n\frac{h}{2\pi m_e r}. \tag{14}$$

4. *Calculation of the Bohr radius:*

   By equaling (6) and (14), we obtain:

   $$v = \frac{e}{\sqrt{4\pi\varepsilon_0 m_e r}} = n\frac{h}{2\pi m_e r} \quad \Rightarrow \quad r = n^2\frac{\varepsilon_0\,h^2}{\pi m_e e^2} = n^2\,r_1. \tag{15}$$

   We deduce the Bohr radius which corresponds to the trajectory of the electron in its fundamental state $n = 1$:

   $$r_1 = \frac{h^2\varepsilon_0}{\pi m_e e^2} = \frac{\left(8.854 \times 10^{-12}\ \text{F.m}^{-1}\right)\left(6.626 \times 10^{-34}\ \text{J.s}\right)^2}{(3.14)\left(9.1 \times 10^{-31}\ \text{kg}\right)\left(1.6 \times 10^{-19}\ \text{C}\right)^2}$$
   $$= 53.1 \times 10^{-12}\ \text{m} = 53.1\ \text{pm}. \tag{16}$$

5. *Quantized total energy of the electron:*

   By starting from (12) and (15), we find:

   $$E = -\frac{e^2}{8\pi\varepsilon_0 r} = -\frac{e^2}{8\pi\varepsilon_0 n^2\dfrac{\varepsilon_0\,h^2}{\pi m_e e^2}} = -\frac{1}{n^2}\frac{m_e e^4}{8\varepsilon_0^2 h^2} = -\frac{E_1}{n^2}. \tag{17}$$

6. *Calculation of the ionization energy of the atom in eV and the speed v_1:*

   • *Ionization energy of the atom in eV:*

   The ionization energy is the energy that must be supplied to tear the electron off its electronic shell:

$$E_{\text{ionization}} = E(n \to \infty) - E(n = 1) = E_1 = \frac{m_e e^4}{8\varepsilon_0^2 h^2}$$

$$= \frac{\left(9.1 \times 10^{-31} \text{ kg}\right) \left(1.6 \times 10^{-19} \text{ C}\right)^4}{8 \left(8.854 \times 10^{-12} \text{ F.m}^{-1}\right)^2 \left(6.626 \times 10^{-34} \text{ J.s}\right)^2} = 2.166 \times 10^{-18} \text{J}$$

$$= \frac{(1 \text{ eV}) \left(2.166 \times 10^{-18} \text{ J}\right)}{1.6 \times 10^{-19} \text{ J}} = 13.6 \text{ eV}.$$

$$(18)$$

- *Speed v_1:*

  According to (15), the speed of the electron in the ground state ($n = 1$) is:

$$v_1 = \frac{h}{2\pi m_e r_1} = \frac{6.626 \times 10^{-34} \text{ J.s}}{(2)(3.14)\left(9.1 \times 10^{-31} \text{ kg}\right)\left(53.1 \times 10^{-12} \text{ m}\right)}$$

$$= 2.2 \times 10^6 \text{ m.s}^{-1}.$$

$$(19)$$

- The ratio of the speed of the electron v_1 to the speed of light c gives:

$$\frac{v_1}{c} = \frac{2.2 \times 10^6 \text{ m.s}^{-1}}{3 \times 10^8 \text{ m.s}^{-1}} = 7.3 \times 10^{-3}.$$

$$(20)$$

This speed is far from the speed of light $\Rightarrow$ the electron is not relativistic.

7. *Literal expression of the constant of Rydberg R_H relative to the hydrogen atom and calculation of its value:*

By starting from (17), when the hydrogen atom in the level n of energy $E_n = -\dfrac{E_1}{n^2}$ de-excites and jumps to the lower level m of energy $E_m = -\dfrac{E_1}{m^2}$, it releases a photon of energy $h\nu_{n \to m}$ such that:

$$h\nu_{n \to m} = E_n - E_m = E_1 \left(\frac{1}{m^2} - \frac{1}{n^2}\right) = h\frac{c}{\lambda_{n \to m}}.$$

$$(21)$$

The inverse of (21) gives:

$$\frac{1}{\lambda_{n \to m}} = \frac{E_1}{hc}\left(\frac{1}{m^2} - \frac{1}{n^2}\right) = R_H \left(\frac{1}{m^2} - \frac{1}{n^2}\right).$$

$$(22)$$

Here R_H is the Rydberg constant, of which the value is:

$$R_H = \frac{E_1}{hc} = \frac{m_e e^4}{8\varepsilon_0^2 h^3 c}$$

$$= \frac{(9.1 \times 10^{-31}\,\text{kg})(1.6 \times 10^{-19}\,\text{C})^4}{8(8.854 \times 10^{-12}\,\text{F.m}^{-1})^2(6.626 \times 10^{-34}\,\text{J.s})^3(3 \times 10^8\,\text{m.s}^{-1})} \tag{23}$$

$$= 1.09 \times 10^7\,\text{m}^{-1}.$$

Exercise 26

1. *What is a hydrogen-like ion?*
2. *Establish for a hydrogen atom the formulas giving:*
   (a) *The radius of the orbit of rank n.*
   (b) *The energy of the electron associated with the orbit n.*
   (c) *Express the radius and the total energy of rank n for the hydrogen-like ion as a function of same quantities relative to the hydrogen atom.*
3. (a) *What is the energy of level 1 of lithium?*
   (b) *What is the energy of level n of lithium as a function of the energy of level 1 of the same ion?*
   (c) *Calculate in electronvolts (eV) and joules (J) the energy of the first four levels of the hydrogen-like ion Li^{2+}.*
4. *What energy must an ion Li^{2+} absorb so that the electron jumps from the fundament level to the first excited level?*
5. *If this energy is supplied in luminous form, what is the wavelength λ_{1-2} of the radiation capable of causing this transition? Which electromagnetic domain does it belong to? We give:*
   • *Charge number of lithium $Z = 3$.*
   • *Planck's constant $h = 6.62 \times 10^{-34}$ J.s.*
   • *Speed of light $c = 3 \times 10^8$ m/s.*
   • *Mass of the electron $m_e = 9.1 \times 10^{-11}$ kg.*
   • *Charge of the electron $e = 1.6 \times 10^{-19}$ C.*
   • *Permittivity of vacuum $\varepsilon_0 = 8.854 \times 10^{-12}$ F/m.*
   • *$1\,eV = 1.6 \times 10^{-19}$ J*

Solution

1. *Definition of the hydrogen-like ion:*

A hydrogen-like ion is a cation that has one electron. It has a structure similar to that of the hydrogen atom, except for the charge of its nucleus Ze instead of e for hydrogen and where Z is the atomic number (charge number) of the chemical element and e the electron charge. It is therefore an atom to which we have torn all its electrons except one.

2. (a) *Radius of the orbit of rank n*:

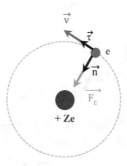

The considered system is the electron which orbits around the nucleus of a uniform circular movement. In any circular movement, the radius of the trajectory r is constant. Let ω be the angular speed of a mobile M. In the Frenet-Serret frame $\left(\vec{n}, \vec{\tau} \right)$, the vector-speed of the circular movement has the following expression:

$$\vec{v} = v\,\vec{\tau}.\tag{1}$$

It is therefore tangent to the trajectory. The vector-acceleration has the following components:

$$\vec{a} = \frac{d\vec{v}}{dt} = a_{\tau}\vec{\tau} + a_n\,\vec{n} = \frac{dv}{dt}\vec{\tau} + \frac{v^2}{r}\vec{n}.\tag{2}$$

The vector-acceleration is always directed towards the interior of the trajectory. The value of the tangential component a_{τ} can be positive, negative, or zero. The normal component a_n can be positive or zero. In a uniform circular movement (the case of our exercise), the speed modulus is constant:

$$\mid \vec{v} \mid = v = \text{Cte} \quad \Rightarrow \quad \frac{dv}{dt} = 0.\tag{3}$$

By injecting (3) into (2), we obtain the expression of the vector-acceleration $\vec{a}$:

$$\vec{a} = \frac{d\vec{v}}{dt} = \frac{v^2}{r}\,\vec{n}.\tag{4}$$

Equation (4) shows that the vector-acceleration is not zero, although the speed modulus is zero. This is due to the change of orientation of the vector-speed over time. The vector-acceleration is directed towards the circle center. During its rotation around the nucleus, the electron is subjected to the Coulomb force. Let us apply the fundamental equation of dynamics:

$$\sum \vec{F} = m_e\,\vec{a} \text{ that is to say: } \frac{Ze^2}{4\pi\varepsilon_0 r^2}\,\vec{n} = \frac{m_e v^2}{r}\,\vec{n}. \tag{5}$$

By the projection of (5) on the Frenet-Serret frame, we obtain:

$$\frac{Ze^2}{4\pi\varepsilon_0 r^2} = \frac{m_e v^2}{r} \quad \Rightarrow \quad \frac{Ze^2}{4\pi\varepsilon_0} = m_e v^2 r. \tag{6}$$

To explain the discontinuous spectrum of the hydrogen atom and to understand why the electron does not fall on the nucleus in spiral by loss of energy, Bohr formulates two hypotheses:

(i) The electron revolves in a permitted orbit.
(ii) There is an energy absorption if $E = E_2 - E_1$ and an energy emission if $E = E_1 - E_2$ $(E_2 > E_1)$. These two hypotheses lead to the quantization of the orbital kinetic momentum:

$$\sigma = m_e v r = n\hbar = n\frac{h}{2\pi} \quad \Rightarrow \quad m_e^2 v^2 r^2 = n^2 \frac{h^2}{4\pi^2}. \tag{7}$$

By injecting (7) into (6), we arrive at the expression of the radius of the orbit of rank n:

$$\frac{Ze^2}{4\pi\varepsilon_0} = \frac{m_e^2 v^2 r^2}{m_e r} = n^2 \frac{h^2}{4\pi^2 m_e r} \quad \Rightarrow \quad r_n = \frac{n^2}{Z}\left(\frac{h^2 \varepsilon_0}{\pi m_e e^2}\right). \tag{8}$$

2. (b) *Energy of the electron associated with the orbit n:*

The mechanical energy E of the electron is the sum of its kinetic energy E_C and its potential energy E_P:

$$E = E_C + E_P = \frac{1}{2}m_e v^2 - \frac{Ze^2}{4\pi\varepsilon_0 r}. \tag{9}$$

By injecting (6) into (9), we find:

$$E = \frac{1}{2}\times\frac{Ze^2}{4\pi\varepsilon_0 r} - \frac{Ze^2}{4\pi\varepsilon_0 r} = -\frac{Ze^2}{8\pi\varepsilon_0 r}. \tag{10}$$

By replacing r_n with its expression (8) in (10), we obtain:

$$E_n = -\frac{Ze^2}{8_0} \times \frac{1}{r_n} = -\frac{Ze^2}{8\pi\varepsilon_0} \times \frac{1}{n^2 \left(\dfrac{h^2\varepsilon_0}{\pi m_e e^2}\right)} = -\left(\frac{Z^2}{n^2}\right)\left(\frac{m_e e^4}{8\varepsilon_0^2 h^2}\right). \tag{11}$$

2. (c) *Expression of the radius and the total energy of rank n for the hydrogen-like ion as a function of same quantities relative to the hydrogen atom:*

- *Radius:*

The radius of the first orbit of the hydrogen atom is deduced from (8) for $n = 1$ and $Z = 1$:

$$r_1^H = \frac{h^2\varepsilon_0}{\pi m_e e^2} = \frac{(6.626 \times 10^{-34} \text{ J.s})^2 (8.854 \times 10^{-12} \text{ F.m}^{-1})}{(3.14)(9.1 \times 10^{-31} \text{ kg})(1.6 \times 10^{-19} \text{ C})^2}$$

$$= 5.31 \times 10^{-11} \text{ m} = 0.53\text{Å}. \tag{12}$$

The radius of the orbit n of the hydrogen-like ion is:

$$r_n = \frac{n^2}{Z} r_1^H = \left(\frac{n^2}{Z}\right) 0.53\text{Å}. \tag{13}$$

- *Energy:*

The energy of the first orbit of the hydrogen atom is deduced from (11) for $n = 1$ and $Z = 1$:

$$E_1^H = -\frac{m_e e^4}{8\varepsilon_0^2 h^2}$$

$$= -\frac{(9.1 \times 10^{-31} \text{ kg})(1.6 \times 10^{-19} \text{ C})^4}{8 (8.854 \times 10^{-12} \text{ F.m}^{-1})^2 (6.626 \times 10^{-34} \text{ J.s})^2} \tag{14}$$

$$= 2.166 \times 10^{-18} \text{ J} = \frac{(1 \text{ eV})(2.166 \times 10^{-18} \text{ J})}{1.6 \times 10^{-19} \text{ J}}$$

$$= -13.6 \text{ eV}.$$

The energy of the orbit n of the hydrogen-like ion is:

$$E_n = \left(\frac{Z^2}{n^2}\right) E_1^H = \left(\frac{Z^2}{n^2}\right)(-13.6 \text{ eV}). \tag{15}$$

3. (a) *Level 1 lithium energy*:

The lithium charge number being $Z = 3$, the energy of lithium of level 1 is, by starting from (15):

$$E_1^{Li^{2+}} = \left(\frac{3^2}{1^2}\right) E_1^H = (-13.6 \text{ eV}) \left(3^2\right) = -122.4 \text{ eV}. \tag{16}$$

3. (b) *Energy of the level n of lithium as a function of the energy of level 1 of the same ion*:

The energy of the level n of lithium as a function of the energy of level 1 of the same ion is:

$$E_n^{Li^{2+}} = \frac{E_1^{Li^{2+}}}{n^2} = -\frac{122.4 \text{ eV}}{n^2}. \tag{17}$$

3. (c) *Calculation, in electronvolts (eV) and joules (J), of the energy of the first four levels of the hydrogen-like ion* Li^{2+}:

- *Orbit* $n = 1$:

By starting from (16), we obtain:

$$E_1^{Li^{2+}} = -122.4 \text{ eV} = -\frac{(122.4 \text{ eV}) \left(1.6 \times 10^{-19} \text{ J}\right)}{1 \text{ eV}} = -19.58 \, 10^{-18} \text{ J}. \tag{18}$$

- *Orbit* $n = 2$:

By starting from (17) we get:

$$E_2^{Li^{2+}} = -\frac{122.4 \text{ eV}}{2^2} = -30.6 \text{ eV}. \tag{19}$$

By starting from (17) and (18), we find:

$$E_2^{Li^{2+}} = -\frac{-19.58 \times 10^{-18} \text{ J}}{2^2} = -4.90 \times 10^{-18} \text{ J}. \tag{20}$$

- *Orbit* $n = 3$:

By starting from (17), we get:

$$E_2^{Li^{2+}} = -\frac{122.4 \text{ eV}}{3^2} = -13.60 \text{ eV}. \tag{21}$$

By starting from (17) and (18), we obtain:

$$E_2^{Li^{2+}} = -\frac{-19.58 \times 10^{-18} \text{ J}}{3^2} = -2.18 \ 10^{-18} \text{ J}. \tag{22}$$

- *Orbit* $n = 4$:

By starting from (17), we obtain:

$$E_2^{Li^{2+}} = -\frac{122.4 \text{ eV}}{4^2} = -7.65 \text{ eV}. \tag{23}$$

By starting from (17) and (18), we get:

$$E_2^{Li^{2+}} = -\frac{-19.58 \times 10^{-18} \text{ J}}{4^2} = -1.22 \times 10^{-18} \text{ J}. \tag{24}$$

4. *Absorbed energy by an ion* Li^{2+} *so that the electron jumps from the fundamental level to the first excited level:*

The absorbed energy so that the electron jumps from the orbit $n = 1$ to the orbit $n = 2$ is:

$$\Delta E_{1 \to 2} = E_2 - E_1 = -30.6 \text{ eV} - (-122.4 \text{ eV}) = +91.8 \text{ eV}. \tag{25}$$

This energy is positive because it is won by the ion.

5. *Wavelength* λ_{1-2} *of the radiation capable of causing the transition* $1 \to 2$:

During the transition, the ion absorbs a light photon, of which the energy is given by the Einstein equation:

$$
\begin{aligned}
h\nu_{1 \to 2} = \Delta E_{1 \to 2} &= \frac{hc}{\lambda_{1 \to 2}} \Rightarrow \lambda_{1 \to 2} = \frac{hc}{\Delta E_{1 \to 2}} \\
&= \frac{\left(6.62 \times 10^{-34} \text{ J.s}\right)\left(3 \times 10^8 \text{ m.s}^{-1}\right)(1 \text{ eV})}{(91.8 \text{ eV})\left(1.6 \times 10^{-19} \text{ J}\right)} = 1.35 \times 10^{-8} \text{ m} \\
&= 13.5 \text{ nm}.
\end{aligned}
\tag{26}
$$

- The incident source emits in the extreme ultraviolet.

Exercise 27

1. *The emission spectrum of the hydrogen atom is composed of several series of lines.*
   (a) *Establish the formula giving* $1/\lambda_{i-j}$ *where* λ_{i-j} *represents the wavelength of the emitted radiation when the electron jumps from level* n_i *to level* n_j *with* $n_i > n_j$.

(b) *Give for each of the first three series the wavelengths of the first line and the limit line.*

(c) *In which spectral domain do you observe each of these lines?*

2. *The first line of the Brackett series of the emission spectrum of the hydrogen atom has a wavelength of 4.052 μm. Calculate the wavelength of the following three lines. We give:*

- *Extreme ultraviolet range: λ from 10 nm to 200 nm.*
- *Visible range: λ from 380 nm to 780 nm.*
- *Near-infrared range: λ from 0.78 μm to 3 μm.*

Solution

1. (a) *Formula giving* $1/\lambda_{i-j}$:

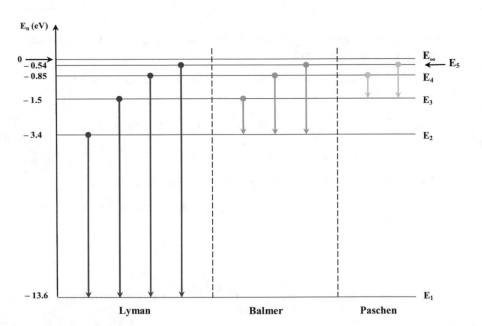

According to Eq. (1.28) established in the course, the emitted energy during the passage from the orbit n_i to the orbit n_j with $n_i > n_j$ is given by:

$$E_{n_i} - E_{n_j} = \frac{m_e\, q_e^4}{8\varepsilon_0^2 h^2}\left(\frac{1}{n_j^2} - \frac{1}{n_i^2}\right) \tag{1}$$

with

$$\lambda_{i \to j} = \frac{c}{\nu_{i \to j}} = \frac{hc}{h\nu_{i \to j}} = \frac{hc}{E_{n_i} - E_{n_j}}. \tag{2}$$

By expressing the inverse of the wave, it becomes:

$$\frac{1}{\lambda_{i \to j}} = \frac{E_{n_i} - E_{n_j}}{hc} = \frac{m_e q_e^4}{8\varepsilon_0^2 h^2} \left(\frac{1}{n_j^2} - \frac{1}{n_i^2} \right) = R_H \left(\frac{1}{n_j^2} - \frac{1}{n_i^2} \right). \tag{3}$$

It is the Ritz formula where R_H is the Rydberg constant. It is about $R_H = 1.097 \times 10^7 \text{ m}^{-1}$. This value is very close to the experimental value obtained from the spectrum of the hydrogen atom, hence the great success of the Bohr model. The hydrogen spectrum is the set of wavelengths that the hydrogen atom is capable of emitting or absorbing. This light spectrum is composed of discrete wavelengths, of which values are given by the formula of Ritz. The main series of spectral lines of the hydrogen atom are the series of Lyman, Balmer, Paschen, Brackett, Pfund, and Humphreys. The first three series are:

- Lyman series: $j = 1$ and $i \geq 2$.
- Balmer series: $j = 2$ and $i \geq 3$.
- Paschen series: $j = 3$ and $i \geq 4$.

1. (b) and (c) *Wavelengths of the first line and the limit line*:

   - The first line of each series is the transition $n_{j+1} \to n_j$: $\lambda_{j+1 \to j}$.
   - The limit line of each series is the transition $n_\infty \to n_j$: $\lambda_{\infty \to j}$.

   By starting from (3), it becomes:

- *For the Lyman series*:

   - *First line*:

$$\frac{1}{\lambda_{2 \to 1}} = \left(1.097 \times 10^7 \text{ m}^{-1} \right) \left(\frac{1}{1^2} - \frac{1}{2^2} \right) = 0.823 \times 10^7 \text{m}^{-1} \quad \Rightarrow \quad \lambda_{2 \to 1}$$

$$= 121.5 \text{ nm}. \tag{4}$$

   - *Limit line*:

$$\frac{1}{\lambda_{\infty \to 1}} = \left(1.097 \times 10^7 \text{m}^{-1} \right) \left(\frac{1}{1^2} - \frac{1}{\infty} \right) = 1.097 \times 10^7 \text{ m}^{-1} \quad \Rightarrow \quad \lambda_{\infty \to 1}$$

$$= 91.2 \text{ nm}. \tag{5}$$

The emission is located in the extreme ultraviolet domain.

- *For the Balmer series*:
  - *First line*:

$$\frac{1}{\lambda_{3\to2}} = \left(1.097 \times 10^7 \text{ m}^{-1}\right)\left(\frac{1}{2^2} - \frac{1}{3^2}\right) = 0.1523 \times 10^7 \text{ m}^{-1} \Rightarrow \lambda_{3\to2}$$

$$= 656.6 \text{ nm}. \tag{6}$$

  - *Limit line*:

$$\frac{1}{\lambda_{\infty\to2}} = \left(1.097 \times 10^7 \text{m}^{-1}\right)\left(\frac{1}{2^2} - \frac{1}{\infty}\right) = 0.2743 \times 10^7 \text{ m}^{-1} \Rightarrow \lambda_{\infty\to2}$$

$$= 364.6 \text{ nm}. \tag{7}$$

The emission is in the visible range.

- *For the Paschen series*:
  - *First line*:

$$\frac{1}{\lambda_{4\to3}} = \left(1.097 \times 10^7 \text{ m}^{-1}\right)\left(\frac{1}{3^2} - \frac{1}{4^2}\right) = 0.0533 \times 10^7 \text{ m}^{-1} \Rightarrow \lambda_{4\to3}$$

$$= 1.876 \text{ μm}. \tag{8}$$

  - *Limit line*:

$$\frac{1}{\lambda_{\infty\to3}} = \left(1.097 \times 10^7 \text{ m}^{-1}\right)\left(\frac{1}{3^2} - \frac{1}{\infty}\right) = 0.1219 \times 10^7 \text{ m}^{-1} \Rightarrow \lambda_{\infty\to3}$$

$$= 0.820 \text{ m}. \tag{9}$$

The emission is in the near-infrared domain.

2. *Wavelength of the following three lines of the Brackett series*:
   - *Brackett's series: $j = 4$ and $i \geq 5$*:
     - *First line*:

$$\frac{1}{\lambda_{5\to4}} = R_H\left(\frac{1}{4^2} - \frac{1}{5^2}\right) = (0.0225\,R_H)\,\mu m^{-1} = \frac{1}{4.052\,\mu m} = 0.2468\,\mu m^{-1}. \quad (10)$$

– *Second line*:

$$\frac{1}{\lambda_{6\to4}} = R_H\left(\frac{1}{4^2} - \frac{1}{6^2}\right) = (0.03472\,R_H)\,\mu m^{-1}. \quad (11)$$

On the basis of the ratio (10)/(11), we obtain:

$$\frac{\dfrac{1}{\lambda_{5\to4}}}{\dfrac{1}{\lambda_{6\to4}}} = \frac{(0.0225\,R_H)\,\mu m^{-1}}{(0.03472\,R_H)\,\mu m^{-1}} = 0.648 \;\Rightarrow\; \lambda_{6\to4} = 0.648\,\lambda_{5\to4}$$
$$= (0.648)\,(4.052\,\mu m) = 2.626\,\mu m. \quad (12)$$

– *Third line*:

$$\frac{1}{\lambda_{7\to4}} = R_H\left(\frac{1}{4^2} - \frac{1}{7^2}\right) = (0.0421\,R_H)\,\mu m^{-1}. \quad (13)$$

On the basis of the ratio (10)/(13), we obtain:

$$\frac{\dfrac{1}{\lambda_{5\to4}}}{\dfrac{1}{\lambda_{7\to4}}} = \frac{(0.0225\,R_H)\,\mu m^{-1}}{(0.0421\,R_H)\,\mu m^{-1}} = 0.0534 \;\Rightarrow\; \lambda_{7\to4} = 0.0534\,\lambda_{5\to4}$$
$$= (0.0534)\,(4.052\,\mu m) = 0.2164\,\mu m. \quad (14)$$

– *Fourth line*:

$$\frac{1}{\lambda_{8\to4}} = R_H\left(\frac{1}{4^2} - \frac{1}{8^2}\right) = (0.0469\,R_H)\,\mu m^{-1}. \quad (15)$$

On the basis of the ratio (10)/(15), we obtain:

$$\frac{\dfrac{1}{\lambda_{5\to4}}}{\dfrac{1}{\lambda_{8\to4}}} = \frac{(0.0225\,R_{\rm H})\,\mu{\rm m}^{-1}}{(0.0469\,R_{\rm H})\,\mu{\rm m}^{-1}} = 0.480 \quad \Rightarrow \quad \lambda_{8\to4} = 0.480\,\lambda_{5\to4} \tag{16}$$

$$= (0.480)(4.052\ \mu{\rm m}) = 1.945\ \mu{\rm m}.$$

Exercise 28

If a hydrogen atom in its fundamental state absorbs a photon of the wavelength λ_1 and then emits a photon of the wavelength λ_2, on what level does the electron end up after this emission? We give $\lambda_1 = 97.28$ nm, $\lambda_2 = 1879$ nm, and $R_H = 1.097 \times 10^7\,m^{-1}$, Rydberg's constant.

Solution

Let us consider the first transition between the fundamental level $n = 1$ and a higher level n_i (absorption) and the successive transition between the levels n_i and n_j with $i > j$ (emission). By starting from Eq. (1.30) of the course that gives the formula of Ritz, it becomes for the case of the absorption:

$$\frac{1}{\lambda_{1\to i}} = \frac{E_1 - E_j}{hc} = R_{\rm H}\left(\frac{1}{1^2} - \frac{1}{n_i^2}\right) = R_{\rm H}\left(1 - \frac{1}{n_i^2}\right) \quad \Rightarrow \quad 1 - \frac{1}{n_i^2} = \frac{1}{\lambda_{1\to n_i} R_{\rm H}}$$

$$= \frac{1}{\left(97.28 \times 10^{-9}{\rm m}\right)\left(1.097 \times 10^7{\rm m}^{-1}\right)} = 0.937. \tag{1}$$

Here $R_{\rm H}$ is the Rydberg constant. It is equal by hypothesis to $1.097 \times 10^7{\rm m}^{-1}$. From (1) we deduce:

$$1 - \frac{1}{n_i^2} = 0.937 \quad \Rightarrow \quad \frac{1}{n_i^2} = 1 - 0.937 = 0.063 \quad \Rightarrow \quad n_i^2 = \frac{1}{0.063} = 15.87$$

$$\Rightarrow \quad n_i = 4. \tag{2}$$

$|E_i - E_1| = hc/\lambda_1$	$|E_i - E_j| = hc/\lambda_2$
The energy of the electron increases by the absorption of a photon.	The energy of the electron decreases by the emission of a photon.

The electron is found at level 4. By starting from Eq. (1.30) of the course which gives the formula of Ritz, we obtain for the case of the emission ($n_i > n_j$):

$$\frac{1}{\lambda_{n_i \to n_j}} = \frac{E_i - j}{hc} = R_H \left(\frac{1}{n_j^2} - \frac{1}{n_i^2} \right) = R_H \left(\frac{1}{n_j^2} - \frac{1}{4^2} \right) \quad \Rightarrow \quad \frac{1}{n_j^2} - \frac{1}{16} = \frac{1}{\lambda_{n_i \to n_j} \, R_H}$$

$$= \frac{1}{(1879 \times 10^{-9} \text{m}) \, (1.097 \times 10^7 \text{m}^{-1})} = 0.0485$$

$$(3)$$

where R_H is Rydberg's constant. It is equal by hypothesis to $1.097 \times 10^7 \text{m}^{-1}$.
From (1) we deduce:

$$\frac{1}{n_j^2} - \frac{1}{16} = 0.0485 \; \Rightarrow \; \frac{1}{n_j^2} = \frac{1}{16} + 0.0485 = 0.111 \; \Rightarrow \; n_j^2 = \frac{1}{0.111} = 9 \; \Rightarrow \; n_j = 3. \quad (4)$$

The electron eventually ends up at level 3.

Exercise 29
In the hydrogen atom, the energy of the electron in its fundamental state is
$E_1 = -13.6 \; eV.$
1. *What is in electronvolts and in joules the smallest amount of energy that it must*
   absorb to jump?
   (a) *To the first excited state?*
   (b) *From the first excited state to the ionized state?*
2. *What are the wavelengths of the emission spectrum lines corresponding to the*
   return?
   (a) *From the ionized state to the first excited state?*
   (b) *From the first excited state to the fundamental state? We give:*
   - *Rydberg's constant: $R_H = 1.097 \times 10^7 m^{-1}$.*
   - *Planck's constant: $h = 6.62 \times 10^{-34} \; J.s.$*
   - *Speed of light: $c = 3 \times 10^8 \; m.s^{-1}$.*
   - *1 eV $= 1.6 \times 10^{-19} \; J.$*

Solution
1. (a) *Quantity of energy absorbed to jump to the first excited state*:

By starting from Eq. (1.26) of the course, the energy associated with the shell n of
the hydrogen atom is expressed by:

$$E_n = \frac{E_1}{n^2}. \tag{1}$$

The energy E_1 is called *Rydberg energy*. Its value is –13.6 eV. It is a negative
energy, meaning that the electron is more stable in the atom than outside. This
energy will have to be supplied to extract it from the atom and make it a free electron.
When an atom initially in the fundamental state $n = 1$ jumps to the first excited state
$n = 2$, it absorbs energy equal to:

$$\Delta E_{1\to2} = \frac{E_2}{2^2} - \frac{E_1}{1^2} \quad\Rightarrow\quad \frac{\Delta E_{1\to2}}{E_1} = \frac{1}{4} - 1 = -\frac{3}{4} \quad\Rightarrow\quad \Delta E_{1\to2} = -\frac{3}{4}$$

$$E_1 = -\frac{3}{4}\,(-13.6\ \text{eV}) = 10.2\ \text{eV}. \tag{2}$$

As $1\ \text{eV} = 1.6 \times 10^{-19}\ \text{J}$, it becomes:

$$|\Delta E_{1\to2}| = \frac{(1.6 \times 10^{-19}\ \text{J})\,(10.2\ \text{eV})}{1\ \text{eV}} = 1.63 \times 10^{-18}\ \text{J}. \tag{3}$$

1. (b) *Quantity of the absorbed energy to jump from the first excited state to the ionized state*:

When an atom initially in the excited state $n = 2$ jumps to the ionized state $n = \infty$, it absorbs energy equal to:

$$\Delta E_{2\to\infty} = \frac{E_\infty}{\infty} - \frac{E_2}{2^2} \quad\Rightarrow\quad \frac{\Delta E_{2\to\infty}}{E_1} = 0 - \frac{1}{4} = -\frac{1}{4} \quad\Rightarrow\quad \Delta E_{2\to\infty} = -\frac{1}{4}E_1$$

$$= -\frac{1}{4}\,(-13.6\ \text{eV}) = 3.4\ \text{eV}. \tag{4}$$

As $1\ \text{eV} = 1.6 \times 10^{-19}\ \text{J}$, we obtain:

$$|\Delta E_{1\to2}| = \frac{(1.6 \times 10^{-19}\ \text{J})\,(3.4\ \text{eV})}{1\ \text{eV}} = 5.44\ 10^{-19}\ \text{J}. \tag{5}$$

2. (a) *Wavelength corresponding to the return from the ionized state to the first excited state*:

When the hydrogen atom jumps from the ionized state $n = \infty$ to the first excited state $n = 2$, it emits a photon of energy:

$$\Delta E_{\infty\to2} = h\nu_{\infty\to2} = \frac{hc}{\lambda_{\infty\to2}}$$

$$= \frac{hc}{\Delta E_{\infty\to2}} = \frac{(6.62 \times 10^{-34}\ \text{J.s})\,(3 \times 10^8\ \text{m.s}^{-1})}{5.44 \times 10^{-19}\ \text{J}} \tag{6}$$

$$= 3.651 \times 10^{-7}\ \text{m} = 365.1\ \text{nm}.$$

2. (b) *Wavelength corresponding to the return from the excited state to the fundamental state*:

When the hydrogen atom jumps from the excited state $n = 2$ to the fundamental state $n = 1$, it emits a photon of energy:

$$\Delta E_{2\to1} = h\nu_{2\to1} = \frac{hc}{\lambda_{2\to1}}$$

$$= \frac{hc}{\Delta E_{2\to1}} = \frac{\left(6.62 \times 10^{-34} \text{ J.s}\right)\left(3 \times 10^8 \text{ m.s}^{-1}\right)}{1.63 \times 10^{-18} \text{ J}} \qquad (7)$$

$$= 1.218 \times 10^{-7} \text{ m} = 121.8 \text{ nm}.$$

Exercise 30

1. *An initially fundamental-state hydrogen atom absorbs energy of 10.2 eV. At what level is the electron of the excited state of the atom?*
2. *The electron of a hydrogen atom initially at the level $n = 3$ emits a radiation of wavelength $\lambda = 1027$ Å. At what level does it end up? We give:*
   - *Rydberg's constant: $R_H = 1.097 \times 10^7 m^{-1}$.*
   - *Rydberg's energy: $E_1 = -13.6$ eV.*
   - *Planck's constant: $h = 6.62 \times 10^{-34}$ J.s.*
   - *Speed of light: $c = 3 \times 10^8 m.s^{-1}$.*
   - *1 eV $= 1.6 \times 10^{-19}$ J*

Solution

1. *Level of the electron of the excited state*:

By starting from Eq. (1.26) of the course, the energy associated with the shell n of the hydrogen atom is expressed by:

$$E_n = \frac{E_1}{n^2}. \qquad (1)$$

The energy E_1 is called *Rydberg energy*. Its value is -13.6 eV. It is a negative energy, meaning that the electron is more stable in the atom than outside. This energy will have to be supplied to extract it from the atom and make it a free electron. When an atom initially in the fundamental state absorbs energy of 10.2 eV, the electron jumps from the level $n = 1$ to a level n_i with $i > 1$:

$$\Delta E_{1\to i} = \frac{E_1}{n_i^2} - \frac{E_1}{1^2} \implies \frac{\Delta E_{1\to i}}{E_1} = \frac{1}{n_i^2} - 1 \implies \frac{1}{n_i^2} = \frac{\Delta E_{1\to n_i}}{E_1} + 1 = \frac{10.2 \text{ eV}}{-13.6 \text{ eV}} + 1$$

$$= 0.25 \implies n_i^2 = \frac{1}{0.25} = 4 \implies n_i = 2.$$

$$(2)$$

The electron is at level 2.

2. *Level of the electron after the emission*:

The energy of the emitted photon is:

$$|\Delta E_{3\to j}| = \frac{hc}{\lambda_{3\to j}} = \frac{\left(6.62 \times 10^{-34} \text{ J.s}\right)\left(3 \times 10^8 \text{ m.s}^{-1}\right)}{1027 \times 10^{-10} \text{ m}} = 1.934 \times 10^{-18} \text{ J}. \qquad (3)$$

As 1 eV $= 1.6 \times 10^{-19}$ J, it becomes:

$$|\Delta E_{3\rightarrow j}| = \frac{(1.934 \times 10^{-18}\ J)\,(1\ eV)}{1.6 \times 10^{-19}\ J} = 12.088\ eV. \tag{4}$$

By starting from Eq. (1.30) of the course that gives the formula of Ritz, we obtain for the case of the emission:

$$\frac{1}{\lambda_{3\rightarrow j}} = \frac{E_3 - E_j}{hc} = R_H \left(\frac{1}{n_j^2} - \frac{1}{3^2}\right) = R_H \left(\frac{1}{n_j^2} - \frac{1}{9}\right) \quad \Rightarrow \quad \frac{1}{n_j^2} - \frac{1}{9} \tag{5}$$

$$= \frac{1}{\lambda_{3\rightarrow j}\,R_H} = \frac{1}{(1027 \times 10^{-10}\ m)\,(1.097 \times 10^7\ m^{-1})} = 0.8876.$$

Here R_H is the Rydberg constant. It is equal by hypothesis to $1.097 \times 10^7\ m^{-1}$. From (5) we deduce:

$$\frac{1}{n_j^2} - \frac{1}{9} = 0.8876 \quad \Rightarrow \quad \frac{1}{n_j^2} = \frac{1}{9} + 0.8876$$

$$= 0.9987 \quad \Rightarrow \quad n_j^2 \tag{6}$$

$$= \frac{1}{0.9987} = 1 \quad \Rightarrow \quad n_j = 1.$$

The electron returns to the fundamental level.

Exercise 31

1. *Calculate the energy required for the ionization of He^+, Li^{2+}, and Be^{3+} ions from their fundamental states.*
2. *What are the wavelengths of the first line and the limit line of Balmer series for He^+? We give:*
   - *Rydberg's energy: $E_1 = -13.6\ eV$.*
   - *Planck's constant: $h = 6.62 \times 10^{-34}\ J.s$.*
   - *Speed of light: $c = 3 \times 10^8\ m.s^{-1}$.*
   - *$1\ eV = 1.6 \times 10^{-19}\ J$*

Solution

The energy associated with the n-shell of a hydrogen-like ion (cation possessing only one electron) has the following expression:

$$E_n = \frac{Z^2}{n^2} E_1. \tag{1}$$

The energy E_1 is called *Rydberg energy*. Its value is –13.6 eV. It is a negative energy, meaning that the electron is more stable in the atom than outside. Z is the charge number of the hydrogen-like ion (number of protons). When a hydrogen-like ion initially in the fundamental state $n = 1$ jumps to the ionized state $n = \infty$, it absorbs energy equal to:

$$\Delta E_{1\to\infty} = \frac{Z^2 E_1}{\infty} - \frac{Z^2 E_1}{1^2} \quad \Rightarrow \quad \frac{\Delta E_{1\to\infty}}{Z^2 E_1} = 0 - 1 = -1 \quad \Rightarrow \quad \Delta E_{1\to 2}$$

$$= -Z^2 E_1. \tag{2}$$

- *Case of the helium cation He⁺:*

The charge number of the helium ion He^+ is $Z = 2$, from which we deduce from (2) the ionization energy:

$$\Delta E_{1\to 2} = -Z^2 E_1 = -(2^2)(-13.6\text{ eV}) = 54.4\text{ eV}. \tag{3}$$

- *Case of the lithium cation Li²⁺:*

The charge number of the lithium cation Li^{2+} is $Z = 3$, from which we deduce from (2) the ionization energy:

$$\Delta E_{1\to 2} = -Z^2 E_1 = -(3^2)(-13.6\text{ eV}) = 122.4\text{ eV}. \tag{4}$$

- *Case of the beryllium cation Be³⁺:*

The charge number of the beryllium ion Be^{3+} is $Z = 4$, from which we deduce from (2) the ionization energy:

$$\Delta E_{1\to 2} = -Z^2 E_1 = -(4^2)(-13.6\text{ eV}) = 217.6\text{ eV}. \tag{5}$$

1. *Wavelengths of the first line and the limit line of Balmer series for He⁺:*

   - Balmer series: $j = 2$ and $i \geq 3$.
   - The first line of each series is the transition $j + 1 \to j$: $\lambda_{j+1\to j}$.
   - The limit line of each series is the transition $\infty \to j$: $\lambda_{\infty\to j}$.

- *Wavelength of the first line:*

The first line of Balmer series is the transition from the state $n = 3$ to the state $n = 2$. This transition is accompanied by the emission of a photon of energy equal to:

$$\Delta E_{3\to 2} = \frac{Z^2 E_1}{3^2} - \frac{Z^2 E_1}{2^2} = Z^2 E_1 \left(\frac{1}{9} - \frac{1}{4}\right) = \frac{-5 Z^2 E_1}{36}$$

$$= \frac{(-5)(2^2)(-13.6\text{ eV})}{36} = 7.556\text{ eV}. \tag{6}$$

As $1\text{ eV} = 1.6 \times 10^{-19}$ J, it becomes:

$$|\Delta E_{3\to2}| = \frac{(1.6 \times 10^{-19}\,\text{J})\,(7.556\,\text{eV})}{1\,\text{eV}} = 1.21 \times 10^{-18}\,\text{J}. \qquad (7)$$

The emitted wavelength during this transition is therefore:

$$\Delta E_{3\to2} = h\nu_{3\to2} = \frac{hc}{\lambda_{3\to2}} \;\Rightarrow\; \lambda_{3\to2} = \frac{hc}{\Delta E_{3\to2}}$$

$$= \frac{(6.62 \times 10^{-34}\,\text{J.s})\,(3 \times 10^{8}\,\text{m.s}^{-1})}{1.21 \times 10^{-18}\,\text{J}} \qquad (8)$$

$$= 1.641 \times 10^{-7}\,\text{m} = 164.1\,\text{nm}.$$

- *Wavelength of the limit line*:

The limit line of Balmer series is the transition from the state $n = \infty$ to the state $n = 2$. This transition is accompanied by the emission of a photon of energy equal to:

$$\Delta E_{\infty\to2} = \frac{Z^2 E_1}{\infty} - \frac{Z^2 E_1}{2^2} = Z^2 E_1 \left(\frac{1}{\infty} - \frac{1}{4}\right) = \frac{-Z^2 E_1}{4}$$

$$= \frac{(-1)(2^2)(-13.6\,\text{eV})}{4} = 13.6\,\text{eV}. \qquad (9)$$

As $1\,\text{eV} = 1.6 \times 10^{-19}$ J, we get:

$$|\Delta E_{\infty\to2}| = \frac{(1.6 \times 10^{-19}\,\text{J})(13.6\,\text{eV})}{1\,\text{eV}} = 2.18 \times 10^{-18}\,\text{J}. \qquad (10)$$

The emitted wavelength during this transition is therefore:

$$\Delta E_{\infty\to2} = h\nu_{\infty\to2} = \frac{hc}{\lambda_{\infty\to2}} \;\Rightarrow\; \lambda_{\infty\to2}$$

$$= \frac{hc}{\Delta E_{\infty\to2}} = \frac{(6.62 \times 10^{-34}\,\text{J.s})\,(3 \times 10^{8}\,\text{m.s}^{-1})}{2.18 \times 10^{-18}\,\text{J}} \qquad (11)$$

$$= 9.11 \times 10^{-8}\,\text{m} = 91.1\,\text{nm}.$$

Exercise 32

Calculate the de Broglie wavelength of one electron moving at 1/137th of the speed of light. The relativistic corrections for the mass are negligible at this speed. We give:

- *Mass of the electron: $m_e = 9.109 \times 10^{-31}$ kg.*
- *Planck's constant: $h = 6.62 \times 10^{-34}$ J.s.*
- *Speed of light: $c = 2.998 \times 10^{8}$ m.s^{-1}.*

Solution

De Broglie wavelength associated with the electron:

The wavelength of de Broglie associated with the electron is defined by (see Eq. (1.35) of the course):

$$\lambda_e = \frac{h}{m_e v_e} = \frac{6.626 \times 10^{-34} \text{ J.s}}{\left(9.109 \times 10^{-31} \text{kg}\right) \left(\dfrac{2.998 \times 10^8 \text{ m.s}^{-1}}{137}\right)} = 3.325 \times 10^{-10} \text{ m}$$

$$= 33.25 \text{ nm.} \tag{1}$$

The wave character of the electron can be revealed because its wavelength is comparable to the scale of lengths which characterizes its movement.

Exercise 33
1. *Calculate the wavelength associated with one proton of mass $m_P = 1.6726 \times 10^{-27}$ kg that moves with a kinetic energy $E_C = 700$ eV. Which domain does it belong to?*
2. *Calculate the proton speed. Conclude. We give:*
   - *Planck's constant: $h = 6.62 \times 10^{-34}$ J.s.*
   - *Speed of light: $c = 2.998 \times 10^8$ m.s^{-1}.*
   - *1 eV $= 1.6 \times 10^{-19}$ J.*

Solution
1. *Wavelength associated with the proton:*

De Broglie wavelength associated with the particle is defined by (see Eq. (1.35) of the course):

$$\lambda = \frac{h}{P} = \frac{h}{m_P v} \tag{1}$$

where $P = m_P v$ is the momentum of the particle. Given the magnitude of Planck's constant which conditions the value of the wavelength, the wave character associated with a particle cannot be manifested for macroscopic objects. On the other hand, at the atomic or subatomic scale, the wave character of material particles (electrons, nucleons) can be revealed provided that their wavelengths are comparable to the scale of lengths which characterize their movement. The kinetic energy of the proton in nonrelativistic mechanics is given by:

$$E_C = \frac{1}{2} m_P v^2. \tag{2}$$

From (1) and (2), we obtain:

$$E_C = \frac{1}{2}\frac{m_P^2}{m_P}v^2 = \frac{1}{2}\frac{P^2}{m_P} \quad \Rightarrow \quad P = \sqrt{2\,m_P\,E_C}. \qquad (3)$$

From (1), we deduce:

$$\lambda = \frac{h}{P} = \frac{h}{\sqrt{2\,m_P\,E_C}}$$

$$= \frac{6.626 \times 10^{-34}\ \text{J.s}}{\sqrt{(2)(1.6726 \times 10^{-27}\ \text{kg})\dfrac{(700\ \text{eV})(1.602 \times 10^{-19}\ \text{J})}{1\ \text{eV}}}} \qquad (4)$$

$$= 1.08 \times 10^{-12}\ \text{m} = 1.08\ \text{pm}.$$

This wavelength belongs to the X-ray domain.

2. *Speed of the proton*:

From (2), we deduce:

$$E_C = \frac{1}{2}m_P v^2$$

$$= \sqrt{\frac{(2)\dfrac{(700\ \text{eV})(1.602 \times 10^{-19}\text{J})}{1\ \text{eV}}}{1.6726 \times 10^{-27}\ \text{kg}}} \qquad (5)$$

$$= 3.7 \times 10^5\ \text{m.s}^{-1}.$$

By evaluating the ratio of the proton speed to the speed of light, we find:

$$\frac{v}{c} = \frac{3.7 \times 10^5\ \text{m}}{3 \times 10^8\ \text{m}} \approx 10^{-3}. \qquad (6)$$

The speed of the proton being 1000 times lower than that of the light, this validates the nonrelativistic study of this exercise.

Exercise 34

1. *What is the dimensional analysis of the quantity h/mv and what is it?*
2. *What is the wavelength associated with:*
   (a) *An electron with a mass $m_e = 9.109 \times 10^{-31}$ kg and a kinetic energy of $E_C = 54$ eV. Conclude.*
   (b) *A ball with a speed $v_b = 300$ m.s^{-1} and a mass of $m_b = 2$ g. Conclude. We give:*
      • *Planck's constant: $h = 6.62 \times 10^{-34}$ J.s.*
      • *Speed of light: $c = 3 \times 10^8$ m/s.*
      • *$1\ eV = 1.602 \times 10^{-19}$ J*

Solution

1. *Dimensional analysis of the quantity h/mv*:

- The Planck constant is expressed by hypothesis in J.s. It is therefore an energy multiplied by a time: $[h]$ = [energy × time] = [force × distance × time] = [mass × distance/time squared × distance × time] = $MLT^{-2}LT$ = ML^2T^{-1}.
- The product mv has the following physical dimension: $[mv] = MLT^{-1}$.
- The physical dimension of the quantity h/mv is therefore: $[h/mv] = ML^2T^{-1}/MLT^{-1} = L$.

This quantity therefore has the physical dimension of a length. It is expressed in meters. This is the wavelength of de Broglie.

2. (a) *Wavelength associated with an electron*:

De Broglie's wavelength associated with the particle is defined by (see Eq. (1.35) of the course):

$$\lambda_e = \frac{h}{P_e} = \frac{h}{m_e v_e} \tag{1}$$

where $P_e = m_e v_e$ is the momentum of the particle. Given the magnitude of Planck's constant that conditions the value of the wavelength, the wave character associated with a particle cannot be manifested for macroscopic objects. On the other hand, at the atomic or subatomic scale, the wave character of material particles (electrons, nucleons) can be revealed provided that their wavelengths are comparable to the scale of lengths which characterize their movement. The kinetic energy of the proton in nonrelativistic mechanics is given by:

$$E_C = \frac{1}{2} m_e v_e^2. \tag{2}$$

From (1) and (2), we obtain:

$$E_C = \frac{1}{2} \frac{m_e^2}{m_e} v_e^2 = \frac{1}{2} \frac{P_e^2}{m_e} \Rightarrow P_e = \sqrt{2 m_e E_C}. \tag{3}$$

From (1), we deduce:

$$\lambda_e = \frac{h}{P_e} = \frac{h}{\sqrt{2\, m_e\, E_C}}$$

$$= \frac{6.626 \times 10^{-34}\ \text{J.s}}{\sqrt{(2)\ (9.109 \times 10^{-31}\ \text{kg})\ \dfrac{(54\,\text{eV})\,(1.602 \times 10^{-19}\ \text{J})}{1\ \text{eV}}}} \tag{4}$$

$$= 0.1668 \times 10^{-9}\ \text{m} = 1.668\ \text{Å}.$$

On the subatomic scale, the wave character of the electron can be revealed because its wavelength is comparable to the scale of lengths that characterize its movement.

2. (b) *Wavelength associated with a ball*:

The wavelength of de Broglie associated with the ball is defined by (see Eq. (1.35) of the course):

$$\lambda_b = \frac{h}{m_b v_b} = \frac{6.626 \times 10^{-34}\ \text{J.s}}{(2 \times 10^{-3}\text{kg})(300\ \text{m.s}^{-1})} = 1.104 \times 10^{-33}\ \text{m}$$

$$= 1.104 \times 10^{-23}\ \text{Å}. \tag{5}$$

The wave nature of the ball cannot be revealed because its wavelength is not comparable to the scale of lengths that characterize its move.

Exercise 35

Is de Broglie's wavelength of one hydrogen molecule move at the v_{H2} average speed of molecules of a gas of hydrogen at the temperature T greater or less than that of a molecule of oxygen moving at the v_{O2} average speed of a gas of oxygen at the same temperature T? We recall that the theory of gases gives:

$$\frac{1}{2}mv^2 = \frac{3}{2}k_B T$$

where k_B is the Boltzmann constant. We give:
- *Atomic molar mass of oxygen $= 16$ g.mol^{-1}.*
- *Atomic molar mass of hydrogen $= 1$ g.mol^{-1}.*

Solution

The ratio of speeds of gas molecules of hydrogen and oxygen is:

$$\frac{\frac{1}{2}\, m_{H_2} v_{H_2}^2}{\frac{1}{2}\, m_{O_2} v_{O_2}^2} = \frac{\frac{3}{2}\, k_B T}{\frac{3}{2}\, k_B T} \quad \Rightarrow \quad \frac{v_{H_2}^2}{v_{O_2}^2} = \frac{m_{O_2}}{m_{H_2}} \quad \Rightarrow \quad \frac{v_{H_2}}{v_{O_2}} = \sqrt{\frac{m_{O_2}}{m_{H_2}}}. \tag{1}$$

The de Broglie wavelength associated with a material particle is defined by (see Eq. (1.35) of the course):

$$\lambda = \frac{h}{mv}. \tag{2}$$

We finally get for 1 mole of gas:

$$\frac{\lambda_{O_2}}{\lambda_{H_2}} = \frac{\dfrac{h}{m_{O_2} v_{O_2}}}{\dfrac{h}{m_{H_2} v_{H_2}}} = \frac{m_{H_2} v_{H_2}}{m_{O_2} v_{O_2}} = \frac{m_{H_2}}{m_{O_2}} \sqrt{\frac{m_{O_2}}{m_{H_2}}} = \sqrt{\frac{m_{H_2}}{m_{O_2}}} \Rightarrow \lambda_{H_2} = \sqrt{\frac{m_{O_2}}{m_{H_2}}} \lambda_{O_2} = \tag{3}$$

$$\sqrt{\frac{(2)\,(16 \times 10^{-3}\ \text{kg})}{(2)\,(1 \times 10^{-3}\ \text{kg})}}\ \lambda_{O_2} = 4\,\lambda_{O_2}\ \text{m}.$$

The wavelength of hydrogen gas molecules is greater than the wavelength of oxygen gas molecules.

Exercise 36

1. *An electron moves in a straight line with an uncertainty $\Delta x = 1$ Å. Calculate the uncertainty on the speed Δv by using the uncertainty principle of Heisenberg. Conclude.*
2. *A ball of mass 10 g moves in a straight line with an uncertainty $\Delta x = 1$ μm. Calculate the uncertainty on the speed Δv by using the uncertainty principle of Heisenberg. We give:*
   - *Planck's constant: $h = 6.62 \times 10^{-34}$ J.s.*
   - *Mass of the electron: $m_e = 9.109 \times 10^{-31}$ kg.*

Solution

A quantum particle, contrary to the classical vision, has a position x defined at Δx near and a momentum P_x along the Ox-axis defined at ΔP_x near. Heisenberg's inequality states that:

$$\Delta P_x . \Delta x \geq \frac{h}{4\pi} \tag{1}$$

where h is the Planck constant.

1. *Uncertainty on the speed Δv of the electron*:

   From (2), we obtain:

$$\Delta v \geq \frac{h}{4\pi m_e \Delta x} = \frac{6.62 \times 10^{-34}\ \text{J.s}}{(4)\,(3.14)\,(9.109 \times 10^{-31}\ \text{kg})\,(10^{-10}\ \text{m})}$$

$$= 0.58 \times 10^6\ \text{m.s}^{-1}. \tag{2}$$

On the atomic scale, the uncertainty on the speed Δv is very important. We cannot precisely measure the position of the electron and its speed at the same time. Thus, the position of an electron with a definite amount of motion will only be defined with some uncertainty. Its presence will therefore be described in a domain of probability of presence and not by its position on an orbit.

2. *Uncertainty about the speed Δv for the ball*:

From (1), it becomes:

$$\Delta v \geq \frac{h}{4\pi m_b \Delta x} = \frac{6.62 \times 10^{-34} \text{ J.s}}{(4)\,(3.14)\,\left(10^{-2} \text{ kg}\right)\,\left(10^{-6} \text{ m}\right)} = 0.53 \times 10^{-26} \text{ m.s}^{-1}. \quad (3)$$

This uncertainty is too small to be measured. The uncertainty principle has no physical meaning at the macroscopic scale because at this scale, the position and the speed of the object are accurately determined.

Exercise 37
Let us consider an electron with a speed v = 0.1 c where c is the speed of light.
1. *Calculate the minimum uncertainty on its momentum ΔP by using the uncertainty principle of Heisenberg.*
2. *Calculate the amount of movement. Conclude. We give:*
   - *$\Delta x = 0.2$ pm (position uncertainty).*
   - *$m_e = 9.11 \times 10^{-31}$ kg (mass of the electron),*
   - *$c = 3 \times 10^8$ m/s (speed of light),*
   - *$h = 6.626 \times 10^{-34}$ J.s (Planck's constant),*

Solution
1. *Minimum uncertainty on the momentum ΔP:*

A quantum particle, contrary to classical vision, has a position x defined at Δx and a momentum P_x along the Ox-axis defined at ΔP_x near. Heisenberg inequality states that:

$$\Delta P_x \cdot \Delta x \geq \frac{h}{4\pi} \quad (1)$$

where h is Planck's constant. The minimum value of ΔP_x is taken from (1):

$$\Delta P_x = \frac{h}{4\pi\,\Delta x} = \frac{6.626 \times 10^{-34} \text{ J.s}}{(4)\,(3.14)\,\left(0.2 \times 10^{-12} \text{ m}\right)} = 2.64 \times 10^{-22} \text{ kg.m.s}^{-1}. \quad (2)$$

2. *Momentum*:

The electron with a speed $v = 0.1$ c being nonrelativistic, its momentum is given by:

$$P_x = m_e v = \left(9.11 \times 10^{-31} \text{ kg}\right)(0.1)\left(3 \times 10^8 \text{ m.s}^{-1}\right)$$
$$= 2.73 \times 10^{-23} \text{ kg.m.s}^{-1} \tag{3}$$

where m_e and v are respectively its mass and its speed. We deduce from the results found in (2) and (3) that in the case of an experimental determination, the minimal uncertainty on the momentum of the electron ΔP_x would be 10 times greater than the momentum itself! This is illogical. In fact, the position of the electron and its momentum cannot be determined simultaneously, precisely as it would be the case in classical mechanics where the values of the mass of the object, the coordinates of its position, its speed, its energy, etc. can be precisely determined. All of these parameters constitute the state of the object and the laws of classical mechanics allow predicting the evolution of this state with time. For the electron, the calculations on its state concern the wave function and the amplitude of the wave associated with it. They end up characterizing the state of this electron by associating with a possible value of its energy the geometry of the space where the probability of finding it is the greatest. Each possible state of the electron is defined by a unique set of four quantum numbers that can only take discrete values. The laws of quantum mechanics specify how the electron moves from one state to another.

Exercise 38

A motorist was taken by a radar system at 120 km/h behind the wheel of his car of the mass 1500 kg. To escape a contravention, he invoked the uncertainty principle of Heisenberg. Is he scientifically right to argue his speeding in this way? We give $h = 6.62 \times 10^{-34}$ J.s (Planck's constant).

Solution

The speed of the motorist is:

$$v = \frac{120 \text{ km}}{\text{hour}} = \frac{120000 \text{ m}}{3600 \text{ s}} = 33.33 \text{ m.s}^{-1}. \tag{1}$$

In one second, the driver travels 33.3 m. Let us suppose that its position is known to be 30 m near $\Rightarrow \Delta x = 30$ m. The Heisenberg inequality states that:

$$\Delta P_x \cdot \Delta x \geq \frac{h}{4\pi} \Rightarrow \Delta v \geq \frac{h}{4\pi m \Delta x} = \frac{6.62 \times 10^{-34} \text{ J.s}}{(4)(3.14)(1500 \text{ kg})(30 \text{ m})}$$
$$= 1.17 \times 10^{-39} \frac{\text{m}}{\text{s}} = \left(1.17 \times 10^{-39}\right)\frac{10^{-3} \text{ km}}{\frac{1}{3600}\text{h}} \tag{2}$$
$$= 4.21 \times 10^{-39} \frac{\text{km}}{\text{h}}$$

where h is the Planck constant. This uncertainty is tiny; the speed of the vehicle is thus known with all the required precision. In the macroscopic scale, the Heisenberg principle plays no role and our motorist has no chance of escaping a contravention on this only quantum motive!

Exercise 39

The orbital 1s of the hydrogen atom has the following expression:

$$\Psi = N_{1s} \exp\left(\frac{-r}{r_0}\right).$$

1. *What is the presence probability of the electron inside a sphere between r and r + dr?*
2. *Define the probability density of the radial presence D_r.*
3. (a) *What is the radius r of the sphere over which the radial presence probability density is maximal?*

(b) *Plot qualitatively the function $D_r = f(r)$.*
(c) *What is the radius r_0?*

Solution

The wave associated with an electron is a stationary wave. Its amplitude at each point of space is independent of time. It is given by a mathematical function Ψ called *wave* or *orbital function*. It is a solution of the Schrödinger equation:

$$H\Psi = E\Psi. \tag{1}$$

Here E is the energy of the system, Ψ its wave function, and H the Hamiltonian operator, which involves the second partial derivatives of the function. The wave function Ψ has no physical meaning. On the other hand, the value at a point of its square Ψ^2 or of the square of its modulus $|\Psi^2|$ in the case of a complex function determines the probability dP of finding the electron in a volume dV around this point such that:

$$dP = |\Psi_{1s}|^2 \, dV. \tag{2}$$

The dP/dV ratio is the density of the presence probability of the electron at the considered point or electronic density.

1. *Presence probability of the electron within a sphere between r and r ± dr*:

The infinitesimal volume in spherical coordinates is given by:

$$dV = r^2 \sin\theta \, d\theta \, d\varphi \, dr \tag{3}$$

with $0 < \theta < \pi$, $0 < \varphi < 2\pi$, and $0 < r < r + dr$. The presence probability of the electron in a space limited by two spheres of r and $r + dr$ radii is:

$$dP = |\Psi_{1s}|^2 dV = |\Psi_{1s}|^2 dV \quad \Rightarrow \quad P_{r \to r+dr}$$

$$= \int_0^\pi \sin\theta \, d\theta \int_0^{2\pi} d\varphi \int_r^{r+dr} \Psi_{1s} \Psi_{1s}^* \, r^2 \, dr \qquad (4)$$

$$= 4\pi \int_r^{r+dr} |\Psi_{1s}|^2 \, r^2 \, dr.$$

2. *Radial density of probability of presence D_r:*

The volume of a sphere of radius r is:

$$V = \frac{4}{3}\pi r^3 = f(r) \quad \Rightarrow \quad dV = \frac{df(r)}{dr} dr = \frac{4}{3} \times 3 \times \pi r^2 \, dr = 4\, r^2 \, dr. \qquad (5)$$

From (2) and (5) and the hypothesis of the exercise, we obtain:

$$D_r = \frac{dP}{dr} = \frac{|\Psi_{1s}|^2 4\pi r^2 \, dr}{dr} = 4\pi r^2 |\Psi_{1s}|^2 = 4\pi r^2 \left| N_{1s} \exp\left(\frac{-r}{r_0}\right) \right|^2$$

$$= 4\pi r^2 \, N_{1s}^2 \exp\left(\frac{-2r}{r_0}\right). \qquad (6)$$

3. *Radius r of the sphere on which the radial density of probability of presence is maximal:*

The radial density of presence probability is a function of radius r only. It is maximal for:

$$\frac{dD_r}{dr} = \frac{d}{dr}\left[4\pi r^2 \, N_{1s}^2 \exp\left(\frac{-2r}{r_0}\right) \right] = 8\pi \, N_{1s}^2 \, r \left(1 - \frac{r}{r_0}\right) \exp\left(\frac{-2r}{r_0}\right) = 0. \qquad (7)$$

According to (7), this derivative is zero for:

$$r = 0 \quad \Rightarrow \quad D_{r=0} = 0. \qquad (8)$$

$$r = r_0 \quad \Rightarrow \quad D_{r=r_0} = 4\pi r_0^2 \, N_{1s}^2 \exp(-2). \qquad (9)$$

Moreover, by using L'Hôpital's rule, it becomes:

$$r \to \infty \quad \Rightarrow \quad D_{r \to \infty} \to 0. \qquad (10)$$

(b) *Qualitative plot of the function $D_r = f(r)$:*

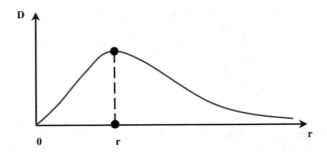

(c) *Radius r_0:*

The radius r_0 is at the maximum of the probability density of the presence of the electron (where we are almost sure to find it). It corresponds to Bohr's radius defined by (see Eq. (1.23) of the course):

$$r_1 = \frac{\hbar^2}{m_e e^2} \tag{11}$$

where $\hbar$ is the reduced Bohr constant, m_e the mass of the electron, and e its charge. Its value is $r_1 = 0.53$ Å.

Exercise 40

The s and p orbitals of the hydrogen atom are described by the following wave functions:

$$\Psi_{1s} = \Psi_{1,0,0} = \frac{2}{a_0^{3/2}} e^{-r/a_0} \times \frac{1}{2\sqrt{\pi}}. \tag{a}$$

$$\Psi_{2p_z} = \Psi_{2,1,0} = \left(\frac{1}{24\,a_0^3}\right)^{1/2} \frac{r}{a_0} e^{-r/2a_0} \times \left(\frac{3}{4\pi}\right)^{1/2} \cos\theta. \tag{b}$$

1. *What do the indices 1.0.0 and 2.1.0 represent?*
2. *What are a nodal point and a nodal surface?*
3. *We propose to study the density of presence of the orbital 1s.*
   (a) *By posing $\rho = r/a_0$, what is the expression of the density of presence of the orbital as a function of π, a_0, and ρ?*
   (b) *Does 1s orbital present a nodal point or a nodal surface?*
   (c) *The orbital 1s has a spherical symmetry centered on the nucleus of the atom. By what element is it illustrated?*
   (d) *Represent by a graph $dP/dV = f(\rho)$.*
   (e) *Represent the density of probability of presence by a scatter cloud.*

4. *We propose to study the density of presence of the $2p_z$ orbital.*
   (a) *We set $\rho = r/a_0$, what is the new expression of the density of presence of this orbital?*
   (b) *Does the $2p_z$ orbital have a nodal point or a nodal surface?*
   (c) *Represent by a graph $dP/dV = f(\rho)$.*
   (d) *Represent the density of probability of presence by a scatter cloud.*

Solution
1. *Meaning of the indices 1.0.0 and 2.1.0*:

The wave function of the electron depends on the three quantum numbers n, l, and m. It is usually referred to as $\Psi_{n,l,m}$ where:

- n is the main quantum number. It characterizes the electronic shell:

  - $n = 1$ for the K-shell,
  - $n = 2$ for the L-shell,
  - $n = 3$ for the M-shell, and so on and so forth,

  This number gives the energy level of the electron.

- l is the secondary or azimuthal quantum number. It is an integer that varies from 0 to $n - 1$. It defines the sublayer s, p, d, f, etc. So:

  - $l = 0$ corresponds to the s-sublayer,
  - $l = 1$ corresponds to the p-sublayer,
  - $l = 2$ corresponds to the d-sublayer,
  - $l = 3$ corresponds to the f-sublayer, and for the excited states g, h, I, etc. for $l = 4, 5, 6$, etc.

  This number tells us about the geometry of the atomic orbital, that is to say the spatial region in which the electron moves. For example:

- For the layer $n = 1$, $l = 0$, the function $\Psi_{1,0,m}$ is called *orbital 1s*.
- For the layer $n = 2$, $l = 1$, the function $\Psi_{2,1,m}$ is called *orbital 2p*.
- m is the magnetic quantum number. It takes values between $-l$ and $+l$ (including the values $-l$ and $+l$). It determines the orientation of orbitals in space. As an example, for $l = 1$, we have $m = -1, 0, +1$ so three orientations corresponding to the three axes of a three-dimensional system, that is to say three orbitals p of the same energy (p_x, p_y, p_z) or three quantum boxes. A more naive way of defining this number is to say that it corresponds to the kinetic moment, which gives the direction of rotation of the electron around the nucleus. Since the electron is an electric charge, its rotational movement creates a magnetic field. For example:

  - For $n = 1$, $l = 0$, $m = 0$: a single orbital of s type: $1s$.
  - For $n = 2$, $l = 0$, $m = 0$: a single orbital of s type: $2s$,
    $l = 1$, $m = -1, 0, +1$: three orbitals of p type: $2p_x$, $2p_y$, $2p_z$.

Note:
These three quantum numbers are not enough to completely determine the movement of the electron in an atom. It is necessary to add a fourth quantum

number, the quantum number of the spin or simply spin. This number can take two values, $m_S = +1/2$ and $m_S = -1/2$. If $m_S = +1/2$, it is customary to represent the electron by an arrow directed upwards ↑ (spin up). If $m_S = -1/2$, it will be represented by a down arrow ↓ (spin down). Spin is an intrinsic form of angular momentum. It is the only quantum observable that has no equivalent in classical physics, unlike the position, momentum, or energy of a particle. Its measurement gives discontinuous values and is subject to the Heisenberg uncertainty principle. Naively, we can consider the two values of the spin give the two directions of rotation of the electron on itself.

2. *Meaning of a nodal point or a nodal surface*:

There may be places around the nucleus where the probability of finding the electron is zero. These are the nodal point and the nodal surface. At this point and at every point on this surface, we have $\Psi^2 = 0$ and therefore $\Psi = 0$.

3. (a) *Expression of the density of presence of the orbital* $1s$:

By putting $\rho = r/a_0$ in the Eq. (a) and then by squaring, we get:

$$\frac{dP}{dV} = \Psi_{1s}^2 = \left(\frac{2}{a_0^{\frac{3}{2}}} \, e^{-\frac{r}{a_0}} \times \frac{1}{2\sqrt{\pi}} \right)^2 = \frac{4}{a_0^3} e^{-\frac{2r}{a_0}} \times \frac{1}{4\pi} = \frac{1}{\pi a_0^3} \, e^{-2\rho}. \tag{1}$$

(b) On the basis of (1) and since an exponential is never zero, dP/dV is always a strictly positive quantity $\forall \rho$. The $1s$ orbital therefore has neither point nor nodal surface.

(c) The spherical symmetry is illustrated by the fact that at ρ constant, dP/dV is also constant.

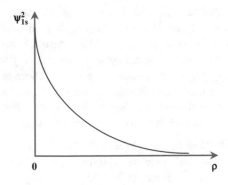

(d) *Graphic representation of dP/dV = f(ρ)*:

The graphical representation of $dP/dV = f(\rho)$ is that of a decreasing exponential of type $k\,e^{-ax}$ with k and α positive constants and $x \geq 0$. The e^{-ax} function is a valuable mathematical tool, in particular to express a probability, because its value is 1 for $x = 0$ and tends asymptotically to 0 when $x \to \infty$.

(e) *Representation of the density of probability of presence by a cloud of points*:

4. (a) *Volume density of the presence of the 2p$_z$-orbital*:

By writing $\rho = r/a_0$ in Eq. (b) and then by squaring, we obtain:

$$\frac{dP}{dV} = \Psi_{1s}^2 = \left[\left(\frac{1}{24\,a_0^3}\right)^{1/2} \frac{r}{a_0} e^{-r/2a_0} \times \left(\frac{3}{4\pi}\right)^{1/2} \cos\theta \right]^2$$

$$= \left(\frac{1}{24\,a_0^3}\right) \frac{r^2}{a_0^2} e^{-2r/2a_0} \times \frac{3}{4\pi} \cos^2\theta = \frac{1}{32\,\pi a_0^3}\, \rho^2\, e^{-\rho} \cos^2\theta. \tag{2}$$

(b) On the basis of (2), the density of the presence of the 2p$_z$-orbital vanishes to:

- $\rho = 0$ that is $r = 0$: the point O is a nodal point.
- $\cos^2\theta$ is $\cos\theta = 0 \Rightarrow \theta = \pi/2$: the plane xOy is a nodal plane.

(c) *Graphic representation of dP/dV = f(ρ)*:

The study of $dP/dV = f(\rho)$ is carried out for a value of fixed θ, that is to say that we place ourselves on a line that passes by the point O. The derivative of dP/dV with respect to ρ has the following expression:

$$\frac{d}{d\rho}\left(\frac{dP}{dV}\right) = \frac{d}{d\rho}\left(\frac{1}{32\,\pi a_0^3}\, \rho^2\, e^{-\rho} \cos^2\theta\right) = \frac{\cos^2\theta}{32\,\pi a_0^3}\left(2\rho\, e^{-\rho} - \rho^2 e^{-\rho}\right) \tag{3}$$

where ρ is the only variable since we have fixed θ. This derivative is zero for $\rho = 0$ and $\rho = 2$ $(r = 2a_0)$. The dP/dV function vanishes for $\rho = 0$, increases until it reaches a maximum for $\rho = 2$ (where the derivative $d/d\rho(dP/dV)$ vanishes), and then decreases to tend to 0 when ρ tends toward ∞.

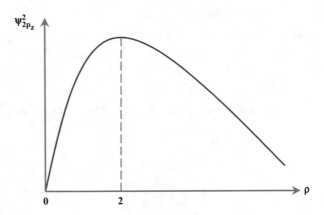

(d) *Representation of the density of probability of presence by a cloud of points*:

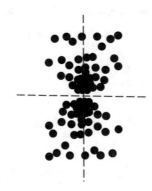

Exercise 41

1. *By using the relationships between the three quantum numbers n, l, and m, determine the number of orbitals in the three energy levels of the hydrogen atom.*
2. *Show that the maximum number of electrons that the n quantum number shell can contain is $2n^2$.*
3. *Give another designations of orbitals $\Psi_{3,0,0}$; $\Psi_{3,2,0}$; and $\Psi_{2,1,-1}$.*

Solution

1. The wave function of the electron depends on the three quantum numbers n, l, and m. It is usually referred to as $\Psi_{n,l,m}$ where:

   • n is the main quantum number. It characterizes the electronic shell:

     – $n = 1$ for the K-shell,
     – $n = 2$ for the L-shell,
     – $n = 3$ for the M-shell, and so on and so forth.

   This number gives the energy level of the electron.

- l is the secondary or azimuthal quantum number. It is an integer that varies from 0 to $n - 1$. It defines the sublayer s, p, d, f, etc. So:

  - $l = 0$ corresponds to the s-sublayer,
  - $l = 1$ corresponds to the p-sublayer,
  - $l = 2$ corresponds to the d-sublayer,
  - $l = 3$ corresponds to the f-sublayer, and for the excited states g, h, I, etc. for $l = 4, 5, 6$, etc.

This number tells us about the geometry of the atomic orbital, that is to say the spatial region in which the electron moves. For example:

- For the layer $n = 1$, $l = 0$, the function $\Psi_{1,0,m}$ is called *orbital 1s*.
- For the layer $n = 2$, $l = 1$, the function $\Psi_{2,1,m}$ is called *orbital 2p*.
- m is the magnetic quantum number. It takes values between $-l$ and $+l$ (including the values $-l$ and $+l$). It determines the orientation of orbitals in space. As an example, for $l = 1$, we have $m = -1, 0, +1$ so three orientations corresponding to the three axes of a three-dimensional system, that is to say three orbitals p of the same energy (p_x, p_y, p_z) or three quantum boxes. A more naive way of defining this number is to say that it corresponds to the kinetic moment, which gives the direction of rotation of the electron around the nucleus. Since the electron is an electric charge, its rotational movement creates a magnetic field. For example:

  - For $n = 1$, $l = 0$, $m = 0$: a single orbital of s type: 1s.
  - For $n = 2$, $l = 0$, $m = 0$: a single orbital of s type: 2s,
    $l = 1$, $m = -1, 0, +1$: three orbitals of p type: $2p_x$, $2p_y$, $2p_z$.

Note:
These three quantum numbers are not enough to completely determine the movement of the electron in an atom. It is necessary to add a fourth quantum number, the quantum number of the spin or simply spin. This number can take two values, $m_S = +1/2$ and $m_S = -1/2$. If $m_S = +1/2$, it is customary to represent the electron by an arrow directed upwards ↑ (spin up). If $m_S = -1/2$, it will be represented by a down arrow ↓ (spin down). Spin is an *intrinsic* form of *angular momentum*. It is the only quantum observable that has no equivalent in classical physics, unlike the position, momentum, or energy of a particle. Its measurement gives discontinuous values and is subject to the Heisenberg uncertainty principle. Naively, we can consider the two values of the spin give the two directions of rotation of the electron on itself.

The number of atomic orbitals for each value of n is n^2. So:

- In the $n = 1$ layer, we have 1 atomic orbital because $n^2 = 1^2 = 1$.
- In the $n = 2$ layer, we have 4 atomic orbitals because $n^2 = 2^2 = 4$.
- In the $n = 3$ layer, we have 9 atomic orbitals because $n^2 = 3^2 = 9$.

The following table gives the atomic orbitals in the three energy levels of the hydrogen atom:

n	l	m	Wave	Orbital
1	0	0	$\Psi_{1,0,0}$	$1s$
2	0	0	$\Psi_{2,0,0}$	$2s$
	1	-1	$\Psi_{2,1,-1}$	$2p_x$
		0	$\Psi_{2,10}$	$2p_y$
		$+1$	$\Psi_{2,1,1}$	$2p_z$
3	0	0	$\Psi_{3,0,0}$	$3s$
	1	-1	$\Psi_{3,1,-1}$	$3p_x$
		0	$\Psi_{3,1,0}$	$3p_y$
		$+1$	$\Psi_{3,1,1}$	$3p_z$
	2	-2	$\Psi_{3,2,-2}$	$3d$
		-1	$\Psi_{3,2,-1}$	$3d$
		0	$\Psi_{3,2,0}$	$3d$
		$+1$	$\Psi_{3,2,1}$	$3d$
		$+2$	$\Psi_{3,2,2}$	$3d$

2. *Maximum number of electrons of the n-shell*:

As we mentioned before, the number of atomic orbitals for each value of the quantum number n is n^2. Now, in each orbital we can have at most two electrons of opposite spins. So the maximum number of electrons that one shell n can hold is $2n^2$. This ceases to be true for $n > 4$.

3. *Other designation of orbitals $\Psi_{3,0,0}$; $\Psi_{3,2,0}$; and $\Psi_{2,1,-1}$*:

The atomic orbitals (abbreviated O.A) are designated by the three quantum numbers n, l, and m.

- $\Psi_{3,0,0} \Rightarrow \Psi_{n,l,m} \Rightarrow n = 3, l = 0$ (s-orbital) and $m = 0 \Rightarrow$ other designation: $3s$.
- $\Psi_{3,2,0} \Rightarrow \Psi_{n,l,m} \Rightarrow n = 3, l = 2$ (d-orbital) and $m = 0 \Rightarrow$ other designation: $3d$.
- $\Psi_{2,2,-1} \Rightarrow \Psi_{n,l,m} \Rightarrow n = 3, l = 2$ (d-orbital) and $m = -1 \Rightarrow$ other designation: $2p$.

Exercise 42
1. *What are the filling electron rules within an atom or an ion?*
2. *Which of the 4s or 3d orbitals fills in first?*
3. *If two sublayers have the same number of $(n + 1)$, which of the two fills first?*

Solution
The filling rules of electrons are:

(i) *Stability rule*:

When the atom is in the fundamental state, electrons occupy the lowest energy levels.

(ii) *Pauli's exclusion principle*:

Two electrons of the same atom cannot have their four quantum numbers n, l, m, and s all identical. This means that in a quantum box, the electron must have opposite spins (up and down).

(iii) *Klechkowski's rule*:

The filling of the sublayers s, p, d, f, etc. is in ascending order of $(n + l)$. The Klechkowski diagram allows finding this sequence by means of a simple construction: we place the s-sublayers in the first column, the p, d, f, etc. sublayers are added in succession on the same line, and the reading is done diagonally, each diagonal representing a given value of $n + l$. Thus, the filling of energy levels is:

1s   2s 2p   3s 3p 4s 3d 4p 5s 4d

K   L   M   N   O

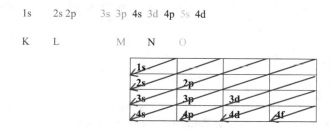

(iv) *Hund's rule*:

About one of five chemical elements has an electronic configuration in the ground state different from that deduced from the Klechkowski rule. This stems from the fact that the latter results from an approximation of electronic subshells energy taking into account only the quantum numbers n (principal) and l (azimuth), whereas the energy of electrons also involves their spin. More precisely, the Hund rule indicates that the lowest energy spectroscopic term is the one with the highest spin multiplicity $2s + 1$, which means that in the same subshell, the most stable configuration is obtained when the electron number with identical spins is maximal. Hund's rule is negligible in front of the Klechkowski rule for the elements of the s-block and the p-block of the Mendeleïev periodic table, so that the Klechkowski rule is always observed; it can, however, be decisive for certain elements of the d-block and f-block, that is to say for transition metals, lanthanides and actinides, because the energy levels of the electronic subshells of the valence shell of these elements are quite similar, so that it may be energetically more favorable to redistribute electrons by observing Hund's rule than to follow the Klechkowski rule.

2. *Filling the 4s and 3d orbitals*:

According to the Klechkowski rule, the 4s-orbital fills before the 2d-orbital because its energy is lower than the energy of the latter, indeed:

- Orbital 3d: $n + l = 3 + 2 = 5$.
- Orbital 4s: $n + l = 4 + 0 = 4$.

The 4s-orbital has the smallest value of $(n + 1) \Rightarrow$ It fills up first.

3. *Filling of the subshells having the same number of* $(n + 1)$:

If for two subshells, this sum is the same, it is the subshell which has the smallest value of n which fills first. For example:
- For orbital $1p$, we have $(n + 1) = 2 + 1 = 3$.
- For the $3s$ orbital, we have $(n + 1) = 3 + 0 = 3$.

In this case, the $2p$-orbital fills before the $3s$-orbital because its number n is smaller than that of the $3s$ subshell.

Exercise 43
Knowing that the electronic configuration of oxygen is, $1s^2$, $2s^2$, and $2p^4$, what are the representations of the quantum boxes and quantum numbers of electrons within the atom?

Solution
- *Level $n = 1$:*

  - $1s^2$ $\boxed{\uparrow\downarrow}$
  - For these two electrons: $n = 1$; $l = 0$ (s-subshell); $m = 0$; $m_S = \pm 1/2$.
  - The quantum numbers of each electron are therefore $[1,0,0,+1/2]$; $[1,0,0,-1/2]$.

- *Level $n = 2$:*

  - $2s^2$ $\boxed{\uparrow\downarrow}$
  - For these two electrons: $n = 2$; $l = 0$ (s-subshell); $m = 0$; $m_S = \pm 1/2$.
  - The quantum numbers of each electron are therefore $[2,0,0,+1/2]$; $[2,0,0,-1/2]$.

- *Level $n = 2$:*

  - $2p^4$ $\boxed{\uparrow\downarrow}\,\boxed{\uparrow}\,\boxed{\uparrow}$
  - For these four electrons: $n = 2$; $l = 1$ (p-subshell); $m = -1$; 0; $+1$; $m_S = \pm 1/2$.
  - The quantum numbers of each electron are therefore $[2,1,-1,+1/2]$; $[2,1,0,+1/2]$; $[2,1,+1,+1/2]$; $[2,1,-1,-1/2]$.

Exercise 44
Are the following statements correct or inaccurate? Justify your answers. An electron for which $n = 4$ and $m = 2$:
1. *Must necessarily have $l = 2$.*
2. *Can have $l = 2$.*
3. *Must necessarily have a spin equal to $+1/2$.*
4. *Is necessarily in a d-subshell.*

Solution

1. *An electron for which* $n = 4$ *and* $m = 2$ *must necessarily have* $l = 2$:

n	l	m
	0	0
		−1
	1	0
		+1
		−2
		−1
	2	0
4		+1
		+2
		−3
		−2
		−1
	3	0
		+1
		+2
		+3

According to the table, so that m to be +2, l must be equal to 2 or 3. So, it is not necessary that m be only equal to 2.

2. *An electron for which* $n = 4$ *and* $m = 2$ *can have* $l = 2$:

Yes, it is more correct to say so because l can also be equal to 3.

3. *An electron for which* $n = 4$ *and* $m = 2$ *must necessarily have a spin equal to* +1/ 2:

No, because a given quantum box of m can contain two electrons of opposite spins and thus have indifferently the spin values +1/2 and −1/2.

4. *An electron for which* $n = 4$ *and* $m = 2$ *is necessarily in an under layer d*:

No, it is not obligatory for the electron to be in a d-subshell because as l can be 3, the electron can be found in the f-subshell.

Exercise 45

Are the following statements correct or inexact? Justify your answers.
1. *If* $l = 1$, *the electron is in a d-subshell.*
2. *If* $n = 4$, *the electron is in the O-shell.*
3. *For a d-electron, m can be equal to 3.*
4. *If* $l = 2$, *the corresponding subshell can receive at most 6 electrons.*
5. *The number n of an electron of an f-subshell can be equal to 3.*
6. *If two atomic buildings have the same electronic configuration, this is necessarily the same element.*
7. *If two atomic structures have different electronic configurations, they are necessarily two different elements.*

Solution

1. *If l = 1, the electron is in a d-subshell:*

   No, because if $l = 1$, the electron is in a p-subshell.

2. *If n = 4, the electron is in the O-shell:*

   No, because if $n = 4$, the electron is in the N-shell.

3. *For an electron d, m can be equal to 3:*

   No, because for an electron d, $l = 2 \Rightarrow m = -2, -1, 0, +1, +2$.

4. *If l = 2, the corresponding subshell can receive at most 6 electrons:*

   No, because if $l = 2$, we have $m = -2, -1, 0, +1, +2$, that is, 5 quantum boxes that can contain each of them 2 electrons that is a total of 10 electrons.

5. *The number n of an electron of an f-subshell can be equal to 3:*

   No, because if $n = 3$, we have $l = 0, 1,$ and 2 corresponding to the s, p, and d sublayers. So, no f-subshell on the 3-shell.

6. *If two atomic buildings have the same electronic configuration, this is necessarily the same element:*

   An atomic building means an atom or an ion. If two atomic buildings have the same electronic configuration, this does not necessarily imply that it is the same element because an ion can have the same electronic configuration as a neutral atom of another element. For example, Na^+, Ne, and O^{2-} have the same electronic configuration.

7. *If two atomic structures have different electronic configurations, they are necessarily two different elements:*

   No, because the ion and the atom of the same chemical element necessarily have different electronic configurations.

Exercise 46

For each of the following elements, give the electronic configuration and abbreviated notation if applicable: Cu(29), He(2), O(8), Cl(17), Na(11), K(19), and Cs(55).

Solution

- Cu (29 e^-): structure: $1s^2 2s^2 2p^6 3s^2 3p^6 3d^{10} 4s^1$, abbreviated notation: [Ar] $3d^{10}4s^1$.
- He (2 e^-): $1s^2$ (no abbreviated notation).
- O (8 e^-): $1s^2 2s^2 2p^4$ (no abbreviated notation).
- Cl (17 e^-): complete structure: $1s^2 2s^2 2p^6 3s^2 3p^5$, abbreviated notation: [Ne] $3s^2 3p^5$.
- Na (11 e^-): complete structure: $1s^2 2s^2 2p^6 3s^1$, abbreviated notation: [Ne] $3s^1$.
- K (19 e^-): complete structure: $1s^2 2s^2 2p^6 3s^2 3p^6 4s^1$, abbreviated notation: [Ar] $4s^1$.

- Cs (55 e^-): complete structure: $1s^2 2s^2 2p^6 3s^2 3p^6 3d^{10} 4s^2 4p^6 4d^{10} 5s^2 5p^6 6s^1$, abbreviated notation: $[Xe]\, 6s^1$.

For elements with an atomic number from 1 to 10, the abbreviated notation does not exist.

Exercise 47

Write in the electronic configuration of K, L, M, etc. type the following elements: Sr (38), V(23), Fe^{2+}(26), Pb(82), Co^{3+}(27), Br(35), S^{2-} (16), and Al^{3+}.

Solution

- Sr(38): $K^2 L^8 M^{18} N^8 O^2$.
- V(23): $K^2 L^8 M^{11} N^2$.
- Fe(26): $K^2 L^8 M^{14} N^2$.
- Pb(82): $K^2 L^8 M^{18} N^{32} O^{18} P^4$.
- Co^{3+}(27): $K^2 L^8 M^{14}$.
- Br(35): $K^2 L^8 M^{18} N^7$.
- S^{2-}(16): $K^2 L^8 M^8$.
- Al^{3+}(13): $K^2 L^8$.

Exercise 48

1. What is the Mendeleïev table?
2. What do "group" and "period" mean in the Mendeleïev periodic table?
3. Explain the method for determining the position of a chemical element in the Mendeleïev periodic table.
4. Give the definition of the valence electron.
5. Boron and barium are elements of which the atomic numbers are, respectively, 5 and 56. They are in the 13th and the second column. Indicate their periods in the periodic table of Mendeleïev and the number of their valence electrons. Justify your answers.

Solution

1. Definition of the Mendeleïev table:

The Mendeleïev table of the name of the Russian chemist Dmitri Mendeleïev who conceived the principle in 1869, also called periodic table of elements or simply periodic table, represents all the chemical elements, ordered by increasing atomic number and organized in accord with their electronic configurations, which underlie their chemical properties.

2. Meaning of "group" and "period" in the periodic table of Mendeleïev:

The Mendeleïev table arranges chemical elements by increasing the atomic number Z in lines called periods and columns called groups. A period has all the elements, of which the s external subshell is at the same energy level (same value of n). The period is numbered with the value of n of this level. Within the period, the elements are arranged iteratively as a function of the successive filling of the ns subshells and then, depending on atoms, $(n-2)f$, and/or $(n-1)d$ and/or np subshells.

A group contains the elements having the same external electronic configuration. To reduce the congestion, there are 18 groups and not 32, and the elements $n = 6$ and $n = 7$ of which the f sublayer is incomplete are carried in two additional lines (lanthanides and actinides).

3. To associate the electronic configuration and the position in the periodic table, it is necessary to follow the following steps:

(i) *Research of the electronic configuration of the concerned atom.*

(ii) *Determination of the period.*

It is a question of locating the value of n of the outer shell: it is the number of the period to which the element belongs.

(iii) *Determination of the group of elements of $Z \leq 18$.*

These are the elements of the first three periods that have no subshell $(n - 1)d$. Two cases may occur:

- The valence electrons are only ns electrons; the number of the group is equal to the number of electrons ns, that is to say 1 or 2.
- The valence electrons are of ns and np type; the number of the group varies from 13 to 18, and the number of units represents the number of electrons of valence $ns + np$.

(iv) *Determination of the group of elements of $Z \geq 18$.*

Look at whether the $(n - 1)d$ subshell is saturated, partially filled, or empty. Three cases can occur:

- If the subshell is saturated, the number of the group varies from 11 to 18, the number of units representing the number of electrons of valence $ns + np$. The only exception is palladium, the configuration of which deviates from Klechkowski's rule ([Kr] $4d^{10}5s^0$), which is in the group 10 and in the fifth period.
- If it is partially filled (the case of transition metals), the number of the group is the number between 3 and 10, equal to the total number of valence electrons $(n - 1)d + ns$ (the case of Co and of Y).
- If it is empty, the number of the group is equal to the number of electrons, that is to say 1 or 2 (the case of Ba).

Note:

This group determination method does not apply to lanthanides or actinides.

4. *Give the definition of the valence electron*:

In chemistry, the valence electrons are the electrons situated in the last shell, that is to say on the outer shell of an atom. The knowledge of the valence electrons of an atom allows better understanding or better anticipating of chemical reactions, since they are the ones that intervene in chemical bonds. The number of the group allows knowing the number of valence electrons. It is possible to determine the number of valence electrons of an atom by knowing which group it belongs to. The number of valence electrons of an atom is the number of units of the number of the group. In other words:

- Column 1: one electron of valence.
- Column 2: two electrons of valence.
- Column 13: three electrons of valence.
- Column 14: four electrons of valence, etc. with few exceptions.

5. *Positions of boron (B) and barium (Ba) in the periodic table of Mendeleïev*:

- *Boron case (B)*:

    (i) *Research for the electronic configuration of the concerned atom.*
        Boron has an atomic number equal to 5. It therefore has five electrons. Its electronic configuration is $1s^2 2s^2 2p^1$.
    (ii) *Determination of the period.*
        The value of n of the outer layer $2p^1$ is 2. It therefore belongs to the second period.
        Determination of the number of electrons of valence of elements of $Z \leq 18$.
        Boron is in the 13th column. It belongs to the 13th group. If the number of the group varies from 13 to 18, the number of units represents the number of valence electrons, i.e., 3 electrons of valence of boron.

- *Case of the barium (Ba)*:
(i) *Research of the electronic configuration of the atom concerned.*

Barium has an atomic number equal to 56. It has 56 electrons. Its electronic configuration is $1s^2 2s^2 2p^6 3s^2 3p^6 3d^{10} 4s^2 4p^6 4d^{10} 5s^2 5p^6 6s^2$.

(ii) *Determination of the period.*
    The value of n of the outer layer $6s^2$ is 6. It therefore belongs to the sixth period.
(iii) *Determination of the number of electrons of valence of elements of $Z \geq 18$.*
    Barium is in the second column. It therefore belongs to the second group. If the $(n-1)d$ subshell is empty, the number of the group is equal to the number of electrons that is 1 or 2, that is to say in the case of barium, two electrons of valence.

1.7.3 Binding Energy

Exercise 49
A nuclide is represented by the following symbol: $_Z^A X$.
1. *What do X, Z, and A mean?*
2. *What is the number of protons, neutrons, and electrons of the following atoms and ions:*

$$^{79}_{35}Br, {}^{16}_{8}O^{2-}, {}^{1}_{1}H^{+}, {}^{35}_{17}Cl^{-}, {}^{70}_{31}Ga^{3+}, {}^{238}_{92}U. {}^{16}_{8}O^{2-}, {}^{1}_{1}H^{+}, {}^{35}_{17}Cl^{-}, {}^{70}_{31}Ga^{3+}, {}^{238}_{92}U.$$

Solution

1. Meaning of X, Z and A:

The resolution of this exercise is based on a simple arithmetic. It is founded on the meaning of the two numbers A (mass number) and Z (number of protons) associated with the symbol of the chemical element $^{A}_{Z}X$ and on the relationship $A = Z + N$ which links the mass number A to the number of protons Z and to the number of neutrons N contained in the nucleus. Regarding the number of electrons which orbit around the nucleus, in an atom it is equal to that of protons Z, but in an ion, it is different from this one, one or more electrons less if it is positively charged (cation), and one or more electrons in addition if it is negatively charged (anion).

2. Number of protons, neutrons, and electrons of the following atoms and ions:

Nuclide	Number of protons	Number of neutrons	Number of electrons
$^{79}_{35}Br$ (atom)	35	44	35
$^{16}_{8}O^{2-}$ (anion)	8	8	10
$^{1}_{1}H^{+}$ (cation	1	0	0
$^{35}_{17}Cl^{-}$ (anion)	17	18	18
$^{70}_{31}Ga^{3+}$ (cation)	31	39	28
$^{238}_{92}U$ (atom)	92	146	92

Exercise 50

1. Define the unified atomic mass unit (u).
2. Convert the masses of proton, neutron, and electron into u.
3. Deduce the sum of nucleon masses of the helium nucleus $^{4}_{2}He$, expressed in u.
4. Knowing that the mass of the particle α is 4.0015 u, calculate:
   (a) The binding energy in unified atomic mass unit (u).
   (b) The binding energy in kilograms.
   (c) The binding energy in joules.
   (d) The binding energy in MeV.
   (e) The binding energy in MeV/nucleon.
   We give:
   - Mass of the proton: $m_P = 1.6726 \times 10^{-27}$ kg.
   - Mass of the neutron: $m_N = 1.6746 \times 10^{-27}$ kg.
   - Mass of the proton: $m_e = 9.11 \times 10^{-31}$ kg.
   - Speed of light: $c = 2.998 \times 10^{8}$ m/s.
   - $1\ u = 1.66054 \times 10^{-27}$ kg
   - $1\ eV = 1.602 \times 10^{-19}$ J

Solution

1. *Definition of the unified atomic mass unit (u):*

The unified atomic mass unit is defined as being one-twelfth (1/12) of the mass of one carbon atom 12. However, 1 mole of this carbon weighs 12 g, i.e.:

$$1\ u = \frac{12 \times 10^{-3}\ kg}{(6.02214 \times 10^{23})\ (12)} = 1.66054 \times 10^{-27}\ kg. \tag{1}$$

This unit is better adapted to the mass expression on the scale of particles as electrons, protons, neutrons, etc.

2. *Proton, neutron, and electron masses in u:*

- *Proton mass in u:*

$$m_P = \frac{(1.6726 \times 10^{-27}\ kg)\ (1\ u)}{1.66054 \times 10^{-27}\ kg} = 1.0073\ u. \tag{2}$$

- *Neutron mass in u:*

$$m_N = \frac{(1.6749 \times 10^{-27}\ kg)\ (1\ u)}{1.66054 \times 10^{-27}\ kg} = 1.0087\ u. \tag{3}$$

- *Electron mass in u:*

$$m_e = \frac{(9.11 \times 10^{-31}\ kg)\ (1\ u)}{1.66054 \times 10^{-27}\ kg} = 5.4862 \times 10^{-4}\ u. \tag{4}$$

3. *The sum of the mass of nucleons of the helium nucleus $_2^4 He$, expressed in u:*

The helium nucleus $_2^4 He$ contains $Z = 2$ protons and $N = A - Z = 4 - 2$ neutrons. The sum of the mass of nucleons is:

$$m_{He} = 2\ m_P + 2\ m_N = (2)\ (1.0073\ u) + (2)\ (1.0087\ u) = 4.032\ u. \tag{5}$$

4. (a) *Binding energy in unified atomic mass unit (u):*

The nucleus of an atom has Z protons and N neutrons. The binding energy, ΔM, is the difference between the sum of masses of all the nucleons of a nucleus (mass of Z protons + mass of N neutrons) and the mass of this same nucleus $M(A,Z)$:

$$\Delta M = (Z\,m_P + N\,m_N) - M(A, Z) = 4.032\ \text{u} - 4.0015\ \text{u} = 0.0305\ \text{u}. \quad (6)$$

This energy can be considered as a potential well in which protons and neutrons are kept locked in the nucleus as the water which gives some energy to the external environment to the point that it freezes. This energy is positive since it corresponds to the energy that the external environment must provide to the nucleus (study system) to separate its nucleons.

(b) *Binding energy in kilograms*:

Knowing that $1\ \text{u} = 1.66054 \times 10^{-27}$ kg, it yields:

$$\Delta M = \frac{(0.0305\ \text{u})\left(1.66054 \times 10^{-27}\ \text{kg}\right)}{1\ \text{u}} = 0.05065 \times 10^{-27}\text{kg}. \quad (7)$$

(c) *Binding energy in joules*:

According to the Einstein mass-energy equivalence, the binding energy in joules is:

$$E_B = \Delta M\,c^2 = \left(0.05065 \times 10^{-27}\ \text{kg}\right)\left(2.998 \times 10^8\ \text{m.s}^1\right)^2$$
$$= 0.44524 \times 10^{-11}\ \text{J}. \quad (8)$$

(d) *Binding energy in MeV*:

Knowing that $1\ \text{eV} = 1.602 \times 10^{-19}$ J, it becomes:

$$E_B = \frac{\left(0.44524 \times 10^{-11}\ \text{J}\right)(1\ \text{eV})}{1.602 \times 10^{-19}\ \text{J}} = 0.2779 \times 10^8\ \text{eV} = 27.79\ \text{MeV}. \quad (9)$$

(e) *Binding energy per nucleon in MeV*:

Since the uranium nucleus contains four nucleons (two protons + two neutrons), the binding energy per nucleon is:

$$\frac{E_B}{A} = \frac{27.79\ \text{MeV}}{4} = 6.95\ \frac{\text{MeV}}{\text{nucleon}}. \quad (10)$$

Exercise 51

1. *What is the number of protons in the uranium nucleus $^{235}_{92}U$?*
2. *What is, in unified atomic mass unit, the mass of protons that form the nucleus?*
3. *What is the number of neutrons in the nucleus?*
4. *What is, in unified atomic mass unit, the mass of neutrons that form the nucleus?*
5. *What is the binding energy in unified atomic mass unit and in kilograms?*
6. *What is the binding energy in joules and in MeV?*

7. *What is the binding energy per nucleon in MeV?*
8. *What is the Aston curve?*
9. *By referring to the Aston curve, how does the uranium nucleus $^{235}_{92}U$ change?*
   We give:
   - *Mass of the proton: $m_P = 1\ 007\ 276$ u.*
   - *Mass of the neutron: $m_N = 1\ 008\ 665$ u.*
   - *Mass of the uranium 235 nucleus: $M(A,Z) = 234\ 994$ u.*
   - *$1\ u = 1.66 \times 10^{-27}$ kg*
   - *$1\ eV = 1.6 \times 10^{-19}$ J*

Solution

1. *Number of protons in the uranium nucleus $^{235}_{92}U$:*

   The number of protons is the charge number Z, that is to say 92 protons.

2. *The mass of protons that form the nucleus in unified atomic mass unit:*

   It is given by:

   $$Z\ m_P = (92 \text{ protons})\ (1.007276 \text{ u}) = 92.669392 \text{ u}. \tag{1}$$

3. *The number of neutrons contained in the nucleus:*

   It is given by the difference between the mass number A and the charge number Z:

   $$N = A - Z = 235 - 92 = 143 \text{ neutrons}. \tag{2}$$

4. *The mass of neutrons which form the nucleus in unified atomic mass unit:*

   It is given by:

   $$N\ m_N = (143 \text{ neutrons})\ (1.008665 \text{ u}) = 144.239095 \text{ u}. \tag{3}$$

5. *Binding energy in unified atomic mass unit and in kilograms:*

   - *Binding energy in unified atomic mass unit:*

   The nucleus of one atom has Z protons and N neutrons. The binding energy, ΔM, is the difference between the sum of the masses of all the nucleons of a nucleus (mass of Z protons + mass of N neutrons) and the mass of this same nucleus $M(A,Z)$:

$$\Delta M = (Z\, m_P + N\, m_N) - M(A, Z)$$
$$= 92.669392 \text{ u} + 144.239095 \text{ u} - 234.994 \text{ u} \tag{4}$$
$$= 1.915 \text{ u}.$$

This energy can be considered as a potential well in which protons and neutrons are kept locked in the nucleus much like the water which gives energy to the external environment to the point where it freezes. This energy is positive since it corresponds to the energy that the external environment must provide to the nucleus to separate its nucleons (studied system: the nucleus of the atom).

- *Binding energy in kilograms*:

  Knowing that $1 \text{ u} = 1.66 \times 10^{-27}$ kg, it becomes:

$$\Delta M = \frac{(1.915 \text{ u})\,(1.66 \times 10^{-27} \text{ kg})}{1 \text{ u}} = 3.1789 \times 10^{-27} \text{ kg}. \tag{5}$$

6. *Binding energy in joules and in MeV*:

   - *Binding energy in joules*:

     According to the Einstein mass-energy equivalence, the binding energy is given by:

$$E_B = \Delta M\, c^2 = (3.1789 \times 10^{-27} \text{ kg})\,(3 \times 10^8 \text{ m.s}^1)^2 = 28.61 \times 10^{-11} \text{ J}. \tag{6}$$

- *Binding energy in MeV*:

  Knowing that $1 \text{ eV} = 1.6 \times 10^{-19}$ J, it yields:

$$E_B = \frac{(28.61 \times 10^{-11} \text{ J})(1 \text{ eV})}{1.6 \times 10^{-19} \text{ eV}} = 17.88 \times 10^8 \text{ eV} = 1788 \text{ MeV}. \tag{7}$$

7. *Binding energy per nucleon in Mev*:

   Since the uranium nucleus contains 235 nucleons (protons + neutrons), the binding energy per nucleon is given by:

$$\frac{E_B}{A} = \frac{1788 \text{ MeV}}{235} = 7.61 \frac{\text{MeV}}{\text{nucleon}}. \tag{8}$$

8. *Definition of the Aston curve*:

   The Aston curve of the name of the English physicist Francis William Aston gives the binding energy per nucleon E_B/A as a function of the number A of nucleons that make up the nucleus. But, it is more telling to plot the curve – $E_B/A = f(A)$

because it allows to compare the stability of different atomic nuclei by proceeding as in an energy diagram where the most stable nuclei occupy the lower part of the curve. This curve reveals that the binding energy per nucleon is minimal for $60 < A < 90$. This corresponds to the most stable nuclei with $E_B/A \approx 9$ MeV. This curve reaches its minimum for a number of nucleons equal to 70. On both sides of this point, nuclei can undergo modifications that bring them closer to this point. The nucleus energy son obtained after modification is greater than the energy of the nucleus father which generated it. This change is manifested by a decrease in the mass of the system and a release of a very great energy towards the external environment. There are two ways to obtain nuclei of high binding energy per nucleon:

• The nucleus is broken into two nuclei, causing a nuclear reaction of fission.
• Two nuclei are assembled into a single nucleus, causing a nuclear reaction of fusion.

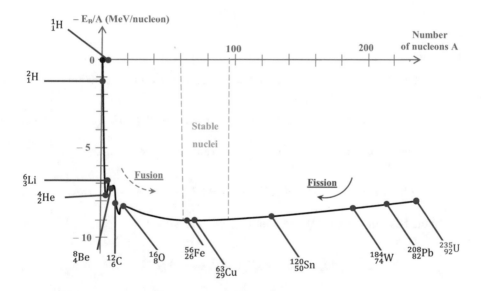

9. *Modification of the uranium nucleus $^{235}_{92}\text{U}$ during a nuclear reaction*:

According to the Aston curve, the uranium nucleus $^{235}_{92}\text{U}$ is modified by fission.

Exercise 52
Let us consider the following fission reaction which describes the collision between a slow neutron and a uranium nucleus:

$$^{235}_{92}U + {}^{1}_{0}n \rightarrow {}^{90}_{36}Kr + {}^{142}_{Z}Ba + y{}^{1}_{0}n.$$

1. *State the conservation laws used to find Z and Y, and rewrite the reaction based on finding results.*
2. *Calculate the binding energy between the reagents and products of the reaction in unified atomic mass unit.*
3. *Calculate in joules and in MeV the energy released during this reaction.*
4. *Calculate in joules and in MeV the energy released per mole of uranium.*
5. *Calculate in joules and in MeV the energy released per kg of uranium.*
6. *A nuclear reactor has a power of 100 megawatts. How long does it take to consume 1 kg of uranium? We give:*
   - *Mass of the uranium nucleus: $m_U = 235.043915$ u.*
   - *Mass of the krypton nucleus: $m_{Kr} = 89.819720$ u.*
   - *Mass of the barium nucleus: $m_N = 1.008665$ u.*
   - *$1\ u = 1.66 \times 10^{-27}$ kg*
   - *$1\ eV = 1.6 \times 10^{-19}$ J*
   - *Number of Avogadro $N_A = 6.02 \times 10^{23}$.*

Solution
1. *Values of Z and y:*

   - Let us write that the charge number is conserved during the reaction:

$$92 + 0 = 36 + Z + y(0) \quad \Rightarrow \quad Z = 92 - 36 = 56. \tag{1}$$

   - Let us write that the mass number is conserved during the reaction:

$$235 + 1 = 90 + 142 + y\,(1) \quad \Rightarrow \quad Z = 235 + 1 - 90 - 142 = 4. \tag{2}$$

   - On the basis of (1) and (2), the fission reaction is written as:

$$^{235}_{92}U + {}^{1}_{0}n \rightarrow {}^{90}_{36}Kr + {}^{142}_{56}Ba + 4\,{}^{1}_{0}n. \tag{3}$$

2. *The binding energy in unified atomic mass unit:*

   The mass of reagents is:

$$m_U + m_N = 235.043915\ \text{u} + 1.008665\ \text{u} = 236.05258\ \text{u}. \tag{4}$$

   The mass of products is:

$$m_{Kr} + m_{Ba} + m_N = 89.819720 \text{ u} + 141.916350 \text{ u} + (4)(1.008665)\text{u}$$
$$= 235.870730 \text{ u.} \tag{5}$$

The binding energy is equal to the difference in the reactants and product masses of the reaction:

$$(m_U + m_N) - (m_{Kr} + m_{Ba} + m_N) = 236.05258 \text{ u} - 235.870730 \text{ u}$$
$$= 0.28185 \text{ u.} \tag{6}$$

3. *Energy released in joules and in MeV*:

   • *Released energy released in joules*:

   Knowing that $1 \text{ u} = 1.66 \times 10^{-27}$ kg and in accordance with the Einstein mass-energy equivalence, the released energy in joules is:

$$E_B = \Delta M \, c^2$$
$$= \frac{(0.28185 \text{ u}) \left(1.66 \times 10^{-27} \text{ kg} \right)}{1 \text{ u}} \left(3 \times 10^8 \text{ m.s}^1 \right)^2 \tag{7}$$
$$= 4.21 \times 10^{-11} \text{ J.}$$

   • *Released energy in MeV*:

   Knowing that $1 \text{ eV} = 1.6 \times 10^{-19}$ J, it becomes:

$$E_B = \frac{\left(4.21 \times 10^{-11} \text{ J}\right) (1 \text{ eV})}{1.6 \times 10^{-19} \text{ J}} = 2.63110^8 \text{ eV} = 263.1 \text{ MeV.} \tag{8}$$

4. *Released energy per mole of uranium expressed in joules and in MeV*:

   • *Released energy per mole of uranium expressed in joules*:

   Knowing that a mole of uranium contains $N_A = 6.02 \times 10^{23}$ atoms, we find:

$$E(J) = \left(4.21 \times 10^{-11} \text{J}\right) \left(6.02 \times 10^{23} \text{ atoms}\right) = 2.5 \times 10^{13} \text{ J.} \tag{9}$$

   • *Released energy per mole of uranium expressed in MeV*:

   Knowing that $1 \text{ eV} = 1.6 \times 10^{-19}$ J, it yields:

$$E(\text{MeV}) = \frac{\left(2.5 \times 10^{13} \text{ J}\right) (1 \text{ eV})}{1.6 \times 10^{-19} \text{ J}} = 1.56 \times 10^{32} \text{ eV} = 1.56 \times 10^{26} \text{ MeV.} \tag{10}$$

5. *Released energy per kg of uranium expressed in joules and in MeV*:

- *Released energy per kilogram of uranium expressed in joules*:

Knowing that 1 mole of uranium weighs 235 g and releases, according to (9), an energy $E(\text{J}) = 2.5 \times 10^{13}$ J, we get:

$$\varepsilon(\text{J}) = \frac{\left(2.5 \times 10^{13} \text{ J}\right) (1000 \text{ g})}{235 \text{ g}} = 1.07 \times 10^{14} \text{ J}. \tag{11}$$

- *Released energy released per kilogram of uranium expressed in MeV*:

Knowing that 1 eV $= 1.6 \times 10^{-19}$ J, we obtain:

$$E(\text{MeV}) = \frac{\left(1.07 \times 10^{13} \text{ J}\right) (1 \text{ eV})}{1.6 \times 10^{-19} \text{ J}} = 0.669 \times 10^{32} \text{ eV}$$

$$= 6.69 \times 10^{25} \text{ MeV}. \tag{12}$$

Exercise 53
1. *Calculate the binding energy in unified atomic mass unit and in kilograms of carbon 12.*
2. *Calculate the binding energy in joules and MeV of carbon 12.*
3. *Calculate the binding energy per nucleon in joules and MeV of carbon 12.*
4. *Calculate the binding energy in unified atomic mass unit and in kilograms of carbon 14.*
5. *Calculate the binding energy in joules and MeV of carbon 14.*
6. *Calculate the binding energy per nucleon in joules and MeV of carbon 14.*
7. *Is there another way to achieve these results more quickly?*
8. *Based on the results found previously, which of the two nuclides is in your opinion the most stable? We give:*
   - *Mass of Carbon 12:* $m\left(^{12}_{6}C\right) = 12.00000$ u.
   - *Mass of Carbon 14:* $m\left(^{14}_{6}C\right) = 14.00320$ u.
   - *Mass of the proton:* $m_P = 1.007276$ u.
   - *Mass of the neutron:* $m_N = 1.008665$ u.
   - *1 u = 1.66 × 10⁻²⁷ kg*
   - *1 eV = 1.6 × 10⁻¹⁹ J*

Solution
1. *Binding energy, in unified atomic mass unit, of carbon 12*:

- *Binding energy in unified atomic mass unit*:

The nucleus of an atom has Z protons and N neutrons. The binding energy, ΔM, is the difference between the sum of the masses of all the nucleons of a nucleus (mass of Z protons + mass of N neutrons) and the mass of this same nucleus $M(A,Z)$:

$$\Delta M = (Z\,m_P + N\,m_N) - M(A,Z) = (6)(1.007276\,u) + (6)(1.008665\,u) - 12.00000\,u$$
$$= 6.043656\,u + 6.05199\,u - 12.00000\,u = 0.095646\,u.$$

(1)

- *Binding energy in kilograms*:

  Knowing that $1\,u = 1.66 \times 10^{-27}$ kg, we obtain:

$$\Delta M = \frac{(0.095646\,u)\,(1.66 \times 10^{-27}\,\text{kg})}{1\,u} = 0.158772 \times 10^{-27}\,\text{kg}.$$

(2)

2. *Calculation of the binding energy in joules and in MeV of carbon 12*:

This energy can be considered as a potential well in which the protons and neutrons are kept locked in the nucleus much like the water which gives energy to the external environment to the point where it freezes. This energy is positive since it corresponds to the energy that the external environment must provide to the nucleus to separate its nucleons (studied system: the nucleus of the atom).

- *Binding energy in joules*:

  It is given by:

$$E_B = \Delta M\,c^2 = (0.158772 \times 10^{-27}\,\text{kg})\,(3 \times 10^8\,\text{m.s}^1)^2$$
$$= 1.4289 \times 10^{-11}\,\text{J}.$$

(3)

- *Binding energy in MeV*:

  Knowing that $1\,eV = 1.6 \times 10^{-19}$ J, we obtain:

$$E_B = \frac{(1.4289 \times 10^{-11}\,\text{J})\,(1\,eV)}{1.6 \times 10^{-19}\,\text{J}} = 0.8931 \times 10^8\,\text{eV} = 89.31\,\text{MeV}.$$

(4)

3. *Binding energy per nucleon in joules and in MeV of carbon 12*:

- *Binding energy per nucleon in joules of carbon 12*:

  Since the uranium nucleus contains 12 nucleons (6 protons + 6 neutrons), the binding energy per nucleon, expressed in joules, is given by:

$$\frac{E_B^{12}}{A_{12}} = \frac{1.4289 \times 10^{-11}\,\text{J}}{12\,\text{nucleons}} = 0.1191 \times 10^{-11}\,\frac{\text{J}}{\text{nucleon}}.$$

(5)

- *Binding energy per nucleus in MeV of carbon 12*:

Since the uranium nucleus contains 12 nucleons (protons + neutrons), the binding energy per nucleon expressed in MeV is:

$$\frac{E_B^{12}}{A_{12}} = \frac{89.31\ \text{MeV}}{12\ \text{nucleons}} = 7.44\ \frac{\text{MeV}}{\text{nucleon}}. \tag{6}$$

4. *Binding energy in unified atomic mass unit of carbon 14*:

- *Binding energy in unified atomic mass unit*:

  The nucleus of an atom has Z protons and N neutrons. The mass defect, ΔM, is the difference between the sum of the masses of all the nucleons of a nucleus (mass of Z protons + mass of N neutrons) and the mass of this same nucleus $M(A,Z)$:

$$\Delta M = (Z\,m_P + N\,m_N) - M(A,Z) = (6)(1.007276\,\text{u}) + (8)\,(1.008665\,\text{u}) - 12.00000\,\text{u}$$
$$= 6.043656\,\text{u} + 8.06932\,\text{u} - 14.00320\,\text{u} = 0.109776\,\text{u}. \tag{7}$$

- *Binding energy in kilograms*:

  Knowing that $1\ \text{u} = 1.66 \times 10^{-27}$ kg, we find:

$$\Delta M = \frac{(0.109776\ \text{u})\ (1.66 \times 10^{-27}\ \text{kg})}{1\ \text{u}} = 0.182228 \times 10^{-27}\ \text{kg}. \tag{8}$$

5. *Calculation of the binding energy in joules and in MeV of carbon 14*:

This energy can be considered as a potential well in which the protons and neutrons are kept locked in the nucleus much like the water which gives energy to the external environment to the point where it freezes. This energy is positive since it corresponds to the energy that the external environment must provide to the nucleus to separate its nucleons (studied system: the nucleus of the atom).

- *Binding energy in joules*:

  It is given by:

$$E_B = \Delta M\ c^2 = (0.182228 \times 10^{-27}\ \text{kg})\ (3 \times 10^8\ \text{m.s}^1)^2$$
$$= 1.64005 \times 10^{-11}\ \text{J}. \tag{9}$$

- *Binding energy in MeV*:

  Knowing that $1\ \text{eV} = 1.6 \times 10^{-19}$ J, we find:

$$E_B = \frac{\left(1.64005 \times 10^{-11} \text{ J}\right) (1 \text{ eV})}{1.6 \times 10^{-19} \text{ J}} = 1.0250 \times 10^8 \text{ eV} = 102.50 \text{ MeV}. \quad (10)$$

6. *Binding energy per nucleon in joules and in MeV of carbon 14*:

• *Binding energy per nucleon in joules of carbon 14*:

Since the uranium nucleus contains 14 nucleons (6 protons + 8 neutrons), the binding energy per nucleon, expressed in joules, is:

$$\frac{E_B^{14}}{A_{14}} = \frac{1.64005 \times 10^{-11} \text{ J}}{14 \text{ nucleons}} = 0.1171 \times 10^{-11} \frac{\text{J}}{\text{nucleon}}. \quad (11)$$

• *Binding energy per nucleus in MeV of carbon 14*:

Since the uranium nucleus contains 14 nucleons (6 protons + 8 neutrons), the binding energy per nucleon expressed in MeV is:

$$\frac{E_B^{14}}{A_{14}} = \frac{102.50 \text{ MeV}}{14 \text{ nucleons}} = 7.32 \frac{\text{MeV}}{\text{nucleon}}. \quad (12)$$

7. *Another method to achieve these results more quickly*:

In special relativity, Einstein has shown that at any mass m_0 corresponds an energy $E = m_0 c^2$ called *rest mass energy* where E is in joules, m_0 in kilograms, and c, the speed of light in m/s. From this equation, we deduce:

$$m_0 = \frac{E}{c^2}. \quad (13)$$

Equation (13) shows that a mass can be expressed by an energy divided by c^2. Although the joule is the SI unit of energy, in nuclear physics, the used unit is the electronvolt (eV) or MeV/c^2 because it is better adapted to nuclear phenomena. As by hypothesis, we have $1 \text{ u} = 1.66 \times 10^{-27} \text{ kg}$ and $1 \text{ eV} = 1.6 \times 10^{-19} \text{ J}$, we obtain:

$$\begin{aligned} 1 \text{ u} &= \frac{\left(1.66 \times 10^{-27} \text{ kg}\right)\left(3 \times 10^8 \text{ m.s}^{-1}\right)^2}{c^2} = \frac{1.494 \times 10^{-10} \text{ J}}{c^2} \\ &= \frac{\left(1.49 \times 10^{-10} \text{ J}\right)(1 \text{ eV})}{\left(1.6 \times 10^{-19}\text{J}\right)c^2} = \frac{0.9313 \times 10^9 \text{ eV}}{c^2} = 931.3 \frac{\text{MeV}}{c^2}. \end{aligned} \quad (14)$$

The binding energy of a nuclide $_Z^A X$ is defined by:

$$E\left({}^A_Z X\right) = \left\{ [Z\, m_P + (A - Z) m_N] - m\left({}^A_Z X\right) \right\} c^2. \qquad (15)$$

- *For carbon ${}^{12}_6 C$, it becomes on the basis of (1) and (17):*

$$E\left({}^A_Z X\right) = (6.043656\ \text{u} + 6.05199\ \text{u} - 12.00000\ \text{u})\, c^2 = (0.095646\ \text{u}) c^2$$

$$= \frac{(0.095646\ \text{u})\left(931.3 \dfrac{\text{MeV}}{c^2}\right) c^2}{1\ \text{u}} = 89.1\ \text{MeV}. \qquad (16)$$

- *For carbon ${}^{14}_6 C$, it becomes on the basis of (7) and (17):*

$$E\left({}^A_Z X\right) = (6.043656\ \text{u} + 8.06932\ \text{u} - 14.00320\ \text{u})\, c^2 = (0.109776\ \text{u}) c^2$$

$$= \frac{(0.109776\ \text{u})\left(931.3 \dfrac{\text{MeV}}{c^2}\right) c^2}{1\ \text{u}} = 102.2\ \text{MeV}. \qquad (17)$$

These two values are, to approximations of calculations near, identical to the found values in (4) and (10).

8. *Stability of the two nuclides*:

By referring to the Aston curve, it appears that between several nuclides, the most stable is the one with the highest binding energy per nucleon. In our case, carbon 12 has the highest binding energy per nucleon (7.44 MeV/nucleon) compared to the binding energy per nucleon of carbon 14, which is 7.32 MeV/nucleon. So, carbon 12 is more stable than carbon 14.

Exercise 54

The fusion of four hydrogen nuclei into a helium nucleus is the result of one of reaction cycles which occur in the Sun:

$$4\,{}^1_1 H \rightarrow {}^4_2 He + 2\,{}^0_x e + 2\,{}^0_0 \nu.$$

1. *Identify x. What does ${}^0_x e$ represent? Rewrite the reaction. What does ${}^0_0 \nu$ represent?*
2. *What is in joules and in MeV the released energy by the formation of a helium nucleus?*
3. *The power radiated by the Sun is 3.9×10^{26} watts. Which is the loss of mass every second?*
4. *The mass of the Sun is 2×10^{30} kg. Its age is 4.6 billion years. What weight has it lost in percentage since it shines? We give:*

- *Mass of hydrogen: $m_H = 1.0073$ u.*
- *Mass of helium: $m_{He} = 4.0015$ u.*
- *Mass of the positron or the electron: $m_e = 0.55 \times 10^{-3}$ u.*

- *Duration of one year: $D = 365.26$ days.*
- *Mass of ${}_{0}^{0}\nu$ negligible.*
- *$1\,u = 1.66054 \times 10^{-27}\,kg$*
- *Speed of light: $c = 2.998 \times 10^{8}\,m/s$.*
- *$1\,eV = 1.602 \times 10^{-19}\,J$*

Solution

1. *Identification of x:*

Let us write that the charge Z is conserved during the reaction:

$$4 = 2 + 2x + 2\,(0) \quad \Rightarrow \quad x = +1. \tag{1}$$

It is a positron. It is the antiparticle associated with the electron. It has an electric charge of $+1$ e, the same spin and the same mass as the electron. The reaction is rewritten as follows:

$$4\,{}_{1}^{1}H \rightarrow {}_{2}^{4}He + 2\,{}_{+1}^{0}e + 2\,{}_{0}^{0}\nu \tag{2}$$

where ${}_{0}^{0}\nu$ is a particle called *neutrino*. It is a fermion of spin 1/2, electrically neutral. It has a very weak mass. This particle is stable and comes from space, from natural radioactive substances, and from man-made nuclear reactions. There are three types: *neutrino-electron*, *neutrino-muon*, and *neutrino-tau*. The neutrino interacts very weakly with matter. This emission is not dangerous for the human species.

2. *Released energy, in MeV, by the formation of a helium nucleus:*

- *Binding energy in unified atomic mass unit:*

The mass of reagents is:

$$4\,m_H = (4)\,(1.0073\,\text{u}) = 4.0292\,\text{u}. \tag{3}$$

The mass of products is:

$$m_{\text{He}} + 2\,m_e = 4.0015\,\text{u} + (2)\,(0.55 \times 10^{-3}\,\text{u}) = 4.0026\,\text{u}. \tag{4}$$

The binding energy is equal to the difference of the masses of the reactants and products of the reaction:

$$\Delta M = 4\,m_H - (m_{\text{He}} + 2\,m_e) = 4.0292\,\text{u} - 4.0026\,\text{u} = 0.0266\,\text{u}. \tag{5}$$

- *Binding energy in kilograms:*

$$\Delta M = \frac{(0.0266\,\text{u})\,(1.66054 \times 10^{-27}\,\text{kg})}{1\,\text{u}} = 4.417 \times 10^{-29}\,\text{kg}. \tag{6}$$

3. *Released energy in joules and in MeV*:

- *Energy released in joules*:

Knowing that 1 u = 1.66054×10^{-27} kg and in accordance with the Einstein mass-energy equivalence, the released energy in joules is:

$$E_B = \Delta M\, c^2$$
$$= \frac{(0.0266\ u)\left(1.66054 \times 10^{-27}\ kg\right)}{1\ u}\left(2.998 \times 10^8\ m.s^1\right)^2 \qquad (7)$$
$$= 3.97 \times 10^{-12} J.$$

- *Released energy in MeV*:

Knowing that 1 eV = 1.602×10^{-19} J, we obtain:

$$E_B = \frac{\left(0.397 \times 10^{-11}\ J\right)(1\ eV)}{1.602 \times 10^{-19}\ J} = 0.2478 \times 10^8\ eV = 24.78\ MeV. \qquad (8)$$

3. *Loss of mass every second*:

Power is the energy released in 1 second, i.e., 3.9×10^{26} J. For a mass loss of 4.417×10^{-29} kg (see Eq. 6), the released energy is 3.97×10^{-12} J (see Eq. 7). The loss of mass every second is:

$$\delta M = \frac{\left(4.417 \times 10^{-29}\ kg\right)\left(3.9 \times 10^{26}\ J\right)}{3.97 \times 10^{-12}\ J} = 4.34 \times 10^9\ kg. \qquad (9)$$

4. *Lost mass of the Sun in percentage since it radiates*:

Let us express the Sun age in seconds:

$$t = \left(4.6 \times 10^9\ years\right)(365.26\ days)\,(24\ h)(3600\ s) = 1.45 \times 10^{17}\ s. \qquad (10)$$

During this period, the mass of the Sun decreased by:

$$dM = \left(4.34 \times 10^9\ kg/s\right)\left(1.45 \times 10^{17}\ s\right) = 6.29 \times 10^{26}\ kg. \qquad (11)$$

The lost mass of the Sun in percentage is:

$$dM\% = \frac{6.29 \times 10^{26}\ kg}{2 \times 10^{30}\ kg} \times 100 = 0.03\%. \qquad (12)$$

1.7.4 Radioactivity

Exercise 55

Complete the following nuclear reaction equations by indicating the nature of particles represented by a point. Explain each time the nuclear reaction.

1. $^{14}_{7}N + ^{4}_{2}He \rightarrow ^{17}_{8}O + \bullet$.
2. $^{7}_{4}Be \rightarrow ^{7}_{3}Li + \bullet + ^{0}_{0}\nu$.
3. $^{6}_{3}Li + \bullet \rightarrow 2\,^{4}_{2}He$.
4. $^{63}_{29}Cu + ^{1}_{1}p \rightarrow ^{63}_{30}Zn + \bullet$.
5. $^{31}_{14}Si \rightarrow ^{31}_{15}P + \bullet + ^{0}_{0}\nu$.
6. $^{2}_{1}H + ^{3}_{1}H \rightarrow ^{1}_{0}n + \bullet$.

Solution

The missing particles are determined by writing that the charge number and the mass number are preserved during the reaction:

1. $^{14}_{7}N + ^{4}_{2}He \rightarrow ^{17}_{8}O + ^{1}_{1}p$.

This is the disintegration of nitrogen 14 by α-bombardment with the emission of one proton.

2. $^{7}_{4}Be \rightarrow ^{7}_{3}Li + ^{0}_{+1}e + ^{0}_{0}\nu$.

This is a spontaneous disintegration of beryllium 7 with the emission of one positron.

3. $^{6}_{3}Li + ^{2}_{1}H \rightarrow 2\,^{4}_{2}He$.

This is the fission of lithium 6 by bombardment with deuterium nuclei also called *deuterons with α-emission*.

4. $^{63}_{29}Cu + ^{1}_{1}p \rightarrow ^{63}_{30}Zn + ^{1}_{0}n$.

This is a disintegration of copper 63 by bombardment with one proton and the emission of one neutron.

5. $^{31}_{14}Si \rightarrow ^{31}_{15}P + ^{0}_{-1}e + ^{0}_{0}\nu$.

This is a spontaneous disintegration of silicon 31 with β^{-}-emission.

6. $^{2}_{1}H + ^{3}_{1}H \rightarrow ^{1}_{0}n + ^{4}_{2}He$.

This is a fusion of the deuterium and the tritium with neutron and α-emissions.

Exercise 56

1. *Define the fission and fusion nuclear reactions.*
2. *Which of the following nuclear reactions is a fission or fusion reaction?*

(i) $\frac{1}{0}n + \frac{235}{92}U \rightarrow \frac{94}{39}Y + \frac{139}{53}I + 3\,\frac{1}{0}n.$

(ii) $\frac{2}{1}H + \frac{3}{1}H \rightarrow \frac{4}{2}He + \frac{1}{0}n.$

(iii) $\frac{131}{53}I + \frac{3}{2}He \rightarrow \frac{4}{2}He + 2\,\frac{1}{1}H.$

Solution

1. *Definition of fission and fusion nuclear reactions*:

   - During a nuclear fission reaction, a slow neutron (thermal neutron) breaks a fissile heavy nucleus into two light nuclei. This reaction releases energy.
   - During a nuclear fusion reaction, two light nuclei unite to form a heavy nucleus. This reaction also releases energy.

2. *Nature of the reaction*:

   (i) $\frac{1}{0}n + \frac{235}{92}U \rightarrow \frac{94}{39}Y + \frac{139}{53}I + 3\frac{1}{0}n$ Fission.

   (ii) $\frac{2}{1}H + \frac{3}{1}H \rightarrow \frac{4}{2}He + \frac{1}{0}n$ Fission.

   (iii) $\frac{131}{53}I + \frac{3}{2}He \rightarrow \frac{4}{2}He + 2\frac{1}{1}H$ Fission.

Exercise 57

Write the fusion reaction equation between two helium 3 nuclei that gives a helium nucleus 4 and two hydrogen nuclei 1.

Solution

The equation of the reaction is:

$$\frac{3}{2}He + \frac{3}{2}He \rightarrow \frac{4}{2}He + 2\frac{1}{1}H. \tag{1}$$

Exercise 58

Write the fission reaction equation of a uranium 235 nucleus that gives a zirconium nucleus of atomic number $Z = 40$, a tellurium nucleus, and 3 neutrons.

Solution

The equation of the reaction is:

$$\frac{1}{0}n + \frac{235}{92}U \rightarrow \frac{A}{40}Zr + \frac{138}{Z}Te + 3\,\frac{1}{0}n. \tag{1}$$

To determine A and Z, let us write that:

- The mass number is conserved:

$$1 + 235 = A + 138 + 3 \quad \Rightarrow \quad A = 95. \tag{2}$$

- The charge number is conserved:

$$0 + 92 = 40 + Z + 0 \quad \Rightarrow \quad Z = 52. \tag{3}$$

In the end, the reaction is written as:

$$_0^1n + _{92}^{235}U \rightarrow _{40}^{95}Zr + _{40}^{138}Te + 3_0^1n. \tag{4}$$

Exercise 59

Radium being a radioactive element, it is transformed into a stable nucleus of lead 206 by a sequence of decays α and β^-.

1. *Give the composition of a nucleus of radium $_{88}^{226}Ra$.*
2. *Define the decays α and β^- by specifying the nature of the emitted particle and give examples of these two decays.*
3. *Write the equation representing the first disintegration of the nucleus $_{88}^{226}Ra$, knowing that it is of α-type.*
4. *Determine the number of decays of α and β^- types which allow to pass from the nucleus $_{88}^{226}Ra$ to the nucleus $_{82}^{206}Pb$.*

Solution

1. *Composition of the nucleus of radium $_{88}^{226}Ra$:*

Since the charge number is equal to $Z = 88$, the radium nucleus contains 88 protons. Since the mass number is $A = 226$, the number of neutrons in this nucleus is $N = A - Z = 226 - 88 = 138$ neutrons.

2. *Definition of the α and β^- decays:*

- *α-Decay:*

  α-Radioactivity corresponds to the helium nucleus emission composed of two protons and two neutrons. The emitted α-particles are not very penetrating and can be stopped by a paper sheet or by the superficial layers of the skin, but these particles may tear away electrons from the material which they cross, making them particularly ionizing. This type of radiation is used for heavy nuclei. The α-decay is written as follows:

$$_Z^AX \rightarrow _{Z-2}^{A-4}Y + _2^4He. \tag{1}$$

  This transformation is not isobaric since the nuclides father and son do not have the same mass number A. Polonium 210, for example, is α-radioactive. Its nucleus is transformed as follows:

$$_{84}^{210}Po \rightarrow _{82}^{206}N_i + _2^4He. \tag{2}$$

- *β^--Decay:*

  The β^--radioactivity corresponds to the emission of electrons. It concerns the nuclei containing an excess of neutrons with respect to the number of protons, both forming the nucleus ($N > Z$). The β^--rays are moderately penetrating: they can cross the superficial layers of the skin, but a few millimeters of aluminum are enough to stop them. The β^--decay is written as follows:

$$\,_Z^A X \rightarrow \,_{Z+1}^A Y + \,_{-1}^0 e^- . \tag{3}$$

This transformation is isobaric since the father and son nuclides have the same mass number A. Carbon 14, for example, is β^--radioactive. Its nucleus is transformed as follows:

$$\,_6^{14} C \rightarrow \,_7^{14} Y + \,_{-1}^0 e^- . \tag{4}$$

3. *Equation representing the first decay of the nucleus* $\,_{88}^{226} Ra$:

$$\,_{88}^{226} Ra \rightarrow \,_{86}^{206} Rn + \,_2^4 He + \,_0^0 \nu . \tag{5}$$

4. *Number of decays of α and β^- types which allow to pass from* $\,_{88}^{226} Ra$ *to* $\,_{82}^{206} Pb$:

The equation of passage from $\,_{88}^{226} Ra$ to $\,_{82}^{206} Pb$ is:

$$\,_{88}^{226} Ra \rightarrow X \,_{-1}^0 e + Y \,_2^4 He + \,_{86}^{206} Pb + \left(X \,_0^0 \nu \right) . \tag{6}$$

- *Number of α-decays*:

Let us write that the mass number A is conserved during these successive reactions:

$$226 = X \times 0 + Y \times 4 + 206 \quad \Rightarrow \quad Y = 5 \ \alpha \ \text{emissions} . \tag{7}$$

- *Number of β^--decays*:

Let us write that the charge number is conserved during these successive reactions:

$$88 = X \times (-1) + Y \times 2 + 82 \quad \Rightarrow \quad X = 2Y - 6 = 4 \,\beta^- \text{emissions} . \tag{8}$$

Exercise 60

1. *Give the expression of the law of radioactive decay of a nuclide by specifying the meaning of all its terms.*
2. *Deduce the expression of the half-life time $t_{1/2}$.*
3. *We consider a sample containing N_0 polonium nuclei $\,_{84}^{210} Po$. The decay constant of polonium 210 is $\lambda = 5.8 \times 10^{-8} \ s^{-1}$.*

   (a) *Calculate its half-life time $t_{1/2}$ in seconds and days.*
   (b) *How many radioactive nuclei remain at times $t_{1/2}$, $2 \ t_{1/2}$, and $3 \ t_{1/2}$?*
   (c) *Give the shape of the decay curve.*

Solution

1. *Expression of the law of radioactive decay of a nuclide by specifying the meaning of all its terms*:

The disintegration of radioactive nuclei is random at the microscopic level, but at the macroscopic level, the average number of nuclides at an instant t follows a well-defined law called *radioactive decay*. For this purpose, let us consider the number of radioactive nuclides $n(t)$ of a given species present at the instant t. Between the instants t and dt, a number of nuclides of this species have disintegrated and changed in nature. Let dn be the variation of the number of nuclides between these two instants. This variation depends on the number $n(t)$, the nature of the nuclide, and the length of time dt:

$$dn = -n(t)\, \lambda\, dt \;\; \Rightarrow \;\; \frac{dn}{n(t)} = -\lambda\, dt \;\; \Rightarrow \;\; n(t) = n_0\, e^{-\lambda t} \tag{1}$$

where λ is a constant which depends on the nuclide nature; it is expressed in s^{-1}. It determines the probability of disintegration of a nucleus per unit of time (not to be confused with the wavelength that is expressed with the same symbol). n_0 is the nuclide number present in the sample at the initial moment $t = 0$. The minus sign indicates a decrease of $n(t)$. The Eq. (1) shows that the decay of radioactive nuclei follows a decreasing exponential law.

2. *Expression of the half-life time $t_{1/2}$*:

The half-life $t_{1/2}$ of a radioactive nucleus is the time necessary for the decay of half of nuclei present in a sample of this species. From (1), we deduce:

$$\frac{n(t)}{n_0} = e^{-t_{1/2}} = \frac{1}{2} \;\; \Rightarrow \;\; t_{1/2} = \frac{ln2}{\lambda} = \frac{0.693}{\lambda}. \tag{2}$$

The half-life $t_{1/2}$ is very different from one nucleus to another. The half-life has only a statistical value. It indicates that a radioactive nucleus has a half chance to disappear after a half-life. It is greater than 10^{30} years for vanadium and less than 2.96×10^{-7} s for polonium 212.

3. (a) *Half-life time $t_{1/2}$ in seconds and days*:

From (2), we deduce:

$$t_{1/2} = \frac{0.693}{\lambda} = \frac{0.693}{5.8 \times 10^{-8}\ s^{-1}} = 1.1948 \times 10^7\ s = 138.29\ \text{days}. \tag{3}$$

(b) *Radioactive nuclei remaining at times $t_{1/2}$, $2\,t_{1/2}$, and $3\,t_{1/2}$:*

The number of remaining radioactive nuclei at the three different dates is summarized in the following table:

t_0	n_0
$t_{1/2}$	$\dfrac{n_0}{2}$
$2\,t_{1/2}$	$\dfrac{n_0}{2^2} = \dfrac{n_0}{4}$
$3\,t_{1/2}$	$\dfrac{n_0}{2^3} = \dfrac{n_0}{8}$

(c) *Shape of the decay curve*:

In this figure, we give the curve of the radioactive decay where the number of radioactive nuclei is divided by two at the end of each radioactive period.

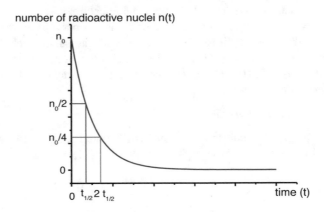

Exercise 61

A radioactive nucleus of radon $^{222}_{86}Ra$ decays by emitting one α-particle. We have a sample of mass $m = 1$ g of this isotope. The life-time of radon 222 is $t_{1/2} = 3.8$ days.

1. *Write the equation of the α-decay by specifying which laws of conservation are used. Specify the nature of the nucleus son.*
2. *Calculate the constant of decay of radon 222.*
3. *How many radioactive nuclei are present in the sample?*
4. (a) *What is the activity of a radioactive source?*
   (b) *What is the activity of this source at the initial moment?*
   (c) *What will it be after 15 days? We give Avogadro's number, $N_A = 6.02 \times 10^{23}$ atoms.*

Solution

1. *Equation of the α-decay and nature of the son nucleus*:

The general decay equation with α-emission is given by:

$$_Z^A X \rightarrow \,_{Z-2}^{A-4} Y + \,_2^4 He + \gamma. \tag{1}$$

In the case of radon 222 decay, we obtain:

$$_{86}^{222} Rn \rightarrow \,_{84}^{218} Po + \,_2^4 He + \gamma. \tag{2}$$

All nuclear reactions verify the laws of conservation of the mass number A and the charge number Z. The nucleus son is polonium $_{84}^{218} Po$.

2. *Calculation of the constant of decay of radon 222*:

The $t_{1/2}$ period of a radioactive nucleus is the time necessary for the decay of half of present nuclei in a sample of this species. It is given by:

$$\frac{n(t)}{n_0} = e^{-\lambda t_{1/2}} = \frac{1}{2} \;\Rightarrow\; t_{1/2} = \frac{\ln 2}{\lambda} = \frac{0.693}{\lambda} \;\Rightarrow\; \lambda = \frac{\ln 2}{t_{1/2}}$$

$$= \frac{0.693}{(3.8 \text{ j}) \, (24 \text{ h}) \, (3600 \text{ s})} = 2.11 \times 10^{-6} \text{ s}^{-1}. \tag{3}$$

3. *Radioactive nuclei present in the sample*:

Knowing that 1 mole of radon 222 weighs 222 g and contains 6.02×10^{23} atoms, the number n_0 of radioactive nuclei initially present is:

$$n_0 = \frac{m \, N_A}{M(\text{Ra})} = \frac{(1 \text{ g}) \, (6.02 \times 10^{23} \text{ atoms})}{222 \text{ g}} = 2.71 \times 10^{21} \text{ atoms}. \tag{4}$$

4. (a) *Definition of the activity of a radioactive source*:

The activity A of a radioactive source at a given instant is equal to the decay rate of the radioactive nuclei constituting it at this moment (not to be confused with the mass number which expresses itself with the same symbol):

$$A(t) = -\frac{dn(t)}{dt} = \lambda n(t) = \lambda n_0 \, e^{-\lambda t}. \tag{5}$$

The unit of the activity is becquerel (Bq) (1 Bq = 1 decay per second). This number depends on the decay constant λ and the average number $n(t)$ of nuclei present in the sample at time t. Just like $n(t)$, the activity A decreases exponentially in time.

(b) *Activity of the radioactive source at the initial time*:

From (3), (4), and (5), we deduce the activity of the radioactive source at $t = 0$:

$$A_0 = \lambda n_0 = \left(2.11 \times 10^{-6}\,\text{s}^{-1}\right)\left(2.71 \times 10^{21}\,\text{atoms}\right) = 5.72 \times 10^{15}\,\text{Bq}. \quad (6)$$

(c) *Activity of the radioactive source A after $t' = 15$ days:*

$$A' = \lambda n_0 e^{-\lambda t'} = A_0\, e^{-\lambda t'}$$
$$= \left(5.72 \times 10^{15}\,\text{Bq}\right) \exp\left[-\left(2.11 \times 10^{-6}\,\text{s}^{-1}\right)(15\,\text{j})(24\,\text{h})(3600\,\text{s})\right] \quad (7)$$
$$= 3.71 \times 10^{15}\,\text{Bq}.$$

Exercise 62

Cosmic neutrons bombard the nitrogen nuclei of the upper atmosphere according to Equation (i). Carbon 14 decays, according to Equation (ii).

$$^{14}_{7}N + ^{1}_{0}n \rightarrow ^{14}_{6}C + ^{b}_{a}X. \quad (i)$$
$$^{14}_{6}C \rightarrow ^{d}_{c}Y + ^{0}_{-1}e + ^{0}_{0}\nu. \quad (ii)$$

1. *Identify X and Y.*
2. *On a prehistoric site, some taken fragments of bone have an activity of 113.75 decays per hour and per gram. On a fragment of bone of a man dead recently, the activity is 911.7 decays per hour and per gram.*
   (a) *Explain the principle of carbon dating.*
   (b) *What is the age of the prehistoric bone? We give the period of carbon 14: $T = 5500$ years.*
   (c) *By assuming that the detection limit corresponds to a residual percentage of 1% of carbon 14, what is the most distant age that can be determined by this method of dating? What solution do you recommend to remedy it?*

Solution
1. *Identification of X and Y:*

   • Let us write that the mass number is conserved during reactions (i) and (ii):

   From (i), we obtain:

   $$14 + d = 14 + b \implies b = 1. \quad (1)$$

   From (ii), we obtain:

   $$14 = d + 0 \implies d = 14. \quad (2)$$

   • Let us write that the charge number is conserved during reactions (i) and (ii):

   From (i), we obtain:

$$7 + 0 = 6 + a \quad \Rightarrow \quad a = 1. \tag{3}$$

From (ii), we obtain:

$$6 = c - 1 \quad \Rightarrow \quad d = 7. \tag{4}$$

We deduce that:

- $_a^b X = {}_1^1 H$: It is the hydrogen atom.
- $_c^d Y = {}_7^{14} N$: It is the nitrogen atom.

(a) *Principle of the carbon 14 dating*:

Carbon 14, produced in the atmosphere by the action of cosmic rays on the nitrogen nuclei, is fixed by living organisms, plants, animals, and humans where its content remains constant as long as they are alive. After their death, carbon 14 is no longer assimilated. The measurement of the content in carbon 14, which decreases in time by radioactive decay, allows dating the death moment.

(b) *Age of the prehistoric bone*:

The temporal evolution of a radioactive decay is given by the following equation:

$$n(t) = n_0 \, e^{-\lambda t} \quad \Rightarrow \quad ln \frac{n_0}{n(t)} = t. \tag{5}$$

The period T is the time after which the initial number of radioactive atoms is divided by 2. The value of T is independent of n_0:

$$T = \frac{ln\,2}{\lambda} \quad \Rightarrow \quad \lambda = \frac{ln\,2}{T}. \tag{6}$$

The activity A of a radioactive source at a given moment is equal to the decay rate of radioactive nuclei which constitute it at this moment:

$$A = -\frac{dn(t)}{dt} = \lambda \, n(t). \tag{7}$$

The unit of A is becquerel (Bq): 1 Bq = 1 decay per second. Let A_0 be the activity of the man recently dead containing n_0 atoms of carbon 14 in 1 gram and A that of the taken bone fragments to date containing N atoms of carbon 14 in 1 gram. According to (7), we have:

$$\begin{aligned} A_0 &= \lambda n_0 \\ A &= \lambda n(t) \end{aligned} \quad \Rightarrow \quad ln \frac{A_0}{A} = ln \frac{n_0}{n(t)}. \tag{8}$$

By injecting (5) into (8), we obtain:

$$\ln\frac{A_0}{A} = \ln\frac{n_0}{n(t)} = \lambda t. \tag{9}$$

By injecting (6) into (9), we find:

$$\ln\frac{A_0}{A} = \ln\frac{n_0}{n(t)} = \lambda t = \frac{\ln 2}{T} \times t \;\Rightarrow\; t = T \times \frac{\ln\frac{A_0}{A}}{\ln 2}$$

$$= (5500 \text{ years}) \; \frac{\ln\left(\dfrac{911.7 \text{ per hour and per gram}}{113.75 \text{ per hour and per gram}}\right)}{\ln 2} \tag{10}$$

$$= 16\,500 \text{ years.}$$

(c) *The most distant age that can be determined by this method of dating and proposed solution*:

- If the detection limit is:

$$\frac{n(t)}{n_0} = 0.01 \;\Rightarrow\; \frac{n_0}{n(t)} = 100. \tag{11}$$

The corresponding value of t is:

$$t = T \times \frac{\ln\frac{n_0}{n(t)}}{\ln 2} = (5500 \text{ years})\frac{\ln 100}{\ln 2} = 36\,548 \text{ years} \approx 36\,600 \text{ years.} \tag{12}$$

- For a dating beyond this value, it will be necessary to use natural isotopes of longer period than that of carbon 14 such as thorium 230 ($T = 75\,000$ years) or uranium 234 ($T = 248\,000$ years).

Exercise 63
1. *The energy production in a pressurized water nuclear reactor is based on the fission of uranium 235. In fact, when a neutron strikes one uranium nucleus $^{235}_{92}U$, one of the possible fissions leads to the formation of one cesium nucleus $^{146}_{58}Ce$ and one selenium nucleus $^{85}_{34}Se$, as well as to the emission of a number x of neutrons.*
   (a) *Write the complete equation of this nuclear reactions.*
   (b) *Deduce the value of x. Justify your answer by expressing the applied laws.*
   (c) *Calculate the mass variation Δm that accompanies the fission of one uranium 235 nucleus in unified atomic mass unit and in kilograms.*
   (d) *Calculate the released energy in joules and in MeV.*

2. *The nuclear power plant that uses uranium 235 provides at most an electrical power $P = 1455$ MW. The combustion of 1 kilogram of oil releases an energy $E = 45 \times 10^6$ J in the form of heat. The yield of the transformation of the thermal energy into the electrical energy is 34.2%.*

   (a) *Deduct the mass of oil in kilograms and in tonnes that would be needed to produce for 1 year the same electrical energy as the nuclear power plant.*

3. *The fission of uranium 235 produces cesium 137 which is a γ-radioactive emitter. An employee of the plant accidentally remains for 1 hour near a source of cesium 137. During this exposure, he absorbs, evenly over his entire body, 5% of emitted γ-rays from this source. One cesium 137 nucleus releases 0.66 MeV. We assume that the activity of this source is equal to $A = 3 \times 10^{12}$ Bq.*

   (a) *Given that this employee weighs 70 kg, calculate the absorbed dose in gray (J/kg).*

   (b) *The concept of dose is not sufficient to explain the dose-response reactions. For this reason, we match it a radiological weighting factor W_R of the incident radiation. So, the product of this factor of the absorbed dose reflects an "equivalent dose" expressed in sievert (Sv); the maximum authorized dose per year is 50 mSv. Calculate the dose received by the employee of the plant, victim of the accident, knowing that W_R is 0.06 in these conditions.*

   (c) *Comment on this result. We give:*
   - *Mass of the uranium nucleus: $m_U = 235.044$ u.*
   - *Mass of the cesium nucleus: $m_{Ce} = 145.910$ u.*
   - *Mass of the selenium nucleus: $m_{Se} = 84.922$ u.*
   - *Mass of the neutron: $m_N = 1.008\,6$ u.*
   - *$1\ u = 1.66 \times 10^{-27}$ kg*
   - *$1\ eV = 1.6 \times 10^{-19}$ J*
   - *Speed of light $c = 3 \times 10^8$ m/s.*
   - *1 year $= 365$ days*

Solution

1. (a) *Complete the equation of the nuclear reaction:*

On the basis of hypotheses, the equation of the nuclear reaction is given by:

$$^{235}_{92}U + ^{1}_{0}n \rightarrow ^{146}_{58}Ce + ^{85}_{34}Se + x\ ^{1}_{0}n. \tag{1}$$

- Let us write that the mass number is conserved during the reaction:

From (1), we obtain:

$$235 + 1 = 146 + 85 + x \quad \Rightarrow \quad x = 5. \tag{2}$$

- Let us write that the charge number is conserved during the reaction:

From (1), we get:

$$92 + 0 = 58 + 34 + (x)(0) \quad \Rightarrow \quad 92 = 92. \tag{3}$$

From this, we deduce the complete equation of the nuclear reaction:

$$_{92}^{235}U + _{0}^{1}n \rightarrow _{58}^{146}Ce + _{34}^{85}Se + 5_{0}^{1}n. \tag{4}$$

(c) *Variation of the mass Δm that accompanies the fission of one nucleus of uranium 235 in unified atomic mass unit and in kilograms*:

- *Variation of the mass Δm in unified atomic mass unit*:

  The mass of reagents is:

$$m_U + m_N = 235.044 \text{ u} + 1.0086 \text{ u} = 236.053 \text{ u}. \tag{5}$$

The mass of products is:

$$m_{Ce} + m_{Se} + 4\,m_N = 145.910 \text{ u} + 84.922 \text{ u} + (4)(1.0086)\text{u} = 235.841 \text{ u}. \tag{6}$$

The mass variation is equal to the difference in the masses of reagents and products of the reaction:

$$\Delta m = (m_U + m_N) - (m_{Ce} + m_{Se} + 4m_N) = 236.053 \text{ u} - 235.841 \text{ u}$$
$$= 0.212 \text{ u}. \tag{7}$$

- *Variation of the mass Δm in kilograms*:

  Knowing that $1 \text{ u} = 1.66 \times 10^{-27}$ kg, we deduce:

$$\Delta m = \frac{(0.212 \text{ u})(1.66 \times 10^{-27} \text{ kg})}{1 \text{ u}} = 0.352 \times 10^{-27} \text{ kg}. \tag{8}$$

(d) *Released energy in joules and in MeV*:

- *Released energy in joules*:

  According to Einstein's mass-energy equivalence, the released energy in joules is:

$$E_B = \Delta M\, c^2 = (0.352 \times 10^{-27} \text{ kg})(3 \times 10^8 \text{ m.s}^1)^2 = 3.128 \times 10^{-11} \text{ J}. \tag{9}$$

- *Released energy in MeV*:

  Knowing that $1 \text{ eV} = 1.6 \times 10^{-19}$ J, we obtain:

$$E_B = \frac{\left(3.128 \times 10^{-11}\ \text{J}\right)(1\ \text{eV})}{1.6 \times 10^{-19}\ \text{J}} = 1.955 \times 10^8\ \text{eV} = 195.5\ \text{MeV}. \tag{10}$$

2. (a) *Oil mass in kilograms and in tonnes*:

The nuclear plant that uses uranium 235 provides a maximum of electrical power of 1455 megawatts (10^6 watts). It releases in 1 year energy equal to:

$$P = \frac{\varepsilon}{t} \quad \Rightarrow \quad \varepsilon = P \times t = \left(1455 \times 10^6\ \text{W}\right)(365\ \text{d})(24\ \text{h})(3600\ \text{s})$$
$$= 4.6 \times 10^{16}\ \text{J}. \tag{11}$$

The combustion of 1 kilogram of oil releases an energy $E = 45 \times 10^6$ J in the form of heat, which is then transformed into electrical energy with a yield of 34.2%. As a result, each kilogram of oil actually provides an electrical energy equal to:

$$E' = \frac{34.2 \times E}{100} = \frac{(34.2)\left(45 \times 10^6\ \text{J}\right)}{100} = 1.54 \times 10^7\ \text{J}. \tag{12}$$

The oil mass in kilograms needed to produce for 1 year the same electrical energy as the nuclear power station is:

$$M = \frac{\left(4.6 \times 10^{16}\ \text{J}\right)(1\ \text{kg})}{1.54 \times 10^7\ \text{J}} = 2.99 \times 10^9\ \text{kg} = 2.99 \times 10^6\ \text{tonnes}. \tag{13}$$

3. (a) *Absorbed dose by the employee in gray (J/kg)*:

The activity of the source of cesium 137 is 3×10^{12} Bq. Becquerel (Bq) is the unit of measurement of the radioactivity of a source. It characterizes the number of spontaneous decays per second. The more this number is important, the more this source is strongly radioactive. So, in 1 second, 3×10^{12} cesium 137 nuclei decayed, releasing energy equal to:

$$E''(\text{in 1 s}) = (0.66\ \text{MeV})\left(3 \times 10^{12}\ \text{nuclei}\right) = 1.98 \times 10^{12}\ \text{MeV}$$
$$= 1.98 \times 10^{18}\ \text{eV}. \tag{14}$$

That is to say in joules, with by hypothesis 1 eV $= 1.6 \times 10^{-19}$ J:

$$E''(\text{in 1 s}) = \frac{\left(1.6 \times 10^{-19}\ \text{J}\right)\left(1.98 \times 10^{18}\ \text{eV}\right)}{1\ \text{eV}} = 0.32\ \text{J}. \tag{15}$$

In 1 hour, the emitted energy from the source is:

$$E''(\text{in } 1\,\text{h}) = \frac{0.32\,\text{J}}{1\,\text{s}}(3600\,\text{s}) = 1152\,\text{J}. \tag{16}$$

As during this exposure, the employee absorbs only 5% of γ-rays emitted by this source, the energy received is therefore:

$$E'''(\text{in } 1\,\text{h}) = \frac{(1152\,\text{J})\,(5)}{100} = 57.6\,\text{J}. \tag{17}$$

That is to say a received energy per kg is equal to:

$$E''''(\text{in } 1\,\text{h and per kg}) = \frac{57.6\,\text{J}}{70\,\text{kg}} = 0.82\,\text{Gr or J/kg}. \tag{18}$$

(b) *Received equivalent dose in sievert by the employee*:

The sievert (Sv) measures the biological effects of ionizing radiations on the living matter. At equal doses, these effects vary as a function of the nature of the ionizing radiation. On equal energy, the impact of α-radiation is double that of β- and γ-radiations. To translate these differences, a radiological weighting factor W_R is introduced to relate the absorbed dose D in gray to the equivalent dose H in sievert:

$$H = D \times W_R. \tag{19}$$

The weighting factor W_R allows taking into account the radiation nature and its effect on the tissues of the living matter. From (19), we deduce the equivalent dose in sievert:

$$H = D \times W_R = (0.82\,\text{Gr})\,(0.06) = 4.92 \times 10^{-2}\,\text{Sv} = 49.2\,\text{mSv}. \tag{20}$$

(c) The allowable dose threshold for a person is assumed to be 50 mSv/year. We are therefore at the limit of the authorized annual maximal dose.

Chapter 2
Matter-Radiation Interaction

Abstract During interaction of the electromagnetic radiation with matter, five phenomena can occur depending on the energy level of the incident photons: the Rayleigh scattering, photoelectric effect, Compton scattering, creation of electron-positron pairs, and nuclear photo-production. In this chapter, only the photoelectric effect and the Compton scattering are studied because, historically, it is these two phenomena that have defeated classical mechanics.

This chapter begins with a precise course, followed by exercises in order of increasing difficulty. These exercises are solved in a very detailed way; they allow students to assimilate the course and help them prepare for exams. It is intended for first-year university students. It can also benefit bachelor's and doctoral students because of the deliberately simplified presentation of the treated concepts.

2.1 Multiple Interaction Phenomena

When the electromagnetic radiation (photons) interacts with the material, five phenomena can occur according to the energy level of incident photons.

2.1.1 Rayleigh's Scattering

The Rayleigh scattering is an elastic diffusion that corresponds to the interaction of a low-energy photon with an atom or a molecule. The photon is deflected without loss of energy. It occurs when the size of scattering particles is much smaller than the wavelength of the radiation. This explains, for example, the blue color of the sky during the day.

2.1.2 Photoelectric Effect (<0.5 MeV)

This phenomenon occurs at energies lower than those of Compton's scattering. It is not predominant in radiotherapy, but is in medical imagery where the used energies are weaker, of the order of keV (1 keV = 10^3 eV). When an incident photon interacts with an atom of the material, it is absorbed by transferring all of its energy to an electron of that atom. An electron of inner shells is then ejected from the electronic shell diagram. Ejected electrons are called *photoelectrons*; but there is nothing that would distinguish them from other electrons. All electrons are identical to one another in mass, charge, spin, and magnetic moment. This causes the excitation of the atom, leading to a rearrangement of the electron shell chart. Electrons of outer shells will fill the inner shells. This modification causes the emission of a second electron from peripheral shells, called *an Auger electron*. An X-ray fluorescence photon can be also emitted.

2.1.3 Compton's Scattering (Between 0.5 and 3 MeV)

Compton's scattering is a commonly employed phenomenon in radiotherapy. The energy range of the used radiation is of the order of MeV (1 MeV = 10^6 eV). Incident photon interacts with an electron of the atom and transmits to it a part of its energy. This electron belongs to outer shells and thus is less bound to the atom. The electron is freed and becomes a secondary electron. It can in turn interact with the atoms of material. The more the incident photon is energetic, the more the secondary electron is emitted forward with an important energy. The incident photon is scattered in a direction different from its initial direction with a reduced energy because a part of the incident photon energy has been transmitted to the electron.

2.1.4 Creation of Electron-Positron Pairs (Between 1 and 10 MeV)

The creation of electron-positron pairs, also called *materialization effect*, occurs in higher energy ranges than those of previous effects. In radiotherapy, its contribution increases when high radiation energies are used. When a photon interacts with the intense electric field around the nucleus, all its energy is transmitted to it and is re-emitted in the form of two particles: an electron and a positron of equal energy. Thus, the incident photon should have energy superior or equal to the cumulative energy of two created particles: an electron of a negative sign (e^-) and a positron of positive sign (e^+). The electron and the positron exhaust all their kinetic energies by producing electronic collisions with the atoms of the material. Once almost all of the energy of the positron is exhausted, this one meets an electron and both particles of opposite signs annihilate each other: two photons are then emitted in two opposite directions. The annihilation between a particle and its antiparticle thus consists in their disappearance, giving rise to a release of energy.

2.1.5   Nuclear Photo-Production (>10 MeV)

In the energy range of radiotherapy, this reaction is very unlikely. It occurs mainly with very high energies, of the order of GeV (1 GeV $= 10^9$ eV). The nuclear photo-production is the absorption of a very energetic photon by the nucleus of the atom. The latter is then in an unstable state and re-emits a neutron to return to the equilibrium.

In what follows only the photoelectric effect and the Compton scattering will be studied in detail.

2.2   Electromagnetic Waves

2.2.1   Introduction

Electromagnetic radiation refers to a transfer of energy from the transmitting source to the interaction point. This radiation has no mass and is electrically neutral. It is characterized only by its energy, the direction of propagation of the wave and its polarization. This energy is associated with a wavelength λ. The white light is, for example, electromagnetic radiation visible to the human eye. However, it is only a small part of a wide spectrum called the *electromagnetic spectrum*. It corresponds to the decomposition of the electromagnetic radiation, according to its characteristics (energy, wavelength, frequency, etc.). The used radiations in medicine and especially in radiotherapy are mainly X-rays and γ-rays. X-rays and γ-rays are of the same nature (also called *photons X and γ*). Only their means of production differ. Their range of energy goes from some keV to several hundred GeV, which makes them usable for medicine and in particular in medical imaging and radiotherapy.

2.2.2   Wave Definition

A wave is the propagation step by step of a disturbance. A pebble that falls into the water creates a local disturbance. This disturbance propagates in all directions, forming concentric circles of the wave. A rope that we agitate at one of its extremities leads to the propagation of a wave. A spring that we compress or stretch creates a wave that propagates. The sound wave causes a compression of air particles that propagates in the air. A wave does not transmit matter, but a disturbance. The matter may eventually move and temporarily move away from its equilibrium position then back to its original position. The wave does not carry matter, but it carries energy! These examples show that there are several kinds of waves that we can group into two categories: waves which propagate through a physical matter and waves that do not require a material medium.

2.2.3 *Electromagnetic Wave*

An electromagnetic wave includes both an electric field $\vec{E}$ and a magnetic field $\vec{B}$ oscillating at the same frequency ν. These two fields, perpendicular to each other, propagate in an environment in an orthogonal direction (Fig. 2.1). This propagation is carried out at a speed that depends on the considered environment. In the vacuum, its speed is equal to 3×10^8 m.s^{-1}.

An electromagnetic wave is characterized by several physical quantities including:

- *The wavelength (λ)*: it expresses the periodic oscillatory character of the wave in space. This is the length of one wave cycle, that is to say the distance that separates two successive peaks. It is measured in meters or in one of its sub-multiples; the electromagnetic waves used in remote sensing space, for example, have very short wavelengths.
- *The period (T):* it represents the time required for the wave to cycle. Its unit is the second.
- *The frequency (ν):* it is the reciprocal of the period; it reflects the number of cycles per unit of time. It is expressed in hertz (Hz) or the multiples of the hertz (one Hz is equivalent to one oscillation per second). The electromagnetic waves used in remote sensing space have very high frequencies.

The wavelength λ and the frequency ν are inversely proportional and linked by the following equation:

Fig. 2.1 Electromagnetic wave: it includes both an electric field $\vec{E}$ and a magnetic field $\vec{B}$ oscillating at the same frequency ν. This wave is created by the vibration of an electric charge

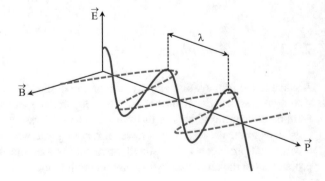

$$\lambda = \frac{c}{\nu} \tag{2.1}$$

where c is the speed of light in the vacuum (3×10^8 m.s^{-1}). Consequently, the more the wavelength is small, the more the frequency is high and vice versa.

Electromagnetic waves are created by the vibration of electric charges: the change of position of the electric charge creates a disturbance of the electric field and the magnetic field, generating a transverse electromagnetic wave. These waves have no material support, so they can propagate in the vacuum. They are classified by their frequencies ν in hertz or by their wavelengths λ in meters. The electromagnetic spectrum is very large, ranging from γ-rays of wavelengths $\lambda \approx$ fm-pm with 1 femtometer (fm) $= 10^{-15}$ m and 1 picometer (pm) $= 10^{-12}$ m to alternating currents of wavelengths $\lambda \approx$ Mm with 1 megameter (Mm) $= 10^6$ m by passing by X-rays of wavelengths $\lambda \approx$ nm with 1 nanometer (nm) $= 10^{-9}$ m, the ultraviolet of wavelengths $\lambda \approx$ nm-μm with 1 micrometer (μm) $= 10^{-6}$ m, the visible of wavelengths $\lambda \approx$ 400–800 nm, the infrared of wavelengths $\lambda \approx \mu$m-mm with 1 millimeter (mm) $= 10^{-3}$ m, microwaves, TV waves of wavelengths $\lambda \approx$ mm-m, and radio waves of wavelengths $\lambda \approx$ m-km with 1 kilometer (km) $= 10^3$ m. The visible domain comprises, in increasing order of wavelengths, violet, blue, green, yellow, orange, and red.

2.2.4 Intensity of an Electromagnetic Wave

The intensity of the electromagnetic field in vacuum is given by:

$$\vec{P} = \frac{1}{\mu_0} \vec{E} \times \vec{B} \tag{2.2}$$

where $\vec{P}$ is the Poynting vector named after the British physicist John Henry Poynting who demonstrated it (it represents the intensity of the electromagnetic field in the vacuum expressed in watts per square meter (W/m^2), $\vec{E}$ is the electric field at the Poynting vector in newtons by coulomb (N/C), $\vec{B}$ is the magnetic field at the Poynting vector in teslas (T), and μ_0 is the permeability of vacuum ($\mu_0 = 4\pi \times 10^{-7}$ N. s^2. C^{-2}). The Poynting vector $\vec{P}$ is, up to a constant, the vectorial product of the electric field $\vec{E}$ and the magnetic field $\vec{B}$.

Let us consider a sinusoidal plane wave propagating in the vacuum:

$$E = E_0 \sin(\omega t + \varphi) \tag{2.3}$$

with E_0, the maximum electric field modulus of the wave in N/C; ω, the angular frequency of the electric field in rad/s; ($\omega t + \varphi$) the phase at time t in radians (rad);

Fig. 2.2 Average intensity $\overline{P}$ of the electromagnetic wave as a function of the modulus of the maximum electric field E_0. This intensity is proportional to the square of the modulus of the electric field in the case of a sinusoidal plane wave

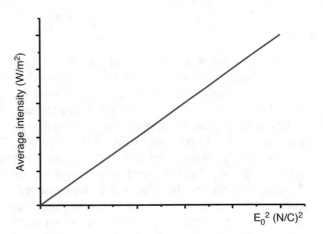

and φ, the phase at $t = 0$ also in rad. The average value of the intensity of the electromagnetic wave $\overline{P}$ in W/m^2 is given in this case by:

$$\overline{P} = \frac{\varepsilon_0 c}{2} E_0^2 \qquad (2.4)$$

where ε_0 is the electric constant $\varepsilon_0 = 8.85 \times 10^{-12}$ C^2.N^{-1}.m^{-2}. Equation (2.4) shows that the average value of the intensity of the electromagnetic wave $\overline{P}$ is proportional to the square of the modulus of the electric field of the wave E_0^2 (Fig. 2.2).

2.3   Photoelectric Effect

2.3.1   Introduction

In 1887, the electromagnetic theory seemed to report all properties of light with the exception of an experiment carried out by Antoine César Becquerel and his son Alexandre Edmond Becquerel in 1839 which revealed a modification of the electrical behavior of electrodes immersed in a liquid by lighting. This effect was considered as paradoxical within the framework of electrodynamics of electrons of the time resulting from the works of Maxwell, Thomson, and Lorentz. Indeed, in this experiment, the energy of electrons did not seem to depend on the intensity of light, in contradiction with what we thought of continuous models of the distribution of light energy and of the first models of the atom. In 1887, Heinrich Rudolf Hertz perceived by realizing an experiment on the production of electric oscillations that a spark sprang more easily between two electrodes when they were negatively polarized and illuminated by an ultraviolet light. This phenomenon was attributed in 1888 by Wilhelm Hallwachs to the emission of negative charges that Philipp Eduard

Fig. 2.3 Leaf electroscope.
When we illuminate the zinc
plate with an ultraviolet light
source, the electroscope leaf
falls down, indicating that
the electroscope discharges

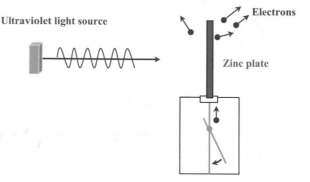

Anton von Lenard identified in 1900 with the electrons that were highlighted. This discovery of the photoelectric emission, and later of its laws, is one of the strongest arguments that led various scientists of that time, in particular Albert Einstein, to take again the corpuscular diagram used by Isaac Newton to describe the light but abandoned after Hertz's works.

2.3.2   Highlighting

Using the apparatus described in Fig. 2.3, Heinrich Rudolf Hertz accidentally discovered in 1887 that negatively charged metals lost negative charges when they are exposed to ultraviolet radiations. Let us have a closer look at this experiment:

(i) Let us consider, for this purpose, a zinc plate placed on a negatively charged electroscope. When this plate is illuminated with an ultraviolet light source, the electroscope leaf falls down, indicating that the electroscope discharges.

(ii) We do the same experiment, but this time the electroscope is positively charged. No modification is observed.

(iii) The electroscope is again negatively charged, but this time we interpose between the zinc plate and the light source a glass slide which has the property to absorb ultraviolet rays. No modification is observed.

These experimental facts are explained by the fact that ultraviolet radiations expelled free electrons from the zinc plate. Excess electrons which were in the electroscope then migrate to the electron-deficient plate and the electroscope discharges.

2.3.3   Definition

The photoelectric effect refers to the emission of electrons by the surface of a metal (or other substance) under the action of light (Fig. 2.4). There are two effects:

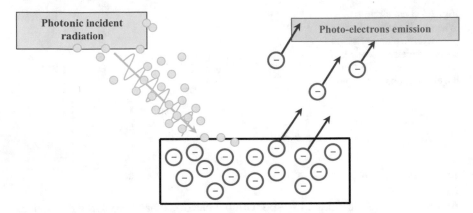

Fig. 2.4 Photoelectric effect. One photon which is absorbed by the metal cedes its energy to one electron. If this energy is sufficiently large to release the electron from the field of attractive forces that binds it to the atom, it can leave the surface of the metal with some kinetic energy

electrons are ejected from the material (photoelectric emission) and a change in the conductivity of the material (photovoltaic effect) that occurs inside a photovoltaic cell.

2.3.4 Observations

The experimental study of the emission of electrons from the surface of a metal under the action of light led to the following conclusions:

  (i) Electrons are emitted only if the frequency of light is sufficiently high and exceeds a limit frequency called *threshold frequency*.
 (ii) This threshold frequency depends on the material. It is directly related to the binding energy of electrons in the electron shells of the atom.
(iii) The number of electrons emitted during the exposure to light, which determines the intensity of the electric current, is proportional to the intensity of the light source.
 (iv) The kinetic energy of emitted electrons is linearly dependent on the frequency of the incident light.
  (v) The photoelectric emission phenomenon is almost instantaneous. It occurs in an extremely short time, less than 10^{-9} second after lighting.

These facts of the experiment are in contradiction with the wave theory of light. Indeed, if the wave nature of the light is admitted, by increasing its intensity, we should be able to supply enough energy to the material to eject electrons. However, the experiment shows this is not the case: the energy transfer causing the emission of electrons can be done only from a certain threshold frequency and no delay is detected between the beginning of illumination and the beginning of the emission of electrons.

2.3.5 *Interpretation*

In 1905, Albert Einstein explained for the first time the photoelectric phenomenon. But, his hypotheses were accepted only in 1916 when Robert Andrews Millikan confirmed it experimentally. Using Max Planck's hypothesis, Einstein postulated that a beam of light consisted of small amounts of energy called *light quanta* or *photons*. The energy E of one photon is equal to its frequency ν multiplied by a constant h:

$$E = h\nu \tag{2.5}$$

where h is Planck's constant ($h = 6.626 \times 10^{-34}$ J.s). The energy E is expressed in joules (J) if the frequency ν is expressed in hertz (Hz). One photon that is absorbed cedes its energy to one electron. If this energy is sufficiently large to release the electron from the field of attractive forces that binds it to the atom, it can leave the surface of the metal. As the probability that one electron absorbs simultaneously two photons is very weak, each electron extracted from the metal takes all its energy from a single photon. These electrons are attached to the metal, and to extract one, it is necessary to supply an energy of extraction W_S that compensates the electron-core binding energy. So that one electron escapes from the metal, the energy of the photon must be greater than the electron-core binding energy W_S, i.e.:

$$h\nu > W_S \quad \text{with} \quad W_S = h\nu_S \tag{2.6}$$

where ν_S is the threshold frequency of one photon that would have energy equal to the energy required to detach one electron from its shell. The photoelectric equation of Einstein is written as:

$$\Delta E = h\nu - h_S = \frac{1}{2} m_e v_{max}^2 \tag{2.7}$$

where $m_e = 9.1 \times 10^{-31}$ kg is the electron mass and v_{max} its maximal speed. The excess energy is used by the detached electron in the form of kinetic energy to move. When the frequency of the incident radiation is equal to the threshold frequency ν_S, the speed of emitted electrons vanishes. The electron is dissociated from the atom, but does not leave the material, and Eq. (2.7) is written as:

$$\Delta E = h\nu - h_S = 0 \Rightarrow h\nu = h\nu_S. \tag{2.8}$$

The determination of the kinetic energy of the emitted electron is done by the method of the retarding field. In this method, the electron crosses an area where there is an electric field which opposes to its move, and of which the intensity is varied until this electron stops. In this case, we obtain, for a measurable stopping potential U_0, the following equation:

$$eU_0 = \frac{1}{2}mv^2_{\text{max}}$$ (2.9)

2.3.6   Electronic Shells Concerned with the Photoelectric Effect

The layered structure of atoms is at the heart of the photoelectric effect. Indeed, one photon extracts one electron from a shell only if its energy is greater than the binding energy of this electron in its shell. Below this value, the probability of extracting one electron from this shell is zero.

Low-energy light photons only extract the most external and least bound electrons of the atoms. This capacity decreases very quickly, until the energy of more energetic photons exceeds the threshold of the binding energy of the first shell of the atom: the photon will then be able to extract deeper electrons than this shell. Beyond this threshold, the probability of extracting of electrons from the new shell adds to that of previous shells. The probability of the photoelectric effect increases accordingly. It is manifested by a jump in the graph giving the probability of interaction as a function of the incident photon energy (Fig. 2.5). The probability of extracting electrons from the new shell decreases in turn until the energy of an even more energetic photon exceeds the threshold of the binding energy of electrons of the next shell which then become the main contributors to the photoelectric effect. As its energy increases, the photon interacts with deeper and deeper shells of the atom. These are the two electrons of the deepest layer K, in direct contact with the nucleus of the atom, which constitute in a way the last cartridge. After an ultimate jump, the probability of interaction decreases inexorably. To be able to tear one electron from the K-layer, the photon must have the energy of an X-photon or even that of a γ-photon. Once the electron of the K-shell is extracted, this atom is in an excited state.

A process of de-excitation engages automatically (Fig. 2.6): an electron of the outer L-shell comes to fill the place left vacant by the extracted electron, which causes either the emission of one X-photon (also called *X-ray* or *X-ray fluorescence*) or a direct transmission of this energy to another electron of any shell, called *Auger electron*, of the name of the French physicist Pierre Victor Auger (1899–1993), which is then ejected with a relatively low kinetic energy. The Auger emission is in competition with the X-emission. Emitted within a dense matter, the X-ray will usually be absorbed after a short course. The photoelectric effect is preponderant for the photons, of which the energy is less than 0.5 MeV.

Fig. 2.5 Typical
probability of the
photoelectric effect as a
function of the incident
photon energy. The jumps
correspond to the binding
energy of electrons in
electronic shells. As soon as
the energy of the incident
photon becomes greater than
or equal to the binding
energy of the electron in the
shell, the photon can interact
with that electron. This
interaction is manifested by
a jump in the graph

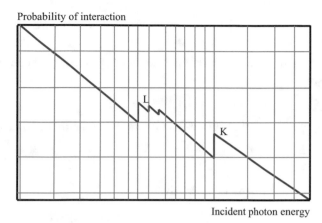

Probability of interaction

Incident photon energy

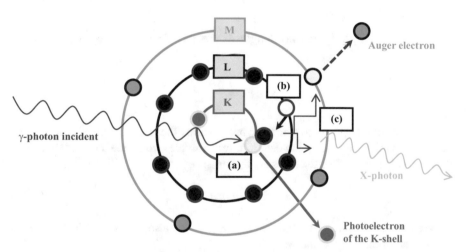

Fig. 2.6 The photoelectric effect occurs in three phases: (a) at first, the γ-photon extracts one
electron from the *K*-shell. (b) The atom that has lost one of its internal electrons is in an excited state.
One electron of the superior *L*-shell comes to fill the vacant place left by the ejected electron.
(c) Either one X-photon is emitted during this transition (X-ray fluorescence) or by a direct
transmission of this energy to another electron of any shell, called Auger electron, which is then
ejected

2.3.7 Wave-Particle Duality

As we have just seen in the previous paragraphs, the light has a dual aspect,
corpuscular and wave, not contradictory but rather complementary. Indeed,
according to the relativity theory, every particle of energy E carries a momentum
P defined by:

$$P = \frac{E}{c} \quad \Rightarrow \quad E = Pc \qquad (2.10)$$

where c is the speed of light. However, the energy of one photon i of frequency ν_i is:

$$E_i = h\nu_i \qquad (2.11)$$

where h denotes Planck's constant. By equaling (2.10) and (2.11), we obtain:

$$P_i c = h\nu_i \qquad (2.12)$$

As by definition, we have:

$$\lambda_i = \frac{c}{\nu_i} \qquad (2.13)$$

the momentum P_i and the wavelength λ_i are therefore linked by the following equation:

$$P_i = \frac{h}{\lambda_i}. \qquad (2.14)$$

This reasoning applies to photons, but it raises the question of whether it also applies to particles of matter (an electron, for example) where the momentum can be defined as the product of the mass m_e by the speed v:

$$P_e = m_e v. \qquad (2.15)$$

Well as unbelievable as it may sound, Eq. (2.14) also applies to electrons, protons, or neutrons and we can write:

$$m_e v = \frac{h}{\lambda_e}. \qquad (2.16)$$

In other words, as soon as an object acquires a certain speed, it can be associated with a wavelength equal to the ratio of Planck's constant by the acquired momentum. In 1924, Louis de Broglie postulated that for every material particle of mass m_e and of speed v was associated a wave of wavelength λ given by the so-called de Broglie's equation:

$$\lambda_e = \frac{h}{P_e}. \qquad (2.17)$$

To fix orders of magnitude, we can, by virtue of what has just been said, give the value of the wavelength of de Broglie associated with the grain of dust of mass $m = 10^{-15}$ kg, of diameter $D = 1$ μm, and of speed $v = 1$ mm/s. We find a wavelength of $\lambda_e = 6.6 \times 10^{-16}$ m, which is negligible on the scale of the size of the dust grain. This suggests that a wave behavior will be very difficult to put into evidence for an object of the macroscopic size. Let us notice that when the speed becomes very low, the thermal agitation is to be taken into account, and the wavelength λ_e of de Broglie is then calculated by the equation:

$$\lambda_e = \frac{h}{\sqrt{3mk_B T}} \tag{2.18}$$

where k_B is Boltzmann's constant ($k_B = 1.38 \times 10^{-23}$ J/K) and T the temperature in kelvins. If we consider this time a thermal neutron, as there is in some reactors of nuclear power stations, of mass $m_N = 1.67 \times 10^{-27}$ kg and of speed given by the following equation:

$$\frac{1}{2} m_N v^2 = \frac{3}{2} k_B T \tag{2.19}$$

with $T = 300$ K, the wavelength of de Broglie is $\lambda_e = 1.4 \times 10^{-10}$ m. It is of the same order of magnitude as the distance between two atoms or two ions of the crystal lattice. We can therefore expect that the wave character of a thermal neutron is revealed when it interacts with a crystal lattice. Today, matter waves are routinely used by physicists and chemists in the electron microscope or the neutron diffraction to probe nanoscale matter and finally visualize these famous atoms and molecules!

2.3.8   Applications

After the photoelectric phenomenon was explained and definitively understood, its practical applications were progressively developed. At first, it allowed measuring or detecting the light. In 1914, the light meter was invented. This was a device used in photography to measure the luminosity of a scene and thus to determine the optimum exposure of a photo shooting. It consisted of a photoelectric cell and a calculator. Then in the 1950s, we began to think about devices which allow producing some electricity: they were the photovoltaic cells. They were used at first to supply power installations located in inaccessible regions, water pumps in Africa, markers in the sea, and of course by the aerospace industry to assure the working of satellites. Gradually, these cells and then photovoltaic panels found their place in the energy network of different countries. Today, in addition to the explosion of the number of photovoltaic production installations, the photoelectric effect is used in several areas: secure closing systems, escalator startup, triggering of alarm systems, adjustment of the public lighting in cities, etc.

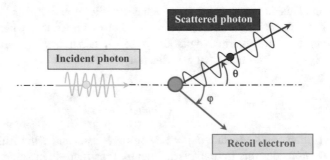

Fig. 2.7 Compton scattering: it links the energy transfer of one photon when it collides with one electron that is free or loosely bound to one atom, this electron is at rest. After the collision, the photon is scattered at an angle θ and the electron acquires a speed and leaves in a direction given by the angle φ

2.4   Compton's Scattering

2.4.1   Photon Scattering on a Target Electron

The corpuscular behavior of the photon and the hypothesis of the quantization of the photon energy by the wavelength were experimentally verified in 1923 by Arthur Holly Compton by the observation of the scattering of the photon on the electron. Compton was awarded the Nobel Prize in Physics in 1927. His discovery showed that electromagnetic waves also behaved like particles.

This experiment called *Compton effect* or *Compton scattering* or *Compton diffusion* links the energy transfer of one photon when it collides with a loosely bound outer-shell orbital electron of an atom. The loss of energy is manifested by an increase in the wavelength of the scattered photon (decrease of its frequency and its energy) (Fig. 2.7). This experiment also shows that there is a deviation of the trajectory of the photon of an angle θ which depends on the difference $\lambda_2 - \lambda_1$ between the wavelengths λ_2 of the scattered photon and λ_1 of the incident photon, both expressed in meters:

$$\lambda_2 - \lambda_1 = \frac{h}{m_e c} \left(1 - cos\theta\right) \tag{2.20}$$

where h is Planck's constant (6.626×10^{-34} J.s), m_e the electron rest mass (9.11×10^{-31} kg), and c the speed of light ($c = 3 \times 10^8$ m/s). To find this expression, let us consider the collision between one photon coming from the left and one supposed motionless electron (Fig. 2.8). The studied system is the photon+electron set. This system is isolated because there is no exchange of energy with the external environment. The collision is thus elastic; it results in a conservation of the momentum and energy of the system. After the collision, the electron acquires a very high speed, and the use of relativistic expressions of energies then proves necessary in this case. The conservation of momentum before and after the collision results in the following vector equation:

$$\vec{P}_1 = \vec{P}_2 + \vec{P}_e. \tag{2.21}$$

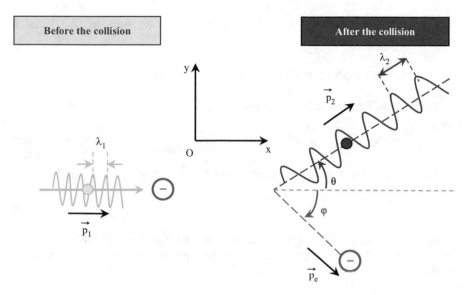

Fig. 2.8 As the photon-electron collision is elastic, the energy E and the linear momentum $\vec{P}$ are conserved

By projection of (2.21) on the x-axis, we obtain (see Fig. 2.8):

$$P_1 = P_2 \cos\theta + P_e \cos\varphi. \tag{2.22}$$

Let us isolate $P_e \cos\varphi$ in a side of (2.22), and it becomes:

$$P_e \cos\varphi = P_1 - P_2 \cos\theta. \tag{2.23}$$

Let us square (2.23), and it becomes:

$$P_e^2 \cos^2\varphi = P_1^2 - 2\,P_1 P_2 \cos\theta + P_2^2 \cos^2\theta. \tag{2.24}$$

By projection of the vector equation (2.21) on the y-axis, we obtain (see Fig. 2.8):

$$0 = P_2 \sin\theta - P_e \sin\varphi. \tag{2.25}$$

Let us isolate $P_e \sin\varphi$ in a side of (2.25), and it becomes:

$$P_e \sin\varphi = P_2 \sin\theta. \tag{2.26}$$

Let us square (2.26), and we get:

$$P_e^2 \sin^2\varphi = P_2^2 \sin^2\theta. \tag{2.27}$$

Let us add to both sides of (2.24) the corresponding sides of (2.27) and it turns out:

$$P_e^2 \cos^2\varphi + P_e^2 \sin^2\varphi = P_1^2 - 2\,P_1P_2 \cos\theta + P_2^2 \cos^2\theta + P_2^2 \sin^2\theta. \qquad (2.28)$$

Let us factorize P_e^2 and P_2^2: we obtain

$$P_e^2 \left(\cos^2\varphi + \sin^2\varphi\right) = P_1^2 - 2\,P_1P_2 \cos\theta + P_2^2 \left(\cos^2\varphi + \sin^2\varphi\right). \qquad (2.29)$$

By application of the trigonometric identity $\cos^2\varphi + \sin^2\varphi = 1$, we find:

$$P_e^2 = P_1^2 - 2\,P_1P_2 \cos\theta + P_2^2. \qquad (2.30)$$

On the other hand, the energy conservation before and after the collision results in the following equation (in relativistic mechanics, the energy at rest of the electron is $m_e c^2$):

$$E_1^P (\text{incident photon}) + E_1^e (\text{electron at rest}) = E_2^P (\text{scattered photon}) + \\ E_2^e (\text{recoil electron}). \qquad (2.31)$$

As the relativistic energy of the photon is $E_i = P_i c = h\nu_i = hc/\nu_i$, we get:

$$P_1 c + E_1^e = P_2 c + E_2^e \quad \text{or} \quad P_2 c + E_2^e = P_1 c + E_1^e. \qquad (2.32)$$

Knowing that the relativistic energy of the electron is $E_e^2 = P_e^2 c^2 + m_e^2 c^4$, we find:

$$P_2 c + \sqrt{P_{e2}^2 c^2 + m_e^2 c^4} = P_1 c + \sqrt{P_{e1}^2 c^2 + m_e^2 c^4}. \qquad (2.33)$$

However, $P_{e1} = 0$ (the electron is at rest $\Rightarrow$ zero momentum) and $P_{e2} = P_e$ (in order not to encumber the writing), and we obtain:

$$P_2 c + \sqrt{P_e^2 c^2 + m_e^2 c^4} = P_1 c + \sqrt{m_e^2 c^4} = P_1 c + m_e c^2. \qquad (2.34)$$

Let us isolate the square root in a side of (2.34), and we find:

$$\sqrt{P_e^2 c^2 + m_e^2 c^4} = P_1 c + m_e c^2 - P_2 c. \qquad (2.35)$$

Let us factorize c in (2.35), and we obtain:

$$\sqrt{P_e^2 c^2 + m_e^2 c^4} = c(P_1 - P_2) + m_e c^2. \qquad (2.36)$$

Let us square (2.36), and it becomes:

$$P_e^2 c^2 + m_e^2 c^4 = \left[c(P_1 - P_2) + m_e c^2\right]^2. \tag{2.37}$$

Let us expand the squared term of the second side of (2.37), and we get:

$$P_e^2 c^2 + m_e^2 c^4 = c^2 (P_1 - P_2)^2 + 2 m_e c^3 (P_1 - P_2) + m_e^2 c^4. \tag{2.38}$$

Let us simplify $m_e{}^2 c^4$ that appears on both equation sides and divide by c^2, and it becomes:

$$P_e^2 = (P_1 - P_2)^2 + 2 m_e c (P_1 - P_2). \tag{2.39}$$

Let us subtract (2.30) and (2.39), and we get:

$$P_e^2 - P_e^2 = \left(P_1^2 - 2\,P_1 P_2 \cos\theta + P_2^2\right) - \left[(P_1 - P_2)^2 + 2 m_e c\,(P_1 - P_2)\right]. \tag{2.40}$$

After simplification, we obtain:

$$0 = P_1^2 - 2\,P_1 P_2 \cos\theta + P_2^2 - (P_1 - P_2)^2 - 2\,m_e c\,(P_1 - P_2). \tag{2.41}$$

Let us expand the squared term $(P_1 - P_2)^2$, and (2.41) transforms into:

$$0 = P_1^2 - 2\,P_1 P_2 \cos\theta + P_2^2 - \left(P_1^2 - 2 P_1 P_2 + P_2^2\right) - 2\,m_e c\,(P_1 - P_2). \tag{2.42}$$

Let us simplify (2.42), and this yields:

$$0 = -2\,P_1 P_2 \cos\theta + 2\,P_1 P_2 - 2\,m_e c\,(P_1 - P_2). \tag{2.43}$$

Let us factorize $2 P_1 P_2$, and we obtain:

$$0 = 2\,P_1 P_2 (1 - \cos\theta) - 2\,m_e c\,(P_1 - P_2). \tag{2.44}$$

By rearranging (2.44), it becomes:

$$2\,m_e c\,(P_1 - P_2) = 2\,P_1 P_2 (1 - \cos\theta). \tag{2.45}$$

Let us divide by $2 m_e c$, and we find:

$$P_1 - P_2 = \frac{P_1 P_2}{m_e c}(1 - \cos\theta). \tag{2.46}$$

Let us substitute P_i by h/λ_i, and it becomes:

$$\frac{h}{\lambda_1} - \frac{h}{\lambda_2} = \frac{\left(\frac{h}{\lambda_1}\right)\left(\frac{h}{\lambda_2}\right)}{m_e c} (1 - cos\theta).$$ (2.47)

Let us simplify by h, and (2.47) transforms into:

$$\frac{1}{\lambda_1} - \frac{1}{\lambda_2} = \frac{\left(\frac{1}{\lambda_1}\right)\left(\frac{1}{\lambda_2}\right)}{m_e c} (1 - cos\theta).$$ (2.48)

By multiplying (2.48) by $\lambda_1\lambda_2$, we finally come back to (2.20):

$$\lambda_2 - \lambda_1 = \frac{h}{m_e c} (1 - cos\theta).$$ (2.49)

Equation (2.49) shows that the ratio $\lambda_C = h/m_e c$ has the dimension of a wavelength. It is called *Compton wavelength of the electron* and has a value of 0.002426 nm (or 2.426×10^{-12} m). This is not, of course, an actual wavelength, but really a proportionality constant for the wavelength shift $\Delta\lambda = \lambda_2 - \lambda_1$. This equation reveals that the wavelength shift does not depend on the density, the atomic number, or any other property of the absorbing material; the Compton scattering is strictly a photon-electron interaction. Looking again at this equation, it becomes clear that the entire shift can be measured purely in terms of the angle at which the photon gets scattered. Everything else on the right side of the equation is constant. Experiments show that this is the case, giving a great support to the photon interpretation of light.

2.4.2 Relationships Between Energies and Deviation Angles

Let E_1 be the energy of the incident photon; E_2, the energy of the scattered photon; $E_C = E_1 - E_2$, the kinetic energy transferred to the electron; θ, the scattering angle of the photon; φ, the angle between the trajectory of the electron after the collision and the initial direction of the photon; and $\alpha = E_1/m_e c^2$, the ratio between the energy of the incident photon and the rest energy of the electron with $m_e c^2 = 511$ keV.

2.4.2.1 Relationship Between E_2, E_1, and θ

The determination of the relationship $E_2 = f(E_1,\theta)$ requires beforehand the knowledge of certain relativistic notions. In particular, in special relativity, Einstein's equation states that at any mass m_e corresponds an energy $m_e c^2$ called *rest mass energy* and that every particle in motion includes both the rest mass energy and the kinetic energy E_c, namely:

$$E_e = mc^2 = E_c + m_e c^2 \qquad (2.50)$$

where $E_e = mc^2$ is the energy of the particle and m its relativistic mass given by:

$$m = \frac{m_e}{\sqrt{1 - \dfrac{v^2}{c^2}}} \Rightarrow m^2 = \frac{m_e^2}{1 - \dfrac{v^2}{c^2}} = \frac{m_e^2}{\dfrac{c^2}{c^2} - \dfrac{v^2}{c^2}} = \frac{m_e^2 c^2}{c^2 - v^2} \Rightarrow m^2 c^2 - m^2 v^2$$

$$= m_e^2 c^2. \qquad (2.51)$$

Moreover, any material particle in motion (one electron, for example) has a momentum:

$$P_e = mv \qquad (2.52)$$

where v is the particle speed. Let us remind that the momentum of the photon P_i has the following expression:

$$E_i = P_i c = h_i = h \frac{c}{\lambda_i} \Rightarrow P_i = \frac{E_i}{c} = \frac{h}{\lambda_i}. \qquad (2.53)$$

From (2.51) and (2.52), we can write:

$$m^2 c^2 - m^2 v^2 = m^2 c^2 - P_e^2 = m_e^2 c^2 \Rightarrow m^2 c^2 = P_e^2 + m_e^2 c^2. \qquad (2.54)$$

Let us multiply (2.54) by c^2, and taking into account that $E_e = mc^2$, we obtain:

$$m^2 c^4 = P_e^2 c^2 + m_e^2 c^4 \text{ that is: } E_e^2 = P_e^2 c^2 + m_e^2 c^4. \qquad (2.55)$$

By injecting (2.50) into (2.55), we find an important expression of the momentum of the electron P_e which we will use later, namely:

$$P_e^2 c^2 = E_e^2 - m_e^2 c^4 = \left(E_c + m_e c^2\right)^2 - m_e^2 c^4 = E_c\left(E_c + 2m_e c^2\right). \qquad (2.56)$$

Now let us go back to the photon-electron interaction described in Sect. 2.4.1. The conservation of energy during the collision allows us to write:

$$h\nu_1 + m_e c^2 = h\nu_2 + \sqrt{P_e^2 c^2 + (m_e c^2)^2}. \qquad (2.57)$$

The first side of (2.57) represents the sum of the energy $h\nu_1$ of the incident photon and the rest mass energy $m_e c^2$ of the electron, supposed motionless. The second side of this equation represents the sum of the energy $h\nu_2$ of the scattered photon and the energy $E_e = mc^2 = \sqrt{P_e^2 c^2 + (m_e c^2)^2}$ of the recoil electron, namely:

$$h\nu_1 + m_e c^2 = h\nu_2 + mc^2 \quad \Rightarrow \quad h_1 = h_2 + (mc^2 - m_e c^2). \tag{2.58}$$

The conservation of momentum allows us to write (see Fig. 2.8):
- By projection on the x-axis, we get:

$$\frac{E_1}{c} = \frac{E_2}{c}\cos\theta + P_e\cos\varphi \quad \Rightarrow \quad P_e\cos\varphi = \frac{E_1}{c} - \frac{E_2}{c}\cos\theta. \tag{2.59}$$

- By projection on the y-axis, we obtain:

$$0 = \frac{E_2}{c}\sin\theta - P_e\sin\varphi \quad \Rightarrow \quad P_e\sin\varphi = \frac{E_2}{c}\sin\theta. \tag{2.60}$$

The squared sum of (2.59) and (2.60) gives us the total momentum:

$$(P_e\cos\varphi)^2 + (P_e\sin\varphi)^2 = \left(\frac{E_1}{c} - \frac{E_2}{c}\cos\theta\right)^2 + \left(\frac{E_2}{c}\sin\theta\right)^2 \Rightarrow P_e^2 = \frac{E_1^2}{c^2} - $$

$$2\frac{E_1 E_2}{c^2}\cos\theta + \frac{E_2^2}{c^2}\cos^2\theta + \frac{E_2^2}{c^2}\sin^2\theta = \frac{E_1^2}{c^2} - 2\frac{E_1 E_2}{c^2}\cos\theta + \frac{E_2^2}{c^2}$$

$$\Rightarrow P_e^2 c^2 = E_1^2 + E_2^2 - 2E_1 E_2 \cos\theta. \tag{2.61}$$

By substituting $P^2 c^2$ by its expression given in (2.56), (2.61) turns into:

$$E_c(E_c + 2m_e c^2) = E_1^2 + E_2^2 - 2E_1 E_2\cos\theta. \tag{2.62}$$

As $E_C = E_1 - E_2$, we finally get the sought equation $E_2 = f(E_1, \theta)$:

$$(E_1 - E_2)(E_1 - E_2 + 2m_e c^2) = E_1^2 + E_2^2 - 2E_1 E_2\cos\theta$$
$$E_1^2 - E_1 E_2 + 2E_1 m_e c^2 - E_1 E_2 + E_2^2 - 2E_2 m_e c^2 = E_1^2 + E_2^2 - 2E_1 E_2\cos\theta$$
$$-2E_1 E_2 + 2E_1 m_e c^2 - 2E_2 m_e c^2 = -2E_1 E_2\cos\theta$$
$$2m_e c^2(E_1 - E_2) = 2E_1 E_2(1 - \cos\theta) \Rightarrow E_1 - E_2 = \frac{E_1 E_2}{m_e c^2}(1 - \cos\theta)$$

$$E_2 + \frac{E_1 E_2}{m_e c^2}(1 - \cos\theta) = E_1 \Rightarrow E_2\left[1 + \frac{E_1}{m_e c^2}(1 - \cos\theta)\right] = E_1 \tag{2.63}$$

$$E_2 = \frac{E_1}{1 + \dfrac{E_1}{m_e c^2}(1 - \cos\theta)} = \frac{E_1}{1 + (1 - \cos\theta)}.$$

Fig. 2.9 The energy E_2 of the scattered photon as a function of the energy E_1 of the incident photon for three values of the angle θ

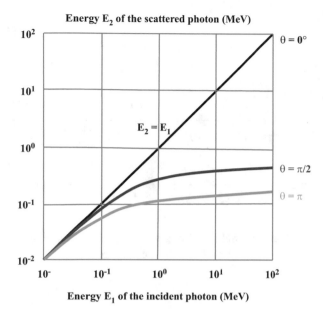

By knowing that $0 < \theta < \pi$, we obtain the expressions of minimal and maximal energies of the scattered photon:

$$(E_2)_{\theta=\pi}^{\min} = \frac{E_1}{1 + 2\alpha} \quad \text{and} \quad (E_2)_{\theta=0}^{\max} = E_1. \tag{2.64}$$

Figure 2.9 shows that the energy E_2 of the scattered photon depends on the energy E_1 of the incident photon. For E_1 small, the scattered photon loses a little energy regardless of the value of the angle θ (the three curves are almost superimposed). In contrast, for high energies of the incident photon E_1, the variation of the energy of the scattered photon E_2 with the angle θ becomes more and more rapid. At $\theta = 0°$, we have $E_2 = E_1$: the scattered photon leaves in the prolongation of the incident photon, taking with it all the energy E_1 of the incident photon. At $\theta = \pi$, the energy of the scattered photon is minimal, meaning that it is the electron that takes most of the energy of the incident photon. The scattered photon follows the same way as that of the incident photon, but in the opposite direction. At $\theta = \pi/2$, the scattered photon carries an intermediate energy between the aforementioned extreme cases and leaves in the direction perpendicular to that of the incident photon.

2.4.2.2 Kinetic Energy Transferred to the Electron as a Function of E_1, α, and θ

The transferred energy to the recoil electron $E_C = E_1 - E_2$ is:

$$E_C = E_1 - E_2 = E_1 - \frac{E_1}{1 + \dfrac{E_1}{m_e c^2}(1 - cos\,\theta)} = \frac{E_1\left[1 + \dfrac{E_1}{m_e c^2}(1 - cos\,\theta)\right]}{1 + \dfrac{E_1}{m_e c^2}(1 - cos\,\theta)}$$

$$- \frac{E_1}{1 + \dfrac{E_1}{m_e c^2}(1 - cos\,\theta)} = \frac{E_1 + \dfrac{E_1^2}{m_e c^2}(1 - cos\,\theta) - E_1}{1 + \dfrac{E_1}{m_e c^2}(1 - cos\,\theta)} = \frac{\dfrac{E_1^2}{m_e c^2}(1 - cos\,\theta)}{1 + \dfrac{E_1}{m_e c^2}(1 - cos\,\theta)}$$

$$= E_1 \frac{\dfrac{E_1}{m_e c^2}(1 - cos\,\theta)}{1 + \dfrac{E_1}{m_e c^2}(1 - cos\,\theta)} = E_1 \frac{\alpha(1 - cos\,\theta)}{1 + \alpha(1 - cos\,\theta)}. \qquad (2.65)$$

Knowing that $0 < \theta < \pi$, we obtain the minimal and maximal expressions of the kinetic energy transferred to the electron:

$$(E_c)_{\theta=0}^{\min} = 0 \text{ and } (E_c)_{\theta=\pi}^{\max} = \frac{2E_1}{1 + 2\alpha}. \qquad (2.66)$$

Equation (2.66) reveals that for a scattering angle of the photon $\theta = 0°$, the kinetic energy of the electron is zero ($E_C = 0$). It is thus the scattered photon which carries all the energy of the incident photon ($E_2 = E_1$), whereas for $\theta = \pi$, the kinetic energy of the electron is maximal. The scattered photon follows the same way as the incident photon, but in the opposite direction, carrying a minimal energy.

2.4.2.3   Relationship Between the Angles φ and θ

By doing the ratio (2.22)/(2.23), we get:

$$\frac{P_e\, sin\varphi}{P_e\, cos\varphi} = \frac{P_2\, sin\,\theta}{P_1 - P_2\, cos\,\theta}. \qquad (2.67)$$

After simplification, we obtain:

$$tan\varphi = \frac{sin\varphi}{cos\varphi} = \frac{sin\,\theta}{\dfrac{P_1}{P_2} - cos\,\theta}. \qquad (2.68)$$

The momentum of the photon is given as a function of its energy by (2.53), and (2.68) can be written as:

$$tan\varphi = \frac{sin\varphi}{cos\varphi} = \frac{sin\,\theta}{\dfrac{E_1}{c}} = \frac{sin\,\theta}{\dfrac{E_1}{E_2} - cos\,\theta}. \qquad (2.69)$$

The ratio of incident and scattered photons energies is given by (2.63):

$$E_2 = \frac{E_1}{1 + \alpha(1 - cos\theta)} \Rightarrow \frac{E_1}{E_2} = 1 + \alpha(1 - cos\theta). \qquad (2.70)$$

By injecting (2.70) into (2.69), we find:

$$tan\,\varphi = \frac{sin\theta}{1 + \alpha(1 - cos\theta) - cos\theta} = \frac{sin\theta}{(1 - cos\theta)(1 + \alpha)}. \qquad (2.71)$$

Using the following trigonometric equation:

$$tan\frac{\theta}{2} = \frac{1 - cos\theta}{sin\theta} \qquad (2.72)$$

we get:

$$tan\,\varphi = \frac{1}{\dfrac{(1 - cos\theta)}{sin\theta}(1 + \alpha)} = \frac{1}{tan\dfrac{\theta}{2}(1 + \alpha)} \qquad (2.73)$$

which we can finally write in the following form:

$$cot\varphi = (1 + \alpha)\,tan\frac{\theta}{2}. \qquad (2.74)$$

Equation (2.74) shows that for $\theta = 0$, we have $\varphi = \pi/2$, and for $\theta = \pi$, we have $\varphi = 0$.

2.4.2.4   Kinetic Energy Transferred to the Electron as a Function of E_1, α, and φ

From (2.63), we obtain:

$$E_2 = \frac{E_1}{1 + \alpha(1 - cos\theta)} \Rightarrow (1 - cos\theta) = \frac{E_1 - E_2}{\alpha E_2}. \qquad (2.75)$$

Returning to the trigonometric Eq. (2.72), we can write:

$$tan\frac{\theta}{2} = \frac{1 - cos\theta}{sin\theta} \Rightarrow 1 - cos\theta = sin\theta\,tan\frac{\theta}{2}. \qquad (2.76)$$

By injecting (2.76) into (2.75), we obtain:

$$sin\,\theta\,tan\frac{\theta}{2} = \frac{E_1 - E_2}{\alpha E_2} \qquad (2.77)$$

From (2.74), we deduce:

$$cot\varphi = (1+\alpha)\,tan\frac{\theta}{2} = \frac{1}{tan\,\varphi} \quad \Rightarrow \quad tan\frac{\theta}{2} = \frac{1}{(1+\alpha)tan\,\varphi}. \tag{2.78}$$

Injecting (2.78) into (2.77), we get:

$$sin\,\theta\,\frac{1}{(1+\alpha)tan\,\varphi} = sin\,\theta\,\frac{1}{(1+\alpha)\dfrac{sin\,\varphi}{cos\,\varphi}} = \frac{sin\,\theta}{sin\,\theta}\,\frac{cos\,\varphi}{(1+\alpha)} = \frac{E_1 - E_2}{\alpha E_2}. \tag{2.79}$$

According to (2.26), we find:

$$P_e\,sin\,\varphi = P_2\,sin\,\theta \quad \Rightarrow \quad \frac{sin\,\theta}{sin\,\varphi} = \frac{P_e}{P_2}. \tag{2.80}$$

By starting from (2.53), the momentum of the photon can be written as:

$$P_i = \frac{E_i}{c} \quad \Rightarrow \quad P_2 = \frac{E_2}{c}. \tag{2.81}$$

Injecting (2.81) into (2.80), we find:

$$\frac{sin\,\theta}{sin\,\varphi} = \frac{P_e}{\dfrac{E_2}{c}} = \frac{P_e c}{E_2}. \tag{2.82}$$

Injecting (2.82) into (2.79), we get:

$$\frac{E_1 - E_2}{\alpha E_2} = \frac{P_e c}{E_2}\,\frac{cos\,\varphi}{(1+\alpha)}. \tag{2.83}$$

Squaring (2.83) and taking into account (2.56) which gives this expression $P_e^2 c^2 = E_c(E_c + 2m_e c^2)$, we arrive at:

$$\frac{(E_1 - E_2)^2}{\alpha^2 E_2^2} = \frac{P_e^2 c^2}{E_2^2}\,\frac{cos^2\varphi}{(1+\alpha)^2} = \frac{E_c(E_c + 2m_e c^2)}{E_2^2}\,\frac{cos^2\varphi}{(1+\alpha)^2}. \tag{2.84}$$

After expansions and simplifications, (2.84) transforms into:

$$E_c = \frac{2\alpha m_e c^2 cos^2\varphi}{(1+\alpha)^2 - \alpha^2 cos^2\varphi}. \tag{2.85}$$

Knowing by hypothesis that $\alpha = E_1/m_e c^2$, we finally get:

$$E_c = E_1 \frac{2a\cos^2 \varphi}{(1+a)^2 - a^2 \cos^2 \varphi}. \tag{2.86}$$

Knowing that $0 < \varphi < \pi/2$, we obtain the minimal and maximal expressions of the kinetic energy transferred to the electron:

$$(E_c)_{\varphi=\frac{\pi}{2}}^{\min} = 0 \text{ and } (E_c)_{\varphi=0}^{\max} = \frac{E_1}{1 + \frac{1}{2a}}. \tag{2.87}$$

Equation (2.87) reveals that for $\varphi = 0$, the electron carries most of the energy of the incident photon, the photon carrying only a small energy of the latter, whereas for $\varphi = \pi/2$, the kinetic energy of the electron is zero and it is therefore the scattered photon that carries all the energy of the incident photon.

2.4.2.5 Conclusion

On the basis of what has just been said, we conclude that:

- For $\theta = 0$, we have $\varphi = \pi/2$: the photon-electron collision is tangential. It corresponds to the limit case where the photon grazes the electron which leaves in the perpendicular direction to that of the incident photon ($\varphi = \pi/2$). There is practically no photon-electron interaction. The scattered photon leaves on the same straight line and in the same direction as the incident photon (Fig. 2.10).
- For $\theta = \pi$, we have $\varphi = 0°$: the photon-electron collision is frontal. It corresponds to the limit case where the photon encounters on its trajectory the electron and strikes it with full force. The interaction is maximal. The scattered photon leaves in the opposite direction of the incident photon ($\theta = \pi$). It has a minimal energy E_2. The electron leaves on the same straight line and in the same direction as the incident photon with a maximum kinetic energy E_C. Thus, we see that even in the extreme case of a frontal collision, E_2 is never zero, whereas E_C can be: so there is always a scattered photon. The incident photon cannot transfer in all cases all of its energy to the recoil electron (Fig. 2.11).

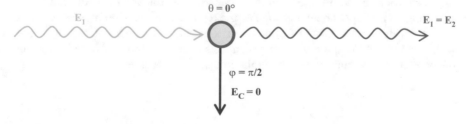

Fig. 2.10 Tangential photon-electron collision

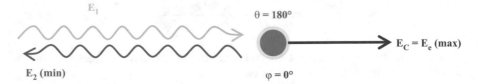

E_1

$\theta = 180°$

$E_C = E_e$ (max)

E_2 (min)

$\varphi = 0°$

Fig. 2.11 Frontal photon-electron collision

2.4.3 *Electronic Shells Involved in the Compton Scattering*

Compton's collisions compete with the photoelectric effect when γ-photons pass through matter. These collisions can be considered as elastic collisions between one photon and one electron. They become preponderant when the energy of the photon becomes large compared to the attraction that holds the electron in an atom. For a light atom such as carbon, the Compton scattering outweighs the photoelectric effect above 20 keV. For copper, it is above 130 keV; for lead, it is from 600 keV. In this energy range, Compton's scattering concerns all the electrons of the atom, while in the photoelectric effect, only the two electrons of the K-shell are involved.

Note
When the energy of the photon γ exceeds 1 MeV, Compton's scattering begins to be competed by a new phenomenon: the transformation of the incident photon into one electron and its antiparticle, a positron. These photons are produced with accelerators of particles.

2.4.4 *Inverse Compton's Scattering*

The inverse Compton scattering is the scattering of electrons onto photons, transferring them a large part of their energy. Its interest is important in astrophysics in situations where high-energy electrons scatter on low-energy photons. This effect, predicted by Rashid Sunyaev and Yakov Zeldovich in 1969, is now accurately observed as a distortion of the Planck spectrum. Since this effect does not depend on the distance, it is now used to detect the existence of distant galaxies. In theory, the description of the inverse Compton scattering is similar to that of Compton's scattering. It is simply about a change of the reference frame: Compton's scattering occurs during the scattering of photons on electrons considered at rest in matter, while the inverse Compton scattering manifests itself in braking of fast electrons by the photons of low energy, present in the interstellar space.

2.5   Solved Exercises

2.5.1   *Photoelectric Effect*

Exercise 1
1. *Do all photons have the same energy?*
2. *When the frequency of light increases, does its wavelength increase or decrease?*
3. *If the violet light with λ = 400 nm does not cause a photoelectric effect in a metal, then is it sure that the red light with λ = 650 nm will not cause the photoelectric effect in this metal?*

Solution
1. No, all photons do not have the same energy because the energy of one photon depends on its emission frequency $E = h\nu$ and the frequency is a characteristic of the emitted electromagnetic wave.
2. When the frequency of light increases, its wavelength decreases because these two physical quantities are inversely proportional:

$$\nu = \frac{c}{\lambda} \tag{1}$$

where c is the speed of light in vacuum. It is constant and has the value $c = 3 \times 10^8$ m/s.

3. True, because the red light with longer wavelength will carry less energy:

$$E = h\nu = h\frac{c}{\lambda}. \tag{2}$$

Exercise 2
Why did the classical description fail to describe the photoelectric effect?

Solution
The classical theory states that as soon as energy is supplied, a current must be measured. In practice, this is not always true. Indeed, the radiation should have a sufficient frequency to reveal an electric current. Below a threshold frequency, you can send in vain a light radiation; no current is detected, even if you increase the power of this radiation.

Exercise 3
When we irradiate the metal surface with a frequency higher than the threshold frequency, explain why the current increases when we increase the power of the radiation.

Solution

Power is the amount of energy consumed per unit of time. The greater the contribution of the incidental energy per unit of time, the greater is the number of photons which strike the metal surface and the greater the number of electrons per unit of time which are able to leave this surface, and the more the current, which is a number of electric charges per unit of time, increases.

Exercise 4

We fix the power of the radiation, and we increase the frequency above the frequency threshold. Explain why the current increases with the frequency.

Solution

The current density j (current per unit area, perpendicular to the current direction) can be written as: $j = \rho v$ where ρ is the concentration of charges per unit volume and v the electron speed. By increasing the frequency, we increase the energy got by one electron in the matter. So this electron is ejected outside the metal with a kinetic energy that grows with the frequency so with a bigger speed, and in the end, the current density j (or simply the electric current) will be larger.

Exercise 5

A helium-neon laser emits a light beam of 0.1 watt with a wavelength of 633 nm. Determine the number of photons emitted by the laser every minute. We give Plank's constant $h = 6.63 \times 10^{-34}$ J.s and the speed of light $c = 3 \times 10^8$ m/s.

Solution

Let us evaluate the energy of one photon:

$$E = h\nu = h\frac{c}{\lambda} = \left(6.63\ 10^{-34}\ \text{J.s}\right)\frac{3 \times 10^8\ \text{m.s}^{-1}}{633 \times 10^{-9}\ \text{m}} = 3.142 \times 10^{-19}\ \text{J}. \quad (1)$$

The light energy released by the laser for 1 minute (60 s) is:

$$W = \text{power} \times \text{time} = (0.1\ \text{watt})\,(60\ \text{s}) = 6\ \text{J}. \quad (2)$$

The number of photons emitted by the laser during 1 minute is equal to:

$$W = NE \quad \Rightarrow \quad N = \frac{W}{E} = \frac{6\ \text{J}}{3.142 \times 10^{-19}\ \text{J}} = 1.19 \times 10^{19}\ \frac{\text{photons}}{\text{minute}}. \quad (3)$$

Exercise 6

Two monochromatic electromagnetic waves of respective wavelengths $\lambda_1 = 500$ nm and $\lambda_2 = 10\ \mu m$ move into the vacuum.
1. *Which part of the spectrum do they belong to?*
2. *What are their respective frequencies?*
3. *What is the energy of one photon associated with these wavelengths? You will express this value in joules and electronvolts.*

4. *We consider two monochromatic laser beams of power 1 watt, the one at λ_1 and the other at λ_2. Which laser beam emits the biggest number of photons per second?*
5. *At what speeds do these two waves move into the vacuum?*
6. *For each wavelength, cite a natural object that emits very large numbers of photons at these wavelengths. We give: $h = 6.626 \times 10^{-34}$ J.s and 1 eV $= 1.6 \times 10^{-19}$ J.*

Solution

1. *Part of the spectrum to which both wavelengths belong:*

The wavelength $\lambda_1 = 500$ nm belongs to the visible range of the electromagnetic spectrum and the wavelength $\lambda_2 = 10$ μm belongs to the infrared domain.

2. *Respective frequencies*:

   – The frequency associated with λ_1 is:

$$\nu_1 = \frac{c}{\lambda_1} = \frac{3 \times 10^8 \text{ m.s}^{-1}}{500 \times 10^{-9} \text{ m}} = 6 \times 10^{14} \text{ Hz}. \tag{1}$$

   – The frequency associated with λ_2 is:

$$\nu_2 = \frac{c}{\lambda_2} = \frac{3 \times 10^8 \text{ m.s}^{-1}}{10 \times 10^{-6} \text{ m}} = 3 \times 10^{13} \text{ Hz}. \tag{2}$$

3. *Energy of one photon associated with these wavelengths*:

   – The energy of one photon of frequency ν_1 is:

$$E_1(\text{J}) = h\nu_1 = \left(6.626 \times 10^{-34}\text{J.s}\right)\left(6 \times 10^{14} \text{ Hz}\right) = 3.978 \times 10^{-19} \text{ J}, \tag{3}$$

namely, in electronvolts:

$$E_1(\text{eV}) = \frac{\left(3.978 \times 10^{-19} \text{ J}\right)\left(1 \text{ eV}\right)}{1.6 \times 10^{-19} \text{ J}} = 2.486 \text{ eV}. \tag{4}$$

   – The energy of one photon of frequency ν_2 is:

$$E_2(\text{J}) = h\nu_2 = \left(6.626 \times 10^{-34} \text{ J.s}\right)\left(3 \times 10^{13} \text{ Hz}\right) = 1.989 \times 10^{-20} \text{ J}, \tag{5}$$

namely, in electronvolts:

$$E_1(\text{eV}) = \frac{(1.989 \times 10^{-20} \text{ J}) (1 \text{ eV})}{1.6 \times 10^{-19} \text{ J}} = 0.1243 \text{ eV}. \qquad (6)$$

4. *A laser beam that emits the biggest number of photons per second*:

The light energy released by the laser for 1 second is:

$$W = \text{power} \times \text{time} = (0.1 \text{ watt}) (1 \text{ s}) = 0.1 \text{ J}. \qquad (7)$$

– The number of photons emitted by the laser at λ_1 (visible) for 1 second is equal to:

$$W = N_1 E_1 \quad \Rightarrow \quad N_1 = \frac{W}{E_1} = \frac{0.1 \text{ J}}{3.978 \times 10^{-19} \text{ J}} = 5.055 \times 10^{18} \frac{\text{photons}}{\text{s}}. \qquad (8)$$

– The number of photons emitted by the laser at λ_2 (infrared) for 1 second is equal to:

$$W = N_2 E_2 \quad \Rightarrow \quad N_2 = \frac{W}{E_2} = \frac{0.1 \text{ J}}{1.989 \times 10^{-20} \text{ J}} = 5.027 \times 10^{19} \frac{\text{photons}}{\text{s}}. \qquad (9)$$

The infrared beam (at λ_2) thus contains more photons than the visible beam (at λ_2) because it is necessary to have more photons of lower energy to reach the same power emitted by the laser (1 J/s). The ratio between the two emitted numbers of photons is:

$$\frac{N_2}{N_1} = \frac{\dfrac{W}{E_2}}{\dfrac{W}{E_1}} = \frac{E_1}{E_2} = \frac{h\nu_1}{h\nu_2} = \frac{\nu_1}{\nu_1} = \frac{\dfrac{c}{\lambda_1}}{\dfrac{c}{\lambda_2}} = \frac{\lambda_2}{\lambda_1} = 20. \qquad (10)$$

5. The Sun emits a lot of photons at λ_1 (visible light to which the human eye is sensitive). Regarding λ_2, any object at room temperature radiates an infrared radiation at this wavelength, especially humans and animals. This radiation is invisible to the naked eye but can be made visible at night by means of special infrared glasses.

Exercise 7

We illuminate a copper plate of which the ionization energy is $W_S = 4.7$ eV with an ultraviolet light at 200 nm. Determine the module of the maximal speed of photo-electrons (the mass of the electron is $m_e = 9.11 \times 10^{-31}$ kg). We give $h = 6.63 \times 10^{-34}$ J.s and 1 eV $= 1.6 \times 10^{-19}$ J.

Solution

Let us evaluate the extraction work for the copper plate in joules:

$$W_s(J) = \frac{(4.7 \text{ eV}) \left(1.6 \times 10^{-19} \text{ J}\right)}{1 \text{ eV}} = 7.52 \times 10^{-19} \text{ J}. \tag{1}$$

Let us evaluate the maximal kinetic energy of the ejected electron which represents the surplus of energy received by the electron and used by it to move:

$$E_c = h\frac{c}{\lambda} - W_s = \frac{\left(6.63 \times 10^{-34} \text{ J.s}\right) \left(3 \times 10^8 \text{ m.s}^{-1}\right)}{\left(200 \times 10^{-9} \text{ m}\right)} - 7.52 \times 10^{-19} \text{ J}$$

$$= 2.425 \times 10^{-19} \text{ J}. \tag{2}$$

From such we deduce the maximal speed of the expelled electron:

$$E_c = \frac{1}{2} m_e v^2 \quad \Rightarrow \quad v = \sqrt{\frac{2E_c}{m_e}} = \sqrt{\frac{(2)\left(2.425 \times 10^{-9} \text{J}\right)}{9.11 \times 10^{-31} \text{kg}}}$$

$$= 7.296 \times 10^5 \text{ m.s.}^{-1}. \tag{3}$$

Exercise 8

The electron extraction energy of the surface of a metal is equal to 2.2 eV. Indicate if there is emission of photoelectrons when this surface is illuminated with the visible light (the visible range goes from 400 nm to 800 nm). We give Planck's constant $h = 6.63 \times 10^{-34}$ J.s and the speed of light $c = 3 \times 10^8$ m/s.

Solution

The threshold wavelength is:

$$\lambda_S = \frac{c}{\nu_S} = \frac{hc}{h\nu_S} = \frac{\left(6.63 \times 10^{-34} \text{ J.s}\right) \left(3 \times 10^8 \text{ m.s}^{-1}\right) (1 \text{ eV})}{(2.2 \text{ eV}) \left(1.6 \times 10^{-19} \text{ J}\right)}$$

$$= 5.66 \times 10^{-7} \text{ m} = 566 \text{ nm}. \tag{1}$$

As the threshold wavelength value is between 400 nm and 800 nm, there is emission of photoelectrons.

Exercise 9

What voltage must be applied to stop the fastest electrons emitted by a nickel surface that receives the ultraviolet of 200 mm? The electron extraction energy of nickel is 5.01 eV. We give Planck's constant $h = 6.626 \times 10^{-34}$ J.s, the speed of light $c = 3 \times 10^8$ m/s, and 1 eV $= 1.602 \times 10^{-19}$ J.

Solution

The energy of one incident photon of the ultraviolet is:

$$E = h\nu = h\frac{c}{\lambda} = \left(6.626 \times 10^{-34} \text{ J.s}\right) \frac{3 \times 10^8 \text{ m.s}^{-1}}{200 \times 10^{-9} \text{ m}} = 9.93 \times 10^{-19} \text{ J}$$

$$= 6.21 \text{ eV}. \tag{1}$$

The received surplus energy is used by the ejected electron in the form of kinetic energy to move:

$$E_c = h\nu - h\nu_S = 6.21 \text{ eV} - 5.01 \text{ eV} = 1.20 \text{ eV}. \tag{2}$$

The determination of the kinetic energy of the emitted electron is done by the method of the retarding field. In this method, the electron crosses an area where there is an electric field which opposes to its move, and of which the intensity is varied until this electron stops. In this case, we get, for a measurable brake voltage U_0, the following equation:

$$e\, U_0 = E_c \;\Rightarrow\; U_0 = \frac{E_c}{e} = \frac{1.20 \text{ eV}}{e} = 1.20 \text{ volts}. \tag{3}$$

Exercise 10
We have a photoelectric cell composed of an anode and a cesium cathode illuminated by a monochromatic light.
1. *The threshold wavelength of cesium is $\lambda_S = 0.66$ μm. Determine the extraction work W_S of one electron in joules and electronvolts.*
2. *The light that illuminates this photocathode has a wavelength $\lambda = 0.44$ μm.*
   (a) *Determine the maximal kinetic energy of an emitted electron by the cathode.*
   (b) *Determine the speed of this electron.*
   (c) *Determine the stopping potential U_0 under these conditions.*
      We give Planck's constant $h = 6.63 \times 10^{-34}$ J.s; $c = 3 \times 10^8$ m/s, the speed of light; $m_e = 9.1 \times 10^{-31}$ kg, the electron mass; $e = 1.6 \times 10^{-19}$ C, its charge; and 1 eV $= 1.6 \times 10^{-19}$ J.

Solution
1. *The extraction work W_S of the electron in joules and electronvolts:*

It is given by:

$$W_S(\text{J}) = h\nu_S = h\frac{c}{\lambda_S} = \left(6.63 \times 10^{-34} \text{ J.s}\right) \frac{3 \times 10^8 \text{ m.s}^{-1}}{0.66 \times 10^{-6} \text{ m}} = 3 \times 10^{-19} \text{ J}. \tag{1}$$

As 1 eV $= 1.6 \times 10^{-19}$ J, we deduce:

$$W_S(\text{eV}) = \frac{\left(3 \times 10^{-19} \text{ J}\right) (1 \text{ eV})}{1.6 \times 10^{-19} \text{ J}} = 1.9 \text{ eV}. \tag{2}$$

2. (a) *Maximal kinetic energy of an emitted electron by the cathode*:

It is given by Einstein's equation:

$$E_c = h\nu - W_S = h\frac{c}{\lambda} - h\nu_S = \left(6.63 \times 10^{-34} \text{ J.s}\right) \frac{3 \times 10^8 \text{ m.s}^{-1}}{0.44 \times 10^{-6} \text{ m}} - 3 \times 10^{-19} \text{ J}$$
$$= 1.5 \times 10^{-19} \text{ J}.$$

(3)

(b) *Speed of the electron*:

It is given by:

$$E_c = \frac{1}{2} m v^2 \Rightarrow v = \sqrt{\frac{2 E_c}{m}} = \sqrt{\frac{2 \left(1.5 \times 10^{-19} \text{ J}\right)}{9.1 \times 10^{-31} \text{ kg}}} = 5.8 \times 10^5 \text{ m.s}^{-1}.$$

(4)

(c) *Stopping potential U_0*:

By application of the kinetic energy theorem between the anode and the cathode, we have:

$$E_C^{\text{Anode}} - E_C^{\text{cathode}} = W.$$

(5)

There is only a work done here: the resistive electrical work that serves to cancel the speed of the electron at its arrival to the anode is:

$$W = eU_0$$

(6)

where U_0 is the stopping potential. This is the voltage that must be applied between electrodes to cancel the electron speed at the anode. Thus, $E_C^{\text{Anode}} = 0$ since on arrival on the anode, the speed of the electron is zero. $E_C^{\text{cathode}} = E_c = 1.5 \times 10^{-19}$ J is the initial energy of the electron at the cathode:

$$0 - E_c = eU_0 \Rightarrow U_0 = \frac{-E_c}{e} = \frac{-1.5 \times 10^{-19} \text{ J}}{1.6 \times 10^{-19} \text{ C}} = -0.94 \text{ V}.$$

(7)

Exercise 11
We have a photoelectric cell of which the extraction threshold is 2.4 eV. It is illuminated by a polychromatic beam composed of two wavelength radiations $\lambda_1 = 430$ nm and $\lambda_2 = 580$ nm.

1. *Define the photoelectric effect.*
2. *Give the diagram of the electrical assembly allowing the study of this effect.*
3. *Represent the variation of the electrical intensity I which passes through the photocell as a function of the voltage U at its terminals in the following three cases:*
   (a) *We fix the cathode metal, the light radiation intensity, and its frequency.*
   (b) *We fix the cathode metal and the light radiation intensity, and we vary the radiation frequency.*
   (c) *We fix the cathode metal and the light radiation frequency, and we vary the radiation intensity.*
   For each case, deduce the appropriate conclusions.
4. *We illuminate the cell with both radiations.*
   (a) *Do the two radiations allow the photoelectric effect?*
   (b) *What is the maximal speed of electrons that are ejected from the photocathode?*
   (c) *Define and calculate the stopping potential. We give Planck's constant $h = 6.63 \times 10^{-34}$ J.s; $c = 3 \times 10^8$ m/s, the speed of light; $m_e = 9.1 \times 10^{-31}$ kg, the electron mass; $e = 1.6 \times 10^{-19}$ C, its charge; and $1\ eV = 1.6 \times 10^{-19}$ J.*

Solution

1. *Definition of the photoelectric effect*:
   The photoelectric effect refers to the emission of electrons from the surface of a metal (or other substance) under the action of light. In this effect, it is not the intensity of the incident radiation that is the dominant factor but rather its frequency.
2. *Diagram of the electrical assembly*:
   The electrical assembly for studying the photoelectric effect consists of a quartz bulb in which a very high vacuum is made of the order of 10^{-7} mm Hg. This bulb, called *photocell*, is composed of two electrodes, a cylinder-shaped cathode and a rectilinear wire-shaped anode placed on the axis of the cylinder. Illuminated by an appropriate light, the cathode (emitter) emits electrons. These electrons are captured by the anode (collector) which is connected to the positive pole of an adjustable generator. It results from this a current detected by a galvanometer incorporated in the circuit. A voltmeter measures the applied voltage to the cell.

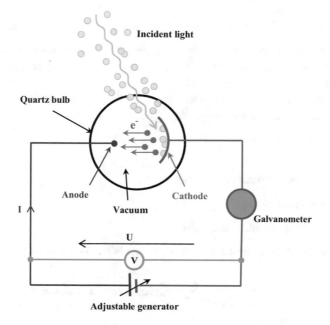

3. *Representation of the variation of the electrical intensity I which crosses the cell as a function of the voltage U at its terminals*:

   (a) *Fixed parameters*:

   - Metal
   - Intensity of the incident radiation
   - Frequency of the incident radiation

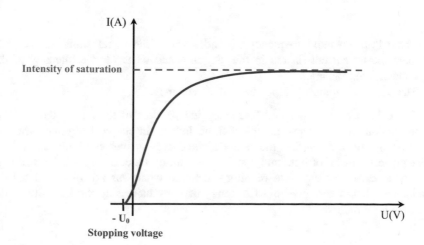

The experiment gives us the curve $I = f(U)$. This curve reveals that:

- The electrical intensity can be canceled by slowing down electrons with an adequate voltage. The voltage $- U_0$ is negative. It is called *stopping voltage*.
- The electrical intensity tends to a maximum when the voltage increases. This maximum is called *saturation intensity*.

(b) *Fixed parameters*:

- Metal
- Intensity of the incident radiation

Studied parameter:

- Frequency of the incident radiation

The experiment gives us the $I = f(U)$ curves. We notice two facts:

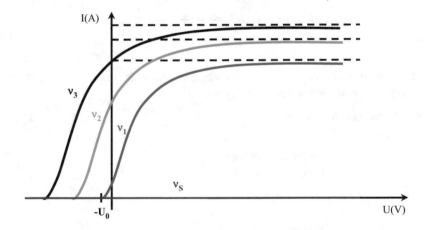

- There is a threshold frequency ν_S under which the metal emits no electron, therefore no current (in red in the diagram) and whatever the intensity of the incident light radiation.
- The saturation intensity depends on the frequency.

The existence of a threshold frequency independent of the light intensity is in contradiction with the wave nature of light. Indeed, in classical physics, when an electromagnetic wave strikes the cathode, the energy arrives continuously. With a very powerful beam of light and a sufficient waiting time, there should be enough energy to cause the emission of photoelectrons; what was not observed for the frequencies below the threshold frequency even by increasing the intensity of the incident light radiation.

(c) *Fixed parameters*:

-   Metal
-   Frequency of the incident radiation

Studied parameter:

•   Intensity of the incident radiation

The experiment gives us the $I = f(U)$ curves. These curves reveal that for a fixed metal and a fixed frequency, the saturation intensity depends on the intensity of the incident light radiation. They also show that the stopping voltage is independent of the intensity of the incident radiation.

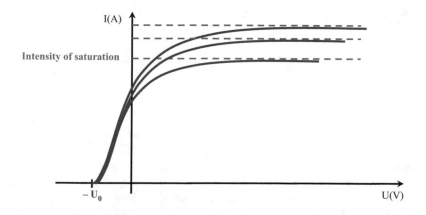

4. (a) *Occurrence of the photoelectric effect for the radiations* $\lambda_1 = 430$ nm *and* $\lambda_2 = 580$ nm:

Let us first determine the threshold wavelength λ_S:

$$W_S = h\frac{c}{\lambda_S} \Rightarrow \lambda_S = \frac{hc}{W_S} = \frac{(6.626 \times 10^{-34} \text{ J.s}) \left(3 \times 10^8 \text{ m.s}^{-1}\right) (1 \text{ eV})}{(2.4 \text{ eV}) \left(1.6 \times 10^{-19} \text{ J}\right)} \tag{1}$$
$$= 517 \times 10^{-9} \text{ m} = 517 \text{ nm}.$$

As the wavelength λ is inversely proportional to the energy W of the incident photon ($\lambda = hc/W$), the greater its value is small, the greater the incidental photon is energetic. From there, we can say that, taken separately, $\lambda_1 = 430$ nm allows the photoelectric effect because $\lambda_1 < \lambda_S = 517$ nm but not $\lambda_2 = 580$ nm which is greater than λ_S. However, if we consider the polychromatic beam composed of the two wavelengths λ_1 and λ_2, this beam obviously allows the emission of photoelectrons.

(b) *Maximal speed* of photoelectrons:

Electrons being ejected only by radiation 1 ($\lambda_1 = 430$ nm), only the speed of these electrons is calculated. The kinetic energy carried by the electron at its exit from the metal is given by Einstein's equation.

$$E_c = h\nu - W_S = h\frac{c}{\lambda_1} - h\nu_S$$
$$= (6.626 \times 10^{-34}\,\text{J.s})\,\frac{3 \times 10^8\,\text{m.s}^{-1}}{430 \times 10^{-9}\,\text{m}} - \frac{(2.4\,\text{eV})\,(1.6 \times 10^{-19}\,\text{J})}{1\,\text{eV}} = 0.76 \times 10^{-19}\,\text{J}.$$

$$(2)$$

The maximal speed of the photoelectron is therefore:

$$E_c = \frac{1}{2}\,m_e\,v^2 \Rightarrow v = \sqrt{\frac{2\,E_c}{m_e}} = \sqrt{\frac{2\,(0.76 \times 10^{-19}\,\text{J})}{9.1 \times 10^{-31}\,\text{kg}}} = 4 \times 10^5\,\text{m.s}^{-1}. \quad (3)$$

(c) *Definition and calculation of the stopping voltage*:

The stopping potential corresponds to the voltage that must be applied between the electrodes in order to cancel the kinetic energy of electrons. This voltage U_0 is negative ($U = V_A - V_C = -U_0$), and it allows to obtain a zero intensity on the curve $I = f(U)$, given above. Indeed, in the absence of speed, electrons do not move anymore, so there is no current. To find the stopping voltage, let us apply the kinetic energy theorem between the anode and the cathode, and it becomes:

$$E_c^A - E_c^C = W_R = 0 - \frac{1}{2}\,m_e\,v^2 = eU_0. \quad (4)$$

Equation (4) shows that the variation of the kinetic energy between the anode E_c^A and the cathode E_c^C is equal to the sum of the works done by the electron. Here, the only work made by the electron is a resistant work which opposes its move. This work, of electrical origin, decelerates it until the cancellation of its speed at its arrival to the anode. By starting from (4), we arrive at the value of the stopping voltage:

$$-E_c^C = eU_0 \Rightarrow U_0 = -\frac{E_c^C}{e} = -\frac{0.76 \times 10^{-19}\,\text{J}}{1.6 \times 10^{-19}\,\text{C}} = -0.48\,\text{V}. \quad (5)$$

Exercise 12
We have three photoemissive cells. The cathodes are respectively covered with cesium (Ce), potassium (K), and lithium (Li). The extraction energies W_S of these metals are, respectively, 1.19 eV, 2.29 eV, and 2.39 eV.
1. *What is the extraction energy?*
2. *We illuminate successively each cell by a monochromatic radiation of wavelength $\lambda = 0.60\ \mu m$.*

(a) *Calculate, in electronvolts, the energy transported by one incident photon.*

(b) *In which of these three cells do we observe the photoelectric effect? Justify your answer.*

(c) *Calculate, in joules, the maximal kinetic energy of the electron at its exit from the cathode.*

3. *Calculate the voltage that must be applied between the anode and the cathode to prevent an electron of the cathode to reach the anode.*

4. *Calculate the maximal speed of one electron at the exit of the cathode. We give:*
   - *Planck's constant: $h = 6.626 \times 10^{-34}$ J.s*
   - *Mass of the electron: $m_e = 9 \times 10^{-31}$ kg*
   - *Speed of light: $c = 3 \times 10^8$ ms^{-1}*
   - *$1 \ eV = 1.6 \times 10^{-19}$ J*
   - *$1 \ \mu m = 10^{-6}$ m*

Solution

1. *Definition of the extraction energy*:

The extraction energy is the minimum energy required to remove one electron from the surface of a given metal. It corresponds to the binding energy of the electron on its electronic shell.

2. (a) *Energy transported by one incident photon*:

The incident photon carries an indivisible energy equal to:

$$W(\text{J}) = h\nu = h\frac{c}{\lambda} = \left(6.626 \times 10^{-34} \ \text{J.s}\right) \frac{3 \times 10^8 \ \text{m.s}^{-1}}{0.60 \times 10^{-6} \ \text{m}} = 3.31 \times 10^{-19} \ \text{J.} \quad (1)$$

As $1 \ \text{eV} = 1.6 \times 10^{-19}$ J, we deduce:

$$W(\text{eV}) = \frac{\left(3.31 \times 10^{-19} \ \text{J}\right)\left(1 \ \text{eV}\right)}{1.6 \times 10^{-19} \ \text{J}} = 2.069 \ \text{eV.} \quad (2)$$

This energy has nothing to do with the nature of the bombarded photocathode metal.

(b) *Cell with which we obtain the photoelectric effect*:

For the emission of photoelectrons, the energy of the incident photon $W = 2.069$ eV must be greater than the extraction energy W_S of the bombarded metal. Only the cathode covered with cesium with an energy of extraction equal to 1.19 eV satisfies this condition.

(c) *Maximal kinetic energy of the electron at its exit of the cathode*:

The kinetic energy carried by the electron at its exit from the metal is given by Einstein's equation:

$$E_c = W - W_S = 3.31 \times 10^{-19}\,\text{J} - \frac{(1.19\,\text{eV})\,(1.6 \times 10^{-19}\,\text{J})}{1\,\text{eV}}$$

$$= 1.406 \times 10^{-19}\,\text{J.} \qquad (3)$$

3. *Stopping voltage U_0:*

The stopping potential corresponds to the voltage that must be applied between the electrodes of the photocell in order to cancel the kinetic energy of electrons. This voltage U_0 is negative ($U = V_A - V_C = -U_0$), and it allows to obtain a zero intensity on the curve $I = f(U)$, given in Exercise 11. Indeed, in the absence of speed, electrons do not move anymore, so there is no current. To find the stopping voltage, let us apply the kinetic energy theorem between the anode and the cathode; it becomes:

$$E_c^A - E_c^C = W_R = 0 - \frac{1}{2}\,m_e\,v^2 = eU_0. \qquad (4)$$

Equation (4) shows that the variation of the kinetic energy between the anode E_c^A and the cathode E_c^C is equal to the sum of the works done by the electron. Here, the only work made by the electron is a resistant work which opposes its move. This work, which is of electrical origin, decelerates it until the cancellation of its speed at its arrival to the anode. By starting from (4), we arrive at the value of the stopping voltage:

$$-E_c^C = eU_0 \quad \Rightarrow \quad U_0 = -\frac{E_c^C}{e} = -\frac{1.406 \times 10^{-19}\,\text{J}}{1.6 \times 10^{-19}\,\text{C}} = -0.88\,\text{V.} \qquad (5)$$

4. *Maximal speed of the electron at the exit of the cathode:*

The maximum speed of the photoelectron is:

$$E_c = \frac{1}{2}\,m_e\,v^2 \quad \Rightarrow \quad v = \sqrt{\frac{2\,E_c}{m}} = \sqrt{\frac{2\,(1.406 \times 10^{-19}\,\text{J})}{9 \times 10^{-31}\,\text{kg}}}$$

$$= 1.77 \times 10^5\,\text{m.s}^{-1}. \qquad (6)$$

Exercise 13
A beam illuminates a cathode of a photocell. The threshold of this cesium cathode is $\nu_S = 4.54 \times 10^{14}$ Hz and its quantum yield is $r = 0.05$. We have $U = V_A - Vc$.
1. *Plot qualitatively the curve $I = f(U)$. What does it reveal?*
2. *Calculate the extraction work W_S in eV and the speed of one electron leaving the cathode if the emitted radiation wavelength is $\lambda = 500$ nm.*
3. *Calculate the speed of one electron when it reaches the anode if $U = 100$ V.*
4. *Establish the relationship between I_S, the saturation intensity, and the luminous power P received by the cathode. Knowing that $P = 10^{-3}$ W, determine I_S. We give:*

- *Planck's constant: h = 6.626 × 10^{-34} J.s*
- *Mass of the electron: m_e = 9 × 10^{-31} kg*
- *Speed of light c = 3 × 10^8 ms^{-1}*
- *1 eV = 1.6 × 10^{-19} J*
- *Charge of the electron e = 1.6 × 10^{-19} C*

Solution

1. *Curve plot I = f(U):*

We fix the metal, the intensity of the incident radiation, and the frequency of the incident radiation, and we study the variation of the intensity of the electric current as a function of the voltage applied across the photocell. We get the following curve:

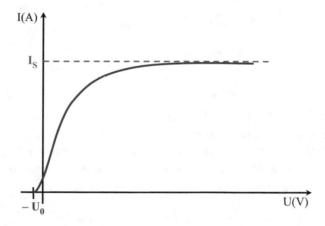

This curve reveals that:

- The electrical intensity can be canceled by slowing down electrons with an adequate voltage. The voltage – U_0 is negative. It is called *stopping voltage.*
- The electrical intensity tends to a maximum when the voltage increases. This maximum, noticed hereafter I_S, is called *saturation intensity.*

2. *Calculation of the extraction work W_S in eV and the speed of one electron leaving the cathode for λ = 500 nm:*

The extraction work is the minimum energy required to remove one electron from the surface of a given metal. It is given by:

$$W_S(\text{J}) = h\nu_S = \left(6.626 \times 10^{-34} \text{ J.s}\right)\left(4.54 \times 10^{14} \text{ Hz}\right) = 3.01 \times 10^{-19} \text{ J}. \quad (1)$$

As 1 eV = 1.6 × 10^{-19} J, we deduce:

$$W_S(\text{eV}) = \frac{\left(3.01 \times 10^{-19} \text{ J}\right)\left(1 \text{ eV}\right)}{1.6 \times 10^{-19} \text{ J}} = 1.88 \text{ eV}. \quad (2)$$

The kinetic energy carried by the electron at its exit from the metal is given by Einstein's equation:

$$E_c = h\nu - W_S = h\frac{c}{\lambda} - h\nu_S$$

$$= \left(6.626 \times 10^{-34} \text{ J.s}\right) \frac{3 \times 10^8 \text{ m.s}^{-1}}{500 \times 10^{-9} \text{ m}} - 3.01 \times 10^{-19} \text{ J} = 0.968 \times 10^{-19} \text{ J}.$$

$$(3)$$

The maximal speed of the photoelectron is therefore:

$$E_c = \frac{1}{2} m_e v^2 \implies v = \sqrt{\frac{2 E_c}{m_e}} = \sqrt{\frac{2 \left(0.968 \times 10^{-19} \text{ J}\right)}{9.1 \times 10^{-31} \text{ kg}}}$$

$$= 4.61 \times 10^5 \text{ m.s}^{-1}.$$

$$(4)$$

3. *Calculation of the speed of one electron when it reaches the anode if U = 100 V*:

The electrical work applied to the electron is $W_e = eU$. The kinetic energy E_c^A of the electron when it reaches the anode is equal to that which it has when it left the cathode E_c^C plus the energy acquired by means of the accelerating voltage $W_e = eU$, taking into account (3) for the numerical value of $E_c^C = W - W_S$, and by application of the kinetic energy theorem, it becomes:

$$E_c^A - E_c^C = eU \implies E_c^A = E_c^C + eU = (W - W_S) + eU$$

$$= 0.968 \times 10^{-19} \text{ J} + \left(1.6 \times 10^{-19} \text{ C}\right) (100 \text{ V}) \approx 1.6 \times 10^{-17} \text{ J}.$$

$$(5)$$

The speed v_A of the electron at the anode is:

$$E_c = \frac{1}{2} m_e v_A^2 \implies v_A = \sqrt{\frac{2 E_c}{m_e}} = \sqrt{\frac{2 \left(1.6 \times 10^{-19} \text{ J}\right)}{9.1 \times 10^{-31} \text{ kg}}} = 5.9 \times 10^6 \text{ m.s}^{-1}. \quad (6)$$

4. *Establishment of the relationship between I_S, the intensity of saturation, and the luminous power P received by the cathode and determination of I_S:*

The quantum yield of the photocell is given by:

$$\alpha = \frac{n}{N}. \qquad (7)$$

It is the ratio between the number n of photons which, arriving at the cathode, pull out electrons and the number N of the total photons received by the cathode. Moreover, the intensity of the electric current I_S is by definition the flow of the electrical charges at a point of the circuit (in a section of the conductor), namely:

$$I_S = \frac{dq(t)}{dt} = \frac{ne}{dt} \quad \Rightarrow \quad n = I_S \frac{dt}{e} \tag{8}$$

where $q(t)$ is the electric charge and dt the short time period. The luminous power P is the radiated energy flux given by the following equation:

$$P = \frac{d\omega}{dt} = \frac{Nh}{dt} \quad \Rightarrow \quad N = P\frac{dt}{h} \tag{9}$$

where ω is the radiated energy in joules. By doing the ratio (8)/(9), we obtain the relationship between the saturation intensity I_S and the luminous power P:

$$\alpha = \frac{n}{N} = \frac{I_S \dfrac{dt}{e}}{P \dfrac{dt}{h\nu}} = \frac{I_S\, h\nu}{P\, e} \quad \Rightarrow \quad I_S = P\frac{\alpha\, e}{h\nu}. \tag{10}$$

The numerical value of the saturation intensity I_S is:

$$I_S = P\frac{\alpha\, e}{h\frac{c}{\lambda}} = (10^{-3}\ \text{W})\,\frac{(0.05)\left(1.6 \times 10^{-19}\ \text{C}\right)}{\left(6.626 \times 10^{-34}\ \text{J.s}\right)\dfrac{3 \times 10^8\ \text{m.s}^{-1}}{500 \times 10^{-9}\ \text{m}}} = 2 \times 10^7\ \text{A}. \tag{11}$$

Exercise 14

1. *Give the definition of the following:*
   - *Photoelectric effect*
   - *Threshold frequency*
   - *Extraction energy*
2. *What experimental setups can be used to observe the photoelectric effect?*
3. *The extraction energy of one electron from a cylindrical sodium plate is $W_S = 2.18\ eV$.*
   This plate is successively illuminated by the following radiations:
   - *Light radiation of wavelength $\lambda = 0.662\ \mu m$*
   - *Frequency light radiation $\nu = 5 \times 10^{14}\ Hz$*
   - *Period luminous radiation $T = 1.3 \times 10^{-15}\ s$*
   Indicate in each case if there is emission of electrons. Justify your answer.
4. *If there is photoelectric effect, calculate:*
   (a) *The maximal speed of electrons emitted from the plate.*
   (b) *The value of the voltage to be applied between the photoemissive metal and the anode in order to cancel the photoelectric current. We give:*
      - *Planck's constant: $h = 6.626 \times 10^{-34}\ J.s$*
      - *Mass of the electron: $m_e = 9 \times 10^{-31}\ kg$*
      - *Speed of light $c = 3 \times 10^8\ ms^{-1}$*
      - *$1\ eV = 1.6 \times 10^{-19}\ J$*
      - *Charge of the electron $e = 1.6 \times 10^{-19}\ C$*

Solution

1. – *Definition of the photoelectric effect*:

The photoelectric effect refers to the emission of electrons from the surface of a metal (or other substance) under the action of light. In this effect it is not the intensity of the incident radiation that is the dominant factor but rather its frequency.

   – *Definition of the threshold frequency*:

This is the minimum frequency that an incident photon must have to extract one electron from a metal.

   – *Definition of the extraction energy*:

The extraction work is the minimum energy required to remove one electron from the surface of a given metal. It corresponds to the binding energy of the electron on its electronic shell.

2. *Highlighting the photoelectric effect*:

To highlight the photoelectric effect, two methods can be used:

   – *First method (Hertz's experiment)*:

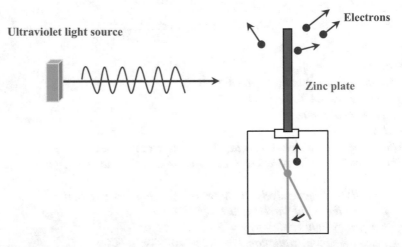

Using the apparatus described in the figure, Heinrich Rudolf Hertz accidentally discovered in 1887 that negatively charged metals lost negative charges when they are exposed to ultraviolet radiation. Let us have a closer look at this experiment:

(i) Let us consider, for this purpose, a zinc plate placed on a negatively charged electroscope. When this plate is illuminated with an ultraviolet light source, the electroscope leaf falls down, indicating that the electroscope discharges.

(ii) We do the same experiment, but this time the electroscope is positively charged. No modification is observed.

(iii) The electroscope is again negatively charged, but this time we interpose between the zinc plate and the light source a glass slide which has the property to absorb ultraviolet rays. No modification is observed.

These experimental facts are explained by the fact that the ultraviolet radiation expelled free electrons from the zinc plate. Excess electrons which were in the electroscope then migrate to the electron-deficient plate and the electroscope discharges.

– *Second method (photocell experiment)*:

The electrical assembly for studying the photoelectric effect consists of a quartz bulb in which a very high vacuum is made of the order of 10^{-7} mm Hg. This bulb, *called photocell*, is composed of two electrodes, a cylinder-shaped cathode and a rectilinear wire-shaped anode placed on the axis of the cylinder. Illuminated by an appropriate light, the cathode (emitter) emits electrons. These electrons are captured by the anode (collector) which is connected to the positive pole of an adjustable generator. It results from this a current detected by a galvanometer incorporated in the circuit. A voltmeter measures the applied voltage to the cell.

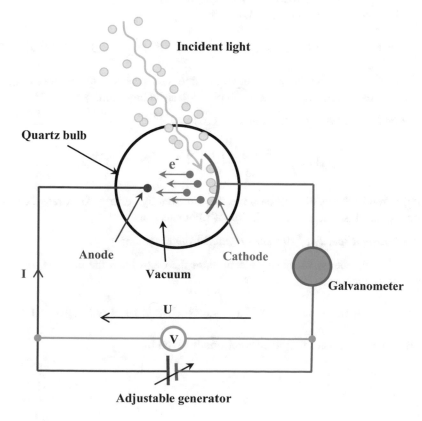

3. *Case where there is emission of electrons*:
    The extraction energy in joules is:

$$W_S(J) = \frac{(2.18 \text{ eV}) \ (1.6 \times 10^{-19} \text{ J})}{1 \text{ eV}} = 3.488 \times 10^{-19} \text{ J}. \tag{1}$$

– *Light radiation of wavelength $\lambda_1 = 0.662 \ \mu m$*:
    The energy of the incident photon is:

$$W_1(J) = h\nu_1 = h\frac{c}{\lambda_1} = (6.626 \times 10^{-34} \text{ J.s}) \ \frac{(3 \times 10^8 \text{ m.s}^{-1})}{0.662 \times 10^{-6} \text{ m}}$$
$$= 0.3 \times 10^{-19} \text{ J}. \tag{2}$$

$W_1(J) < W_S(J)$: the incident photon does not have enough energy to extract electrons from the metal, so there is no photoelectron emission.

– *Light radiation of the frequency $\nu = 5 \times 10^{14}$ Hz*:
    The energy of the incident photon is:

$$W_2(J) = h\nu_2 = (6.626 \times 10^{-34} \text{ J.s}) \ (5 \times 10^{14} \text{ Hz}) = 3.31 \times 10^{-19} \text{ J}. \tag{3}$$

$W_2(J) < W_S(J)$: here too, the incident photon does not have enough energy to extract electrons from the metal, so there is no photoelectron emission.

– *Periodic radiation of the period $T = 1.3 \times 10^{-15}$ s*:

$$W_3(J) = \frac{h}{T} = \frac{6.626 \times 10^{-34} \text{ J.s}}{1.3 \times 10^{-15} \text{ s}} = 5.09 \times 10^{-19} \text{ J}. \tag{4}$$

$W_3(J) > W_S(J)$: the incident photon has this time enough energy to extract electrons from the metal, so there is an emission of photoelectrons.

4. (a) *Maximal speedy of electrons emitted from the plate*:

The kinetic energy carried by electrons at their exit from the metal is given by Einstein's equation:

$$E_c = W_3 - W_S = 5.09 \times 10^{-19} \text{ J} - 3.488 \times 10^{-19} \text{ J} = 1.602 \times 10^{-19} \text{ J}. \tag{5}$$

The maximal speed of emitted electrons is given by:

$$E_c = \frac{1}{2} m_e v^2 \ \Rightarrow \ v = \sqrt{\frac{2\,E_c}{m_e}} = \sqrt{\frac{2\,(1.602 \times 10^{-19}\ \text{J})}{9.1 \times 10^{-31}\ \text{kg}}}$$

$$= 5.933 \times 10^5\ \text{m.s}^{-1}. \tag{6}$$

(b) *Stopping voltage*:

The stopping potential corresponds to the voltage that must be applied between the electrodes of the photocell in order to cancel the kinetic energy of electrons. This voltage U_0 is negative ($U = V_A - V_C = -U_0$), and it allows to obtain a zero intensity on the curve $I = f(U)$, given in Exercise 11. Indeed, in the absence of speed, electrons do not move anymore, so there is no current. To find the stopping voltage, let us apply the theorem of kinetic energy between the anode and the cathode, and we obtain:

$$E_c^A - E_c^C = W_R = 0 - \frac{1}{2}\,m\,v^2 = eU_0. \tag{7}$$

Equation (7) shows that the variation of the kinetic energy between the anode E_c^A and the cathode E_c^C is equal to the sum of the works done by the electron. Here, the only work made by the electron is a resistant work which opposes its move. This work, which is of electrical origin, decelerates it until the cancellation of its speed at its arrival to the anode. By starting from (7), we arrive at the value of the stopping voltage:

$$-E_c^C = eU_0 \ \Rightarrow \ U_0 = -\frac{E_c^C}{e} = -\frac{1.602 \times 10^{-19}\ \text{J}}{1.6 \times 10^{-19}\ \text{C}} = -1.001\ \text{V}. \tag{8}$$

Exercise 15
We have a potassium photocell, of which the extraction work is $W_S = 2.2$ eV. For this cell, we determine the stopping voltage as a function of various illumination frequencies. We obtain the following results (we have indicated in the table the absolute value of the stopping voltage):

ν (Hz)	$7\ 10^{14}$	$8\ 10^{14}$	$9\ 10^{14}$
$|U_0|$ (V)	0.69	1.10	1.52

1. *Plot the curve $\nu = f(|U_0|)$. We expect to find a linear function of the form $\nu = a|U_0| + b$. Deduct the values of the coefficients a and b and what do they mean physically?*
2. *Deduce the value of the photoelectric threshold of this photocathode. Is this result consistent with the value of W_S?*
3. *Determine the value of Planck's constant h from the realized curve. Does this value correspond to the admitted value in the literature? We give:*

 - *Planck's constant: h = 6.626 × 10⁻³⁴ J.s*
 - *Speed of light: c = 3 × 10⁸ ms⁻¹*
 - *1 eV = 1.6 × 10⁻¹⁹ J*
 - *Charge of the electron: e = 1.6 × 10⁻¹⁹ C*

Solution

1. The plot of the curve $\nu = f(|U_0|)$ gives a straight line, of which the equation is:

$$\nu = 2.41 \times 10^{14} \, |U_0| + 5.34 \times 10^{14}. \tag{1}$$

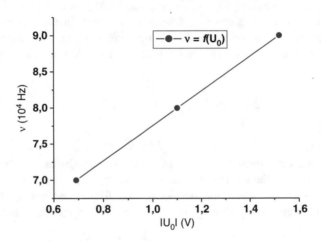

From a theoretical point of view, the stopping potential corresponds to the voltage that must be applied between the electrodes of the photocell in order to cancel the kinetic energy of electrons. This voltage U_0 is negative ($U = V_A - V_C = -U_0$), and it allows to obtain a zero intensity on the curve $I = f(U)$, given in Exercise 11. Indeed, in the absence of speed, electrons do not move anymore, so there is no current. To find the stopping voltage, let us apply the kinetic energy theorem between the anode and the cathode; we obtain:

$$E_c^A - E_c^C = W_R = 0 - \frac{1}{2} \, m \, v^2 = e U_0. \tag{2}$$

Equation (2) shows that the variation of the kinetic energy between the anode E_c^A and the cathode E_c^C is equal to the sum of the works done by the electron. Here, the only work made by the electron is a resistant work which opposes its move. This work, which is of electrical origin, decelerates it until the cancellation of its speed at its arrival to the anode. By starting from (2), we arrive at the value of the stopping voltage:

$$-E_\text{c}^\text{C} = eU_0 \quad \Rightarrow \quad U_0 = -\frac{E_\text{c}^\text{C}}{e} \quad \text{namely: } |U_0| = \frac{E_\text{c}^\text{C}}{e} \quad \text{or} \quad E_\text{c}^\text{C} = e|U_0|. \qquad (3)$$

Moreover, the kinetic energy carried by the electron at its exit of the cathode is given by Einstein's equation:

$$E_\text{c}^\text{C} = h\nu - W_\text{S} = h\nu - h_\text{S} = e|U_0|. \qquad (4)$$

From (4), we deduce a relationship between the frequency of the incident photon ν and the absolute value of the stopping potential $|U_0|$:

$$\nu = \frac{e}{h}|U_0| + \nu_\text{S}. \qquad (5)$$

We retrieve Eq. (1) deduced from the curve $\nu = f(|U_0|)$ where $a = e/h$ (the charge of the electron divided by Planck's constant) and $b = \nu_\text{S}$ (the threshold frequency).

2. *Value of the photoelectric threshold*:

By comparing the experimental (1) and theoretical (5) equations, we deduce the value of the threshold frequency ν_S:

$$\nu_\text{S} = 5.34 \times 10^{14} \text{ Hz}. \qquad (6)$$

The extraction energy W_S is given by:

$$W_\text{S}(\text{eV}) = h\nu_\text{S} = \frac{\left(6.626 \times 10^{-34} \text{ J.s}\right)\left(5.34 \times 10^{14} \text{ Hz}\right)(1 \text{ eV})}{1.6 \times 10^{-19} \text{ J}} = 2.2 \text{ eV}. \qquad (7)$$

This value of the extraction energy, deduced from the curve, is compatible with the given value in this exercise.

3. *Determination of the value of Planck's constant h from the experimental curve*:

According to Eqs. (1) and (5), the directional coefficient of the straight line is:

$$\frac{e}{h} = 2.41 \times 10^{14} \text{ Hz.V}^{-1} \quad \Rightarrow \quad h = \frac{1.6 \times 10^{-19} \text{ C}}{2.41 \times 10^{14} \text{ Hz.V}^{-1}}$$
$$= 6.64 \times 10^{-34} \text{ C.V.Hz}^{-1} \text{ or J.s}. \qquad (8)$$

This value of Planck's constant is very close to the admitted value in the literature.

Exercise 16
The ionization energy of a very pure sodium is 2.75 eV, where 1 eV = 1.602 × 10^{-19} J.

1. *Calculate in electronvolts the maximal kinetic energy that can have photoelectrons emitted by sodium exposed to an ultraviolet radiation of 200 nm.*
2. *Calculate the longest wavelength that can cause a photoelectric effect in pure sodium.*
3. *The sodium extraction energy that has not been thoroughly purified is significantly lower than 2.75 eV because of the absorption of sulfur atoms and other substances derived from atmospheric gases. When impure sodium is exposed to a radiation of 200 nm, will the maximal kinetic energy of a photoelectron be smaller or greater than that of pure sodium exposed to the same radiation of 200 nm? We give:*
   - *Planck's constant: $h = 6.626 \times 10^{-34}$ J.s*
   - *Speed of light: $c = 3 \times 10^8$ ms^{-1}*
   - *1 eV = 1.6×10^{-19} J*
   - *Charge of the electron: $e = 1.6 \times 10^{-19}$ C*

Solution
1. The surplus of the received energy is used by the ejected electron in the form of the kinetic energy to move:

$$E_c = h\nu - h\nu_S = h\frac{c}{\lambda} - h\nu_S$$
$$= \left(6.626 \times 10^{-34} \text{ J.s}\right) \left(\frac{3 \times 10^8 \text{ m.s}^{-1}}{200 \times 10^{-9} \text{ m}}\right) \left(\frac{1 \text{ eV}}{1.6 \times 10^{-19} \text{ J}}\right) - 2.75 \text{ eV} = 3.46 \text{ eV}.$$

$$(1)$$

where $h\nu_S$ is the extraction energy with ν_S, the frequency of extraction.

2. The longest wavelength that can cause a photoelectric effect in pure sodium is:

$$h\nu_S = h\frac{c}{\lambda_S} \Rightarrow \lambda_S = \frac{hc}{h\nu_S} = \frac{\left(6.626 \times 10^{-34} \text{ J.s}\right) \left(3 \times 10^8 \text{ m.s}^{-1}\right) (1 \text{ eV})}{(2.75 \text{ eV}) \left(1.6 \times 10^{-19} \text{ J}\right)}$$
$$= 4.51 \times 10^{-7} \text{m},$$

$$(2)$$

namely, $\lambda_S = 4.51$ nm.

3. The surplus of the received energy is used by the ejected electron in the form of the kinetic energy to move:

$$E_c = h\nu - h\nu_S.$$

$$(3)$$

   Pure sodium and impure sodium are exposed to the same ultraviolet radiation of the wavelength 200 nm $\Rightarrow$ the energy of the incident photon $h\nu$ which strikes both surfaces is the same. So, we reason about the value of the energy of extraction $h\nu_S$ by using Eq. (3): when the extraction energy of impure sodium is lower than the extraction energy of pure sodium, then the maximal kinetic energy of a photoelectron of impure sodium is greater than the maximum kinetic energy of a photoelectron of pure sodium.

Exercise 17

A polychromatic light comprising three radiations (λ_1 = 450 nm, λ_2 = 610 nm, λ_3 = 750 nm) irradiates a potassium sample. The energy of ionization is 2.14 eV.
1. *Establish the relationship $E(eV) = 1241/\lambda$ (nm).*
2. *Which radiation gives rise to the photoelectric effect?*
3. *What is the speed of electrons expelled from the metal?*
    We give:
    – *Planck's constant: h = 6.626 × 10^{-34} J.s*
    – *Speed of light: c = 3 × 10^8 ms^{-1}*
    – *1 eV = 1.6 × 10^{-19} J*
    – *Charge of the electron: e = 1.6 × 10^{-19} C*
    – *Mass of the electron: m_e = 9.1 × 10^{-31} kg*

Solution

1. *Establishment of the relationship $E(eV) = 1241/\lambda$ (nm):*

The energy of the incident photon is given by:

$$E\left(J\right) = h\frac{c}{\lambda} = \left(6.626 \times 10^{-34} \text{ J.s}\right) \frac{3 \times 10^8 \text{ m.s}^{-1}}{\lambda \left(\text{m}\right)}. \tag{1}$$

Knowing that 1 eV = 1.6 × 10^{-19} J and 1 nm = 10^{-9} m, let us divide (1) by (1.6 × 10^{-19} J) to obtain the energy E in electronvolts and let us multiply the wavelength λ by (10^{-9} m) to get a wavelength in nm, namely:

$$E\left(eV\right) = h\frac{c}{\lambda} = \left(6.626 \times 10^{-34} \text{ J.s}\right) \frac{3 \times 10^8 \text{ m.s}^{-1}}{\left(10^{-9}\lambda \text{ nm}\right)\left(1.6 \times 10^{-19} \text{ J}\right)} = \frac{1241}{\lambda \left(\text{nm}\right)}. \tag{2}$$

This equation is useful for calculating the energy E of one photon directly in electronvolts if the wavelength λ is expressed in nm and vice versa.

2. *Radiation that gives rise to the photoelectric effect:*

Using Eq. (2), we obtain the energies of ionization corresponding to the three wavelengths:

$$E_1 = \frac{1241}{\lambda_1} = \frac{1241}{450} = 2.76 \text{ eV}$$

$$E_2 = \frac{1241}{\lambda_2} = \frac{1241}{610} = 2.03 \text{ eV}$$

$$E_3 = \frac{1241}{\lambda_3} = \frac{1241}{750} = 1.65 \text{ eV.} \tag{3}$$

Only the radiation λ_1 gives rise to the photoelectric effect because the energy carried by the photon (1), $E_1 = 2.76$ eV, is greater than the energy of extraction, $E_S = 2.14$ eV, of electrons of the potassium atom.

3. *Speed of electrons expelled from the metal*:

The excess of the received energy by potassium is used by the ejected electron in the form of kinetic energy to move:

$$E_c = E_1 - E_S = 2.76 \text{ eV} - 2.14 \text{ eV} = 0.62 \text{ eV}$$

$$= \frac{(0.62 \text{ eV}) \ (1.6 \times 10^{-19} \text{ J})}{1 \text{ eV}} = 10^{-19} \text{ J.} \tag{4}$$

From such we deduce the maximal speed of the expelled electrons:

$$E_c = \frac{1}{2} m_e v^2 \quad \Rightarrow \quad v = \sqrt{\frac{2 E_c}{m_e}} = \sqrt{\frac{(2) \ (10^{-19} \text{ J})}{9.1 \times 10^{-31} \text{ kg}}} = 4.7 \times 10^5 \text{ m.s}^{-1}. \tag{5}$$

Exercise 18
1. *What is the wavelength of one photoelectron of mass $m_e = 9 \times 10^{-31}$ kg emitted from the material then braked by a potential $U_0 = 10$ kV?*
2. *Describe a general formula between the wavelength λ and the energy E of a particle of mass m.*
3. *Calculate the wavelength λ associated with a man of mass $m_h = 70$ kg moving with a speed $v_h = 5$ km/h.*
4. *Will the wave effects be important for this man in everyday life? We give:*
   - *Planck's constant: $h = 6.626 \times 10^{-34}$ J.s*
   - *Electron mass: $m_e = 9 \times 10^{-31}$ kg*
   - *Electron charge: $e = 1.6 \times 10^{-19}$ C*

Solution
1. *Wavelength of one electron of mass m_e*:

The determination of the kinetic energy of the emitted electron is done by the method of the retarding field. In this method, the electron crosses an area where there is an electric field which opposes to its move, and of which the intensity is varied until this electron stops. In this case, we obtain, for a measurable braking voltage U_0, Eq. (4.9) given in the course:

$$E = eU_0 = \frac{1}{2} m_e v^2 = \frac{m_e^2 v^2}{2m_e} = \frac{P_e^2}{2m_e} \Rightarrow P_e = \sqrt{2m_e eU_0}. \qquad (1)$$

According to Eq. (4.16) of the course which gives the equation of de Broglie, we obtain:

$$P_e = \sqrt{2m_e eU_0} = \frac{h}{\lambda_e} \Rightarrow \lambda_e = \frac{h}{\sqrt{2m_e eU_0}}$$

$$= \frac{6.63 \times 10^{-34} \text{ J.s}}{\sqrt{2 \left(9 \times 10^{-31} \text{kg}\right) \left(1.6022 \times 10^{-19} \text{ C}\right) \left(10 \times 10^3 \text{ volts}\right)}} = 1.2 \times 10^{-11} \text{ m.}$$

$$(2)$$

2. *General formula between the wavelength λ and the energy E of one particle of mass m:*

According to (1) and (2), we deduce a general relationship of the form $\lambda = f(E,m)$:

$$\lambda = \frac{h}{P} = \frac{h}{\sqrt{2mE}}. \qquad (3)$$

3. *Wavelength λ associated with a man of mass m_h and of speed v_h:*

The wavelength associated with a man of mass $m_h = 70$ kg moving with a speed $v_h = 5$ km/h $= 5 \times 10^3$ m/3600 s $= 1.4$ m.s^{-1} is:

$$\lambda_h = \frac{h}{P_h} = \frac{h}{m_h v_h} = \frac{6.63 \times 10^{-34} \text{ J.s}}{(70 \text{ kg}) (1.4 \text{ m.s}^{-1})} = 6.8 \times 10^{-36} \text{ m.} \qquad (4)$$

This wavelength is too small to play any role in the life of this man.

Exercise 19

1. *What does one call the grains of light about which Einstein and Planck speak?*
2. *It is found experimentally that the photoelectric effect only occurs from a minimal frequency v_S of the incoming radiation. We call the extraction work the minimal energy to be supplied to eject one electron from the metal. Write the relationship between the extraction work of the electron W_S and v_S.*
3. *If the used frequency of light satisfies $v > v_S$, give the expression of the kinetic energy of one emitted electron as a function of v and W_S.*
4. *We bombard a metal surface, of which the extraction work is $W_S = 2.2$ eV, by a laser with a wavelength $\lambda = 355$ nm and a power $P = 10$ mW.*
   (a) *Calculate the number of incident photons that strike the metal plate in 1 second.*
   (b) *Calculate the energy of one photon and show that we must observe in this case the photoelectric effect.*

(c) *We assume that all emitted electrons are collected by means of a positively charged electrode placed in front of the metal surface. Knowing that the electric current is defined as the electric charge crossing the inter-electrode area per second, calculate in absolute value the intensity of the current called current of saturation I_S of the photocell. You express this current as a function of e, λ, W, h, and c.*

(d) *Plot the current of saturation as a function of the incident radiation wavelength, for the constant radiation energy W. Interpret this curve. How can we deduce the value of Planck's constant if we know W?*

(e) *Calculate the kinetic energy of each emitted electron in eV.*

5. *We repeat the same experiment with a laser of the same energy, but of the wavelength $\lambda' = 1$ μm. What is going on? Why? We give:*
   - *Planck's constant: $h = 6.626 \times 10^{-34}$ J.s*
   - *Speed of light: $c = 3 \times 10^8$ m.s^{-1}*
   - *1 eV = 1.6×10^{-19} J*
   - *Electron charge: $e = 1.6 \times 10^{-19}$ C*

Solution

1. The grains of energy of which Einstein and Planck speak are called *photons*.

2. *Relationship between the work extraction W_S and ν_S:*

To eject one electron from the metal with a certain speed, the energy of the incident photon $h\nu$ must be greater than the extraction work W_S, namely:

$$h\nu > W_S. \tag{1}$$

For a minimal incident photon energy $h\nu_S$ such that:

$$h\nu_S = W_S. \tag{2}$$

The electron is ejected from the metal without leaving it.

3. *Expression of the kinetic energy of one electron emitted as a function of ν and W_S:*

If the incident photon has an energy $h\nu > W_S$, one part of its energy equal to $h\nu_S$ is used to extract one electron from the metal, and the remaining part of its energy is communicated to this electron in the form of the kinetic energy to move, i.e.:

$$E_C = h\nu - W_S = h\nu - h\nu_S = h(\nu - \nu_S). \tag{3}$$

4. (a) *Number of incident photons that strike the metal surface in 1 second*:

Let us evaluate the energy of one incident photon:

$$E = h\nu = h\frac{c}{\lambda} = \left(6.626 \times 10^{-34} \text{ J.s}^{-1}\right) \frac{3 \times 10^8 \text{ m.s}^{-1}}{355 \times 10^{-9} \text{ m}} = 5.59 \times 10^{-19} \text{ J.} \quad (4)$$

The energy of light released by the laser for 1 second is:

$$W = \text{Power} \times \text{Time} = P \times t = (0.01 \text{ watt}) \, (1 \text{ s}) = 0.01 \text{ J.} \quad (5)$$

The number of photons emitted by the laser during 1 second is therefore equal to:

$$W = NE \implies N = \frac{W}{E} = \frac{0.01 \text{ J}}{5.59 \times 10^{-19} \text{ J}} = 1.79 \times 10^{16} \frac{\text{photons}}{\text{seconds}}. \quad (6)$$

(b) *Energy of a photon in electronvolts*:

Knowing that $1 \text{ eV} = 1.6 \times 10^{-19}$ J, it becomes:

$$E\,(\text{eV}) = \frac{\left(5.59 \times 10^{-18} \text{ J}\right)\,(1 \text{ eV})}{1.6 \times 10^{-19} \text{ J}} = 3.49 \text{ eV.} \quad (7)$$

As the energy of the incident photon ($E = 3.49$ eV) is superior to the extraction work ($W_S = 2.2$ eV), we must observe the photoelectric effect.

(c) *Intensity, in absolute value, of the current of saturation I_S of the photocell*:

Every incident photon causes the emission of one and only one electron. There are thus so many emitted electrons per second and crossing both plates that there are incident photons per second, that is to say the number N calculated previously. The electric charge Q transported by N electrons in 1 second is therefore $Q/s = -Ne$ which is none other than the intensity I_S of the current of saturation, namely, in absolute value:

$$I_S = Ne = \frac{W}{E}e = \frac{W}{h}e = \frac{W}{h\frac{c}{\lambda}}e = \frac{eW}{hc}\lambda. \quad (8)$$

(d) *Curve of the current of saturation as a function of the wavelength of the incident radiation for W constant*:

For a constant laser radiation energy W, Eq. (8) shows that I_S varies linearly with wavelength λ (h, c, and e are constant). However, we have seen that for incident photons of the frequency ν less than ν_S or equivalently for the wavelengths λ greater than $\lambda_S = c/\nu_S$, the photoelectric effect does not occur, so $I_S = 0$. In conclusion, the equation of the curve is:

$$I_S(\lambda) = \begin{cases} \dfrac{eW}{hc}\,\lambda & \text{if } \lambda < \lambda_S \\ 0 & \text{if } \lambda < \lambda_S \end{cases} \tag{9}$$

with after (2):

$$h\nu_S = h\frac{c}{\lambda_S} = W_S \;\Rightarrow\; \lambda_S = \frac{hc}{W_S}$$

$$= \frac{\left(6.626 \times 10^{-34}\text{ J.s}\right)\left(3 \times 10^8\text{ m.s}^{-1}\right)\left(1\text{ eV}\right)}{(2.2\text{ eV})\left(1.6 \times 10^{-19}\text{ J}\right)} = 564.5 \times 10^{-9}\text{ m}. \tag{10}$$

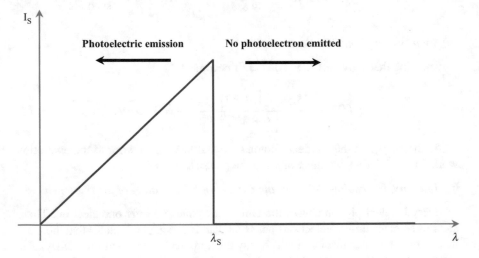

This curve reveals that the intensity of emitted electrons I_S increases when the wavelength of incident photons λ increases because to maintain the energy W, emitted by the laser at the same level (assumed here constant), we necessarily need more and more photons as their wavelength increases, namely, their energy decreases. The slope of the straight line being equal to eW/hc, we can by means of this experiment, knowing W (easy to measure), c, and e, deduce the value of Planck's constant.

(e) *Calculation of the kinetic energy of each emitted electron in eV*:

After (3), we deduce:

$$E_C = h\nu - W_S = 3.5\text{ eV} - 2.2\text{ eV} = 1.3\text{ eV}. \tag{11}$$

5. The energy associated with one photon of wavelength $\lambda' = 1\ \mu m = 10^{-6}\ m$ is:

$$E'(eV) = h\nu' = h\frac{c}{\lambda'} = \left(6.626 \times 10^{-34}\ J.s\right)\frac{\left(3 \times 10^{8}\ m.s^{-1}\right)(1\ eV)}{\left(10^{-6}\ m\right)\left(1.6 \times 10^{-19}\ J\right)}$$

$$= 1.24\ eV. \tag{12}$$

It is less than the extraction work $W_S = 2.2\ eV$, so there is no photoelectric emission in this case.

Exercise 20

We have a quartz bulb in which a very high vacuum is made of the order of 10^{-7} mm Hg. This bulb, called photocell, is formed of two electrodes, a cylinder-shaped cathode and a rectilinear wire-shaped anode placed on the axis of the cylinder. Illuminated by an appropriate light, the cathode (emitter) emits electrons. These electrons are captured by the anode (collector) which is connected to the positive pole of an adjustable generator. It results from this a current detected by a galvanometer incorporated in the circuit. A voltmeter measures the applied voltage to the cell.

1. *From a quantitative point of view, why did the photoelectric effect raise a fundamental problem for scientists of the time?*
2. *Let us consider the case where the frequency ν of the incident light is below the threshold frequency ν_S. Does the application of a voltage produce an electric current?*
3. *We are now considering the case where the frequency ν of the incident light is greater than the threshold frequency ν_S. We apply a positive voltage at the bulb terminals, namely, V+ on the anode and V− on the cathode. Do we observe a current?*
4. *In the case where the frequency ν of the incident light is greater than the threshold frequency ν_S, we are applying now a negative voltage, namely, V− on the anode and V+ on the cathode. Do we observe a current?*
5. *Write a relationship between the kinetic energy of the electron E_c^A arriving on the anode, its initial kinetic energy E_c^C at the exit of the cathode, and the voltage U applied at bulb terminals.*
6. *From which voltage does the current cancel?*
7. *Let us consider now Einstein's corpuscular theory, according to which the light consists of photons of energy $h\nu$. By striking the metal, these photons can extract electrons if their energy is greater than the binding energy of the electron on the electron shell. Let W_S be this extraction energy. Give the relationship between the kinetic energy E_c^C of the electron when it leaves the cathode, the incident light energy $h\nu$, and the extraction energy W_S.*

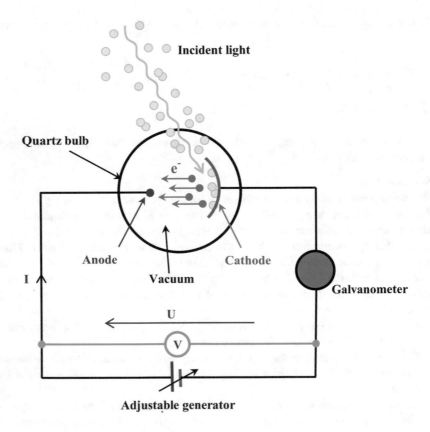

8. *Give the threshold frequency ν_S at zero potential difference.*
9. *Then write the kinetic energy E_c^A of electrons arriving at the anode as a function of frequencies ν and ν_S and the voltage U.*
10. *At a fixed frequency above the threshold, give the value of the voltage U_0 at the terminals of the bulb which allows canceling the current (the stopping voltage).*
11. *Deduce an experimental protocol that allows verifying Einstein's theory.*
12. *If several metals are used as the emitting electrode, what should we observe?*
13. *How can we determine from this experiment the value of Planck's constant h and the extraction work W_S of the used metal? Plot qualitatively the stopping voltage U_0 as a function of the frequency ν of the incident light.*

Solution

1. Qualitatively, it was well understood why a light-illuminated metal surface ejected electrons which were then collected under vacuum and put into circulation in a circuit in the form of the detectable electrical current. Indeed, it was known that light was an electromagnetic wave characterized by an electric field that could exert a force on the electrons of the material. On the other hand, from the quantitative point of view, what we could not understand is why, below a certain frequency of the incident radiation, no electron was ejected, even for a

high intensity of the incident radiation. Moreover, it was also not understood why, above a certain frequency for which a current appears, when the intensity of the radiation was increased by fixing the frequency, the number of detected electrons increased but their kinetic energy remained constant.

2. No, there is not enough light energy to extract electrons from the material, and the applied voltage does not change anything.

3. Yes, a current is always observed because the voltage between the anode and the cathode accelerates the electrons that are in the inter-electrode zone.

4. It depends on the value of the applied voltage. Indeed, a negative voltage slows electrons which leave the emitting electrode (cathode) before reaching the collector electrode (anode). If the voltage is sufficient to slow electrons to zero until they reach the collector electrode, there is no current. In the opposite case, we observe a current even very low.

5. Let be E_c^C and E_c^A the respective kinetic energies of the electron leaving the cathode (emitter), and arriving at the anode (collector), and $U = V_A - V_C$ the voltage at terminals of these two electrodes. The total energy is the sum of the kinetic energy and the potential energy, and the conservation of the energy of the electron of charge q gives:

$$E_c^C + qV_C = E_c^A + qV_A. \tag{1}$$

We thus obtain the kinetic energy at the level of the anode:

$$E_c^A = E_c^C + q(V_A - V_C) = E_c^C + qU. \tag{2}$$

This equation shows that for a positive voltage U, electrons are always accelerated and arrive at the anode with a non-zero kinetic energy E_c^A and there is always a current in the circuit, whereas for a negative voltage, it depends on the value of qU (hence of U) with respect to E_c^C C as we explained it in Question (4). More precisely, if $U < -U_0$ where U_0 is the stopping voltage (voltage that stops the electrons at the anode), electrons do not reach the anode so there is no current; now if $U > -U_0$, electrons reach the anode with a non-zero kinetic energy E_c^A higher or lower depending on the value of U and a current is always detected.

6. On the basis of (2), the current is canceled if:

$$E_c^A = E_c^C + qU = 0 \Rightarrow -qU = E_c^C. \tag{3}$$

In this case, electrons do not have enough kinetic energy to reach the anode: they turn back towards the cathode (a little like an object which is thrown upwards and which finally falls back).

7. *The relationship between the energy of the electron E_c^C leaving the cathode, the incident light energy $h\nu$, and the extraction energy W_S:*

This relationship is given by:

$$E_c^C = h - W_S. \tag{4}$$

This equation indicates that the incident photon of energy $h\nu$ gives one part of its energy to extract one electron from the material. This energy is equal to the binding energy W_S of the electron on the electron shell, and the rest of its energy is carried by the electron in the form of kinetic energy E_c^C.

8. *Threshold frequency ν_S:*

The electron is ejected from the metal but does not leave it because it has no kinetic energy at the level of the cathode $\left(E_c^C = 0\right)$; that is to say by starting from (4):

$$E_c^C = h\nu_S - W_S = 0 \;\Rightarrow\; h\nu_S = W_S \;\Rightarrow\; \nu_S = \frac{W_S}{h}. \tag{5}$$

9. *Kinetic energy E_c^A of electrons arriving at the anode as a function of frequencies ν and ν_S and the voltage U:*

We deduce from (2):

$$E_c^A = E_c^C + qU = (h\nu - W_S) + qU = h(\nu - \nu_S) + qU. \tag{6}$$

10. The current is canceled when the kinetic energy of electrons arriving at the level of the anode becomes null, that is to say starting from (6):

$$E_c^A = 0 \;\Rightarrow\; -qU_0 = h\nu - W_S = h\nu - h\nu_S. \tag{7}$$

11. We send a radiation at a certain frequency on the cathode of the photocell described above at a zero voltage to ensure that it is above the threshold frequency. Then, we apply a negative voltage $- U_0$ (U_0 is the positive stopping voltage) until this current is zero ($U = -U_0$). We thus obtain a pair of values (ν, U_0). Indeed, the kinetic energy of the electron at the exit of the cathode and which does not arrive at the anode $\left(E_c^C = 0\right)$ is, starting from (3):

$$E_c^C = -qU_0 = -(-e)U_0 = eU_0. \tag{8}$$

We repeat this experiment for a series of frequencies, and we plot the stopping voltage U_0 as a function of the frequency ν of the incident light. If Einstein's theory is correct, we must observe under (7) an affine function of the frequency:

$$-qU_0 = eU_0 = h\nu - h\nu_S \quad \Rightarrow \quad U_0 = \frac{h}{e}\nu - \frac{h}{e}\nu_S. \tag{9}$$

12. According to (9), the slope of the straight line $U_0 = f(\nu)$ depends on Planck's constant h and the electron charge, so it does not depend on the type of choosing metal. We must therefore observe for several metals parallel lines of the same inclination. On the other hand, the point at the origin depends on it through the extraction work which is a characteristic of the used material.
13. From the slope of the straight line, we deduce:

$$\frac{h}{e} = \text{measured slope} \quad \Rightarrow \quad h = \text{measured slope} \times \text{charge of the electron.} \tag{10}$$

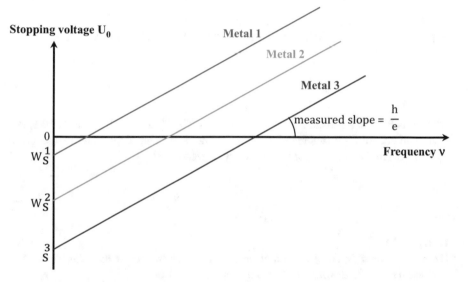

With regard to the extraction work W_S, we write $U_0 = 0$ in (9):

$$U_0 = \frac{h}{e}\nu_S - \frac{W_S}{e} = 0 \quad \Rightarrow \quad W_S = h\nu_S. \tag{11}$$

The stopping voltage U_0 is by definition positive. The straight lines have been intentionally extended to highlight the extraction energy W_S. As we can see, these lines are all parallel (the same slope h/e), but the extraction energy (W_S^1, W_S^2, W_S^3) depends on the metal.

2.5.2   Compton's Scattering

Exercise 21
What is the difference between the photon scattering experiment and the Compton scattering?

Solution
In the diffraction experiment, the scattered photon preserves its energy, whereas in the Compton scattering, the scattered photon transfers a part of its energy to one electron.

Exercise 22
What are the energy and the momentum of one photon as a function of its frequency ν and its wavelength λ?

Solution
A photon carries an indivisible energy proposed by Einstein to explain the photo-electric effect:

$$E = h\nu = h\frac{c}{\lambda} \tag{1}$$

where h is Planck's constant. Its value is 6.626×10^{-34} J.s and c represents the speed of light in vacuum—its value is 3×10^8 m/s. The photon has a momentum also given by Einstein in the framework of the theory of relativity:

$$P = \frac{E}{c} = \frac{h}{c} = \frac{h\frac{c}{\lambda}}{c} = \frac{h}{\lambda}. \tag{2}$$

Exercise 23
What is the energy E_e of one electron in the relativistic case as a function of its momentum P_e. We denote by m_e its mass at rest.

Solution
In special relativity, Einstein's equation states that at any mass m_e corresponds energy $m_e c^2$ called *rest mass energy* and that every particle in motion includes both the rest mass energy and the kinetic energy E_C, namely:

$$E_e = mc^2 = E_c + m_e c^2 \tag{1}$$

where $E_e = mc^2$ is the energy of the particle and m its relativistic mass given by:

$$m = \frac{m_e}{\sqrt{1 - \frac{v^2}{c^2}}} \Rightarrow m^2 = \frac{m_e^2}{1 - \frac{v^2}{c^2}} = \frac{m_e^2}{\frac{c^2}{c^2} - \frac{v^2}{c^2}} = \frac{m_e^2 c^2}{c^2 - v^2} \Rightarrow m^2 c^2 - m^2 v^2$$

$$= m_e^2 c^2. \tag{2}$$

Moreover, any material particle in motion (one electron, for example) has a momentum:

$$P_e = mv \tag{3}$$

where v is the speed of the particle. From (2) and (3), we obtain:

$$m^2c^2 - m^2v^2 = m^2c^2 - P_e^2 = m_e^2c^2 \implies m^2c^2 = P_e^2 + m_e^2c^2. \tag{4}$$

By multiplying (4) by c^2, we find:

$$m^2c^4 = P_e^2c^2 + m_e^2c^4 \text{ namely: } E^2 = P_e^2c^2 + m_e^2c^4. \tag{5}$$

Exercise 24
In the case of a Compton collision between one photon and one electron at rest, write the energies and momenta involved before and after the collision.

Solution
Let us consider the collision between one photon coming from the left and one electron at rest. The energies and momenta involved before and after the collision are summarized in the following table:

	Before the collision		After the collision	
	Incident photon	Electron at rest	Scattered photon	Recoil electron
Energy	$h\nu_1$	m_ec^2	$h\nu_2$	$\sqrt{P_e^2c^2 + m_e^2c^4}$
Momentum	$\dfrac{h\nu_1}{c}\begin{pmatrix} 1 \\ 0 \end{pmatrix}$	0	$\dfrac{h\nu_2}{c}\begin{pmatrix} \cos\theta \\ \sin\theta \end{pmatrix}$	$P_e\begin{pmatrix} \cos\varphi \\ -\sin\omega \end{pmatrix}$

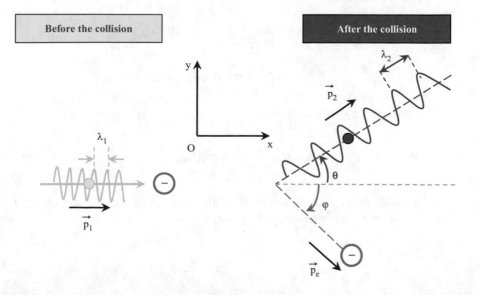

Exercise 25
1. *Write the conservation laws of the momentum and energy in Compton's scattering.*
2. *Find a relationship between the frequencies of the incident photon ν_1, the scattered photon ν_2, and the scattering angle of the photon θ.*
3. *Deduce Compton's equation. What conclusions can you deduce from this equation?*

Solution
1. *Conservation laws of the momentum and energy in Compton's scattering*:

The studied system is the photon+electron set. This system is isolated because there is no exchange of energy with the external environment. The collision is thus elastic; it results in a conservation of the momentum and energy of the system.

A. *Energy conservation*:

$$E_1^P(\text{ incident photon}) + E_1^e(\text{electron at rest}) = E_2^P(\text{scattered photon}) + E_2^e(\text{recoil electron}). \tag{1}$$

According to the table of Exercise 24, we have:

$$h\nu_1 + m_e c^2 = h\nu_2 + \sqrt{P_e^2 c^2 + m_e^2 c^4}. \tag{2}$$

A. *Momentum conservation*:

$$P_1^P(\text{ incident photon}) + P_1^e(\text{electron at rest}) = P_2^P(\text{scattered photon}) + P_2^e(\text{recoil electron}). \tag{3}$$

– *According to the x-axis, we obtain*:

$$\frac{h\nu_1}{c} + 0 = \frac{h\nu_2}{c} \cos\theta + P_e \cos\varphi. \tag{4}$$

– *According to the y-axis, we get*:

$$0 + 0 = \frac{h_2}{c} \sin\theta - P_e \sin\varphi. \tag{5}$$

2. *Relationship between the frequencies of the incident photon ν_1, the scattered photon ν_2, and the scattering angle of the photon θ*:

From (2), we get:

$$P_e^2 c^2 + m_e^2 c^4 = \left(\Delta\nu + m_e c^2\right)^2 \tag{6}$$

with

$$\Delta\nu = \nu_1 - \nu_2. \tag{7}$$

From (4), we obtain:

$$P_e c \cos\varphi = h\nu_1 - h\nu_2 \cos\theta. \tag{8}$$

From (5), we find:

$$P_e c \sin\varphi = h\nu_2 \sin\theta. \tag{9}$$

By summing the squares of (8) and (9) and by injecting them into (7), we eliminate P_e and φ and we get an equation with ν_1, ν_2, h, m_0, c, and θ:

$$(\nu_1 - \nu_2 \cos\theta)^2 + (\nu_2 \sin\theta)^2 + \left(\frac{m_e c^2}{h}\right)^2 = \left(\Delta\nu + \frac{m_e c^2}{h}\right)^2. \tag{10}$$

By writing:

$$\nu_c = \frac{m_e c^2}{h} \tag{11}$$

and expanding (10), we arrive at:

$$\nu_1^2 - 2\nu_1\nu_2 \cos\theta + \nu_2^2 + \nu_c^2 = \nu_1^2 - 2\nu_1\nu_2 + \nu_2^2 + 2\nu_c\Delta\nu + \nu_c^2. \tag{12}$$

After simplification, we get an equation between the frequencies of the incident photon ν_1, scattered photon ν_2, and scattering angle of the photon θ:

$$\frac{\Delta\nu}{\nu_1\nu_2} = \frac{1}{\nu_c}(1 - \cos\theta). \tag{13}$$

3. *Compton's equation*:

   By noticing that:

$$\frac{\Delta \nu}{\nu_1 \nu_2} = \frac{\nu_1 - \nu_2}{\nu_1 \nu_2} = \frac{1}{\nu_2} - \frac{1}{\nu_1} \tag{14}$$

we incorporate (14) in (13), and we obtain:

$$\frac{1}{\nu_2} - \frac{1}{\nu_1} = \frac{1}{\nu_c}(1 - cos\theta). \tag{15}$$

Knowing that frequencies and wavelengths are inversely proportional, we deduce Compton's equation:

$$\frac{1}{\frac{c}{\lambda_2}} - \frac{1}{\frac{c}{\lambda_1}} = \frac{\lambda_2}{c} - \frac{\lambda_1}{c} = \frac{\nu_c}{c} (1 - cos\theta) \text{ that is to say: } \lambda_2 - \lambda_1 = \lambda_c(1 - cos\theta). \tag{16}$$

Equation (16) shows that the ratio $\lambda_C = h/m_e c$ has the dimension of a wavelength. It is called *Compton's wavelength of the electron* and has a value of 0.002426 nm (or 2.426×10^{-12} m). This is not, of course, an actual wavelength, but really a proportionality constant for the wavelength shift $\Delta \lambda = \lambda_2 - \lambda_1$. This equation reveals that the wavelength shift $\Delta \lambda$ does not depend on the density, the atomic number, or any other property of the absorbing material; Compton's scattering is strictly a photon-electron interaction. Looking again at this equation, it becomes clear that the entire shift can be measured purely in terms of the angle at which the photon gets scattered. Everything else on the right side of the equation is constant. Experiments show that this is the case, giving great support to the photon interpretation of light.

Exercise 26
1. *Show by using nonrelativistic mechanics that one electron cannot absorb one photon.*
2. *Does special relativity solve this problem?*
   We suppose for the needs for both demonstrations that once all the energy of the incident photon is absorbed, the electron leaves on the same straight line and in the same direction as the incident photon.

Solution

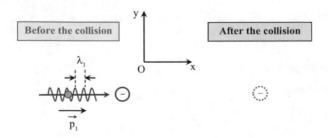

1. *Nonrelativistic study*:

Let us take place in a reference frame where the electron is at rest before the photon-electron interaction, and write the energies and momenta involved before and after the collision:

	Before the collision		After the collision	
	Incident photon	Electron at rest	Scattered photon	Recoil electron
Energy	$h\nu_1$	0	0	$\frac{1}{2}m_e v^2$
Momentum	$\frac{h\nu_1}{c}$	0	0	$m_e v$

The studied system is the photon+electron set. This system is isolated because there is no exchange of energy with the external environment. The collision is thus elastic; it results in a conservation of the momentum and energy of the system.

A. *Energy conservation*:

$$h\nu_1 + 0 = 0 + \frac{1}{2}m_e v^2. \tag{1}$$

A. *Momentum conservation*:

  − *According to the x-axis only, we obtain*:

$$\frac{h\nu_1}{c} + 0 = 0 + m_e v. \tag{2}$$

From (1), we find:

$$m_e v^2 = 2\,h\nu_1. \tag{3}$$

From (2), we get:

$$m_e v = \frac{h\nu_1}{c}. \tag{4}$$

Doing the ratio (3)/(4), it becomes:

$$\frac{m_e v^2}{m_e v} = \frac{2\,h_1}{\dfrac{h\nu_1}{c}} \quad \Rightarrow \quad v = 2c. \tag{5}$$

After (5), we see that the speed of the election would be twice the speed of light, which is absurd. So, in the nonrelativistic framework, one electron cannot absorb all the energy of the incident photon.

2. *Relativistic study*:

Let us see if special relativity solves the problem. For this purpose, let us write the energies and momenta involved before and after the collision:

	Before the collision		After the collision	
	Incident photon	Electron at rest	Scattered photon	Recoil electron
Energy	$h\nu_1$	$m_e c^2$	0	$\sqrt{P_e^2 c^2 + m_e^2 c^4}$
Momentum	$\dfrac{h\nu_1}{c}$	0	0	P_e

The studied system is always the photon+electron set. This system is isolated because there is no exchange of energy with the external environment. The collision is thus elastic; it results in a conservation of the momentum and energy of the system.

A. *Energy conservation*:

$$h\nu_1 + m_e c^2 = \sqrt{P_e^2 c^2 + m_e^2 c^4}. \tag{6}$$

B. *Momentum conservation according to the x-axis only*:

$$\frac{h\nu_1}{c} + 0 = P_e. \tag{7}$$

Let us square (6), and we get:

$$\left(h\nu_1 + m_e c^2\right)^2 = (h\nu_1)^2 + m_e^2 c^4 + 2 m_e c^2 h\nu_1 = P_e^2 c^2 + m_e^2 c^4. \tag{8}$$

After simplification, it becomes:

$$(h\nu_1)^2 + 2 m_e c^2 h\nu_1 = P_e^2 c^2. \tag{9}$$

Let us square (7), and it turns into:

$$\left(\frac{h\nu_1}{c}\right)^2 = P_e^2. \tag{10}$$

By doing (10)/(9), we also arrive at an aberrant result:

$$\frac{2m_ec^2}{h\nu_1} = 0. \tag{11}$$

This result is aberrant because m_e and/or c cannot be zero. In reality, it is the hypothesis of the exercise that is false. Indeed, during the collision of one photon with one electron at rest, the photon cedes only a part of its energy to the recoil electron and scatters with a lesser energy.

Exercise 27

A photon of wavelength $\lambda_1 = 0.71$ pm collides with an immobile electron and deviates at an angle $\theta = 70°$ with regard to its initial trajectory.
1. *Determine the wavelength of the scattered photon.*
2. *Calculate the electron speed.*
3. *Determine the orientation of the recoil electron given by the angle φ. We give:*
   - *Planck's constant: $h = 6.63 \times 10^{-34}$ J.s*
   - *Electron mass: $m_e = 9 \times 10^{-31}$ kg*
   - *Speed of light: $c = 3 \times 10^8$ ms^{-1}*

Solution

1. *Wavelength of the scattered photon*:

By starting from Compton's equation, we get:

$$\lambda_2 - \lambda_1 = \frac{h}{m_ec}\left(1 - cos\,\theta\right) \quad \Rightarrow \quad \lambda_2 = \frac{h}{m_ec}\left(1 - cos\,\theta\right) + \lambda_1. \tag{1}$$

By substituting the numerical values of each physical quantity, we find the value of the wavelength λ_2 of the scattered photon:

$$\lambda_2 = \frac{6.63 \times 10^{-34} \text{ J.s}}{\left(9.11 \times 10^{-31} \text{ kg}\right)\left(3 \times 10^8 \text{ m.s}^{-1}\right)}\left(1 - cos\,70°\right) + 0.71 \times 10^{-12} \text{ m}$$

$$= 7.2596 \times 10^{-11} \text{ m}. \tag{2}$$

2. *Electron speed*:

Let us evaluate the energy loss of the photon which is transferred to the electron in the form of kinetic energy:

$$\varepsilon = -\Delta E = -(E_2 - E_1) = E_1 - E_2 = h\nu_1 - h\nu_2 = h\left(\frac{c}{\lambda_1}\right) - h\left(\frac{c}{\lambda_2}\right)$$

$$= hc\left(\frac{1}{\lambda_1}\right) - hc\left(\frac{1}{\lambda_2}\right) = hc\left[\left(\frac{1}{\lambda_1}\right) - \left(\frac{1}{\lambda_2}\right)\right]. \tag{3}$$

By substituting the numerical values of each physical quantity, we find:

$$\varepsilon = \left(6.63 \times 10^{-34}\,\text{J.s}\right)\left(3 \times 10^{8}\,\text{m.s}^{-1}\right)\left[\left(\frac{1}{7.1 \times 10^{-11}\,\text{m}}\right) - \left(\frac{1}{7.2596 \times 10^{-11}\,\text{m}}\right)\right]$$
$$= 6.1588 \times 10^{-17}\,\text{J}.$$

(4)

Let us suppose that our electron is not relativist. We can determine its speed by means of the classical expression of the kinetic energy:

$$\varepsilon = \frac{1}{2}\,m_e v^2 \quad \Rightarrow \quad v = \sqrt{\frac{2}{m_e}} = \sqrt{\frac{2\left(6.1588 \times 10^{-17}\,\text{J}\right)}{\left(9.11 \times 10^{-31}\,\text{kg}\right)}}$$
$$= 1.1628 \times 10^{7}\,\text{m.s}^{-1}.$$

(5)

By doing the ratio between the speed v of the electron and the speed c of light, we obtain:

$$\frac{v}{c} = \frac{1.1628 \times 10^{7}\,\text{m.s}^{-1}}{3 \times 10^{8}\,\text{m.s}^{-1}} = 0.03876.$$

(6)

Equation (6) reveals that the speed of the electron is very low compared to the speed of light, which fully justifies the nonrelativistic approach of our method by considering in particular that the mass of the electron remains constant ($m = m_e$) during the collision and by using the classical expression of the kinetic energy of the electron after the scattering. Moreover, we started our development by using $\varepsilon = -\Delta E$ rather than the variation of the kinetic energy ΔE in order to avoid the minus sign that inevitably appears in the calculation because the final energy of the scattered photon E_2 is always lower than its initial energy E_1 before the collision given that the incident photon must yield a part from its energy to the electron.

3. *Orientation of the recoil electron*:

Let us write the energies and momenta involved before and after the collision knowing that the electron is nonrelativistic:

	Before the collision		After collision	
	Incident photon	Electron at rest	Scattered photon	Recoil electron
Energy	$h\nu_1$	0	$h\nu_2$	$\frac{1}{2}\,m_e v^2$
Momentum	$\dfrac{h\nu_1}{c}\begin{pmatrix}1\\0\end{pmatrix}$	0	$\dfrac{h\nu_2}{c}\begin{pmatrix}\cos\theta\\\sin\theta\end{pmatrix}$	$m_0 v\begin{pmatrix}\cos\varphi\\-\sin\omega\end{pmatrix}$

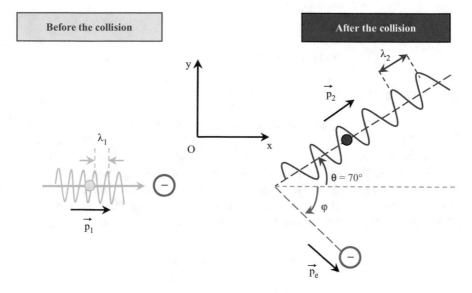

The studied system is the photon+electron set. This system is isolated because there is no exchange of energy with the external environment. The collision is thus elastic. Let us write that the momentum is conserved during the collision:

$$P_1^{\mathrm{P}}(\text{incident photon}) + P_1^{\mathrm{e}}(\text{electron at rest}) = P_2^{\mathrm{P}}(\text{scattered photon}) \\ + P_2^{\mathrm{e}}(\text{recoil electron}). \tag{7}$$

– *According to the x-axis, we get*:

$$\frac{h\nu_1}{c} + 0 = \frac{h\nu_2}{c} \cos\theta + m_e v \cos\varphi. \tag{8}$$

– *According to the y-axis, we find*:

$$0 + 0 = \frac{h\nu_2}{c} \sin\theta - m_e v \sin\varphi. \tag{9}$$

Knowing that $\nu_2 = c/\lambda_2$ and starting from (9), we arrive at:

$$0 = \frac{h}{\lambda_2} \sin\theta - m_e v \sin\varphi. \tag{10}$$

By substituting the numerical values of each physical quantity in (10), it becomes:

$$0 = \frac{6.63 \times 10^{-34} \text{ J.s}}{7.2596 \times 10^{-11} \text{ m}} \sin 70°$$
$$- \left(9.11 \times 10^{-31} \text{ kg}\right) \left(1.1628 \times 10^7 \text{ m.s}^{-1}\right) \sin\varphi. \qquad (11)$$

We finally get the value of the recoil angle φ of the electron:

$$\sin\varphi = 0.8101 \quad \Rightarrow \quad \varphi = 54.106°. \qquad (12)$$

Exercise 28

During an elastic collision of one photon with a target material, we observe a scattered radiation with a frequency different from that of the incident radiation. This is Compton's scattering (1923). Experimentally, Arthur Holly Compton irradiated graphite with X-rays and observed that:

$$\lambda_2 - \lambda_1 = 4.8 \, sin^2\left(\frac{\theta}{2}\right)$$

where λ_1 and λ_2 represented respectively the wavelengths of the incident photon (directed along the x-axis) and the scattered photon; θ is the angle formed by the unit vectors $\vec{e}_P$ and $\vec{i}$. To model this phenomenon, Compton considered that there was a collision of the incident photon (projectile) with one electron (target at rest, mass m_e) of graphite.

1. *Schematize the equation of the X-photon-electron collision process.*
2. *Give the relationships that express the energy E of one photon as a function of its frequency ν, its wavelength λ, and its momentum P, respectively.*
3. *By using appropriate conservation laws, derive the equation obtained by Compton.*
4. *Express the numerical coefficient 4.8 as a function of variables h, m_e, and c where h is Planck's constant; m_e, the mass of the electron; and c, the speed of light.*

Solution

1. *Schematization of the collision process:*

The X-photon+electron system is isolated because there is no energy exchange with the outside environment during the collision. The X-photon-electron collision is therefore elastic. It can be schematized as follows:

$$\text{Photon X} + \text{electron} \rightarrow \text{Photon X}' + \text{electron}. \qquad (1)$$

2. *Relationships that express the energy E of one photon as a function of its frequency ν, its wavelength λ, and its momentum P, respectively:*

One photon carries an indivisible energy proposed by Einstein to explain the photoelectric effect:

$$E = h\nu = h\,\frac{c}{\lambda} \tag{2}$$

where h is Planck's constant. Its value is 6.63×10^{-34} J.s and c represents the speed of light in vacuum, its value is 3×10^8 m/s. The photon also possesses a momentum P; it is given by Einstein in the framework of the theory of relativity:

$$P = \frac{E}{c} \quad \Rightarrow \quad E = Pc. \tag{3}$$

3. *Derivation of the equation obtained by Compton*:

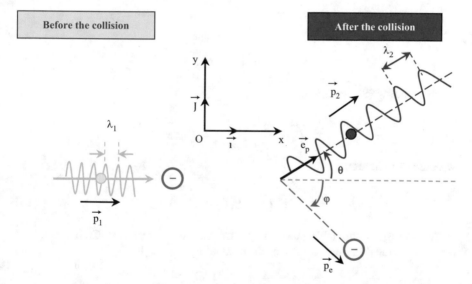

Let us write the energies and momenta involved before and after the collision:

	Before the collision		After the collision	
	Incident photon	Electron at rest	Scattered photon	Recoil electron
Energy	$P_1 c$	$m_e c^2$	$P_2 c$	$E_e = \sqrt{P_e^2 c^2 + m_e^2 c^4}$
Momentum	$\vec{P_1}$	$\vec{0}$	$\vec{P_2}$	$\vec{P_e}$

Let us write that the energy is conserved during the collision:

$$P_1 c + m_e c^2 = P_2 c + E_e. \tag{4}$$

Let us write that the momentum is also conserved during the collision:

$$\vec{P}_1 + \vec{0} = \vec{P}_2 + \vec{P}_e. \tag{5}$$

Once this step is exceeded, by examining more closely the formula obtained by Compton (see the statement of the exercise), we notice that there is no contribution due to the recoil electron. We therefore use the obtained Eqs. (4) and (5), and we eliminate this contribution. Thus, from (4), we obtain:

$$E_e = P_1 c - P_2 c + m_e c^2 = c(P_1 - P_2) + m_e c^2. \tag{6}$$

Let us square (6), and we find:

$$E_e^2 = c^2 (P_1 - P_2)^2 + m_e^2 c^4 + 2c (P_1 - P_2) m_e c^2. \tag{7}$$

From (5), we get:

$$\vec{P}_e = \vec{P}_1 - \vec{P}_2. \tag{8}$$

By squaring (8), it becomes:

$$\vec{P}_e^2 = \vec{P}_1^2 + \vec{P}_2^2 - 2 \vec{P}_1 \vec{P}_2 \tag{9}$$

with after the figure:

$$\vec{P}_1 \vec{P}_2 = P_1 P_2 \, cos\theta. \tag{10}$$

According to the table, the energy of the recoil electron E_e is expressed (as demonstrated in Sect. 2.4.2.1 of the course) by:

$$E_e = \sqrt{P_e^2 c^2 + m_e^2 c^4} \quad \Rightarrow \quad E_e^2 = P_e^2 c^2 + m_e^2 c^4. \tag{11}$$

By incorporating (7) and (9) and (10) in (11), that equation transforms into:

$$c^2 (P_1 - P_2)^2 + m_e^2 c^4 + 2c (P_1 - P_2) m_e c^2 P_1^2 c^2 + P_2^2 \vec{P}_2^2 c^2$$
$$- 2 P_1 P_2 \, c^2 cos\theta + m_e^2 c^4. \tag{12}$$

After simplification, it becomes:

$$-P_1P_2 \cos\theta + P_1P_2 - m_e c \, (P_1 - P_2) = 0 \quad \Rightarrow \quad 1 - \cos\theta = m_e c \left(\frac{1}{P_2} - \frac{1}{P_1}\right)$$
$$= \frac{m_0 c}{h} \, (\lambda_2 - \lambda_1).$$

$$(13)$$

To write the momentum P of the photon as a function of its wavelength λ, we used Eqs. (2) and (3). Indeed, by incorporating (2) in (3), we arrive at:

$$P = \frac{E}{c} = \frac{h \frac{c}{\lambda}}{c} = \frac{h}{\lambda}. \tag{14}$$

By using the following trigonometric equation:

$$1 - \cos\theta = 2 \, \sin^2\left(\frac{\theta}{2}\right) \tag{15}$$

we finally obtain Compton's equation:

$$\lambda_2 - \lambda_1 = \frac{h}{m_e c} \, (1 - \cos\theta) = 2 \, \frac{h}{m_e c} \, \sin^2\left(\frac{\theta}{2}\right). \tag{16}$$

4. *Expression the numerical coefficient 4.8 as a function of variables h, m_e, and c:*

By comparing (15) with the equation established by Arthur Holly Compton which is given in the statement of this exercise, we find:

$$2\frac{h}{m_e c} = 4.8. \tag{17}$$

According to (16), this coefficient has the physical dimension of a length.

Exercise 29
1. (a) *Describe briefly what happens when an X-photon collides with an electron that is free or loosely bound to an atom; this electron is assumed to be at rest.*
    (b) *Illustrate your explanations by a diagram.*
    (c) *Give the energy balance.*
2. *Two extreme cases may occur: the case of a frontal collision and the case of a tangential collision. In which of these two cases is the energy received by the electron maximal?*
3. *The scattering angle of the photon θ and the angle of deviation of the recoil electron φ are linked by the following equation:*

$$cot\,\varphi = (1 + \alpha)\,tan\,\frac{\theta}{2}.$$

In what directions do the scattered photon and the recoil electron leave?

4. *In which of the two cases mentioned in 2 is the received energy by the electron minimal?*
5. *What are the directions of the scattered photon and the recoil electron?*
6. *Knowing that the energy of the incident photon E_1 is 500 keV, calculate its wavelength.*
7. *Calculate the wavelength λ_2 associated with the scattered photon in the case of a frontal collision if $\Delta\lambda = 4.85\ 10^{-12}$ m.*
8. *Calculate the energy E_2 of the scattered photon. We give:*
   - *Planck's constant: $h = 6.63 \times 10^{-34}$ J.s*
   - *Electron mass: $m_e = 9 \times 10^{-31}$ kg*
   - *Speed of light: $c = 3 \times 10^8$ ms^{-1}*
   - *$1\ eV = 1.6 \times 10^{-19}$ J*
   - *$1\ keV = 10^3\ eV$*

Solution

1. (a) *Brief description of the X-photon-electron interaction*:

When an X-photon collides with a peripheral electron that is poorly bound to the atom, the energy of the photon is partly transmitted to the electron which is ejected from the atom. The electron leaves the atom with some kinetic energy. The rest of the energy is found in the form of an X-photon of different direction and lower energy.

(b) *Illustration by a diagram and energy balance*:

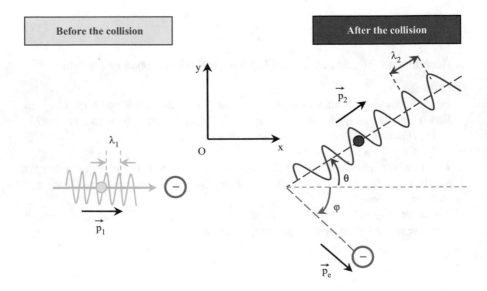

This interaction is called *Compton scattering*. It can be illustrated by the figure: during the collision, the X-photon of energy E_1 strikes one electron at rest. It projects it in a direction given by the angle φ by communicating it a kinetic energy E_C; the rest of the energy is carried by an X-photon of energy E_2 which scatters in a direction given by the angle θ.

(c) *Energy balance*:

The X-photon+electron system is isolated because there is no energy exchange with the outside environment during the collision. There is therefore conservation of energy:

$$E_1 = E_c + E_2. \tag{1}$$

During this interaction, the difference between the wavelength λ_2 of the scattered photon and the wavelength λ_1 of the incident photon is given by Compton's equation:

$$\lambda_2 - \lambda_1 = \frac{h}{m_e c} \, (1 - cos\,\theta). \tag{2}$$

This wavelength shift $\Delta\lambda$ depends on the scattered angle of the photon θ, the mass of the electron m_e, Planck's constant h, and the speed of light c. Surprising facts, this equation reveals that $\Delta\lambda$ does not depend on the density, the atomic number, or any other property of absorbing material, Compton's scattering is strictly a photon-electron interaction. Looking again at this equation, it becomes clear that the entire shift can be measured purely in terms of the angle at which the photon gets scattered. Everything else on the right side of the equation is constant. The experiments show that this is the case, giving great support to the photon interpretation of light.

2. *Maximal received energy*:

According to (1), the energy received by the recoil electron is the difference between the energy E_1 of the incident photon and the energy E_2 of the scattered photon:

$$E_c = E_1 - E_2 = h\,(\nu_1 - \nu_2) = h\left(\frac{c}{\lambda_1} - \frac{c}{\lambda_2}\right) = hc\left(\frac{\lambda_2 - \lambda_1}{\lambda_1 \lambda_2}\right). \tag{3}$$

Equation (3) shows that the energy E_C received by the electron is maximal if the wavelength shift $\Delta\lambda = \lambda_2 - \lambda_1$ is maximal, that is to say by returning to Eq. (2) if $(1 - cos\,\theta)$ is maximal. This comes true if $1 - cos\,\theta = 2 \Rightarrow cos\,\theta = -1 \Rightarrow \theta = 180° \Rightarrow$ in this frontal photon-electron collision, the photon turns back carrying with it a minimal energy given that the energy of the electron is maximal.

3. *Scattered photon and recoil electron directions (frontal case)*:

From the relationship between the angles θ and φ (given in the statement), we deduce that for $\theta = \pi$, the angle of deviation of the recoil electron is $\varphi = 0$: the photon-electron collision is frontal. It corresponds to the limit case where the photon encounters on its trajectory the electron and strikes it with full force. The interaction is maximal. The electron leaves on the same straight line and in the same direction as the incident photon with a maximal kinetic energy E_C. The scattered photon leaves in the opposite direction of the incident photon ($\theta = \pi$). It has a minimal energy E_2. given that the energy of the electron is maximal.

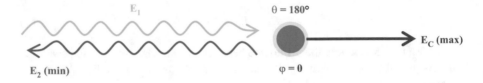

4. *Minimal energy of the electron*:

According to (1), the energy E_C received by the electron is the difference between the energy E_1 of the incident photon and the energy E_2 of the scattered photon. Equation (3) shows that the energy E_C received by the electron is minimal if the wavelength displacement $\Delta\lambda = \lambda_2 - \lambda_1$ is minimal, that is to say by returning to Eq. (2) if $(1 - \cos\theta)$ is minimal. This comes true if $1 - \cos\theta = 0 \Rightarrow \cos\theta = 1 \Rightarrow \theta = 0° \Rightarrow \Delta\lambda = \lambda_2 - \lambda_1 = 0 \Rightarrow \lambda_2 = \lambda_1 \Rightarrow E_2 = E_1$. The photon-electron collision is tangential and the scattered photon carries all the energy of the incident photon.

5. *Directions of the scattered photon and the recoil electron (the tangential case)*:

From the relationship between the angles θ and φ (given in the statement), we deduce that for $\theta = 0$, the angle of deviation of the recoil electron is $\varphi = \pi/2$: the photon-electron collision is tangential. It corresponds to the limit case where the photon grazes the electron which leaves in the perpendicular direction to that of the incident photon ($\varphi = \pi/2$). There is practically no photon-electron interaction. The scattered photon leaves on the same straight line and in the same direction as the incident photon ($\theta = 0$). The two photons have the same wavelength and the same energy ($\lambda_1 = \lambda_2$ and $E_1 = E_2$), which means that the electron does not carry any energy ($E_C = 0$).

$\theta = 0$

E_1 $E_1 = E_2$

$\varphi = \pi/2$
$E_C = 0$

6. *Wavelength of the incident photon*:

This wavelength is given by:

$$E_1 = \frac{hc}{\lambda_1} \quad \Rightarrow \quad \lambda_1 = \frac{hc}{E_1} = \frac{(6.63 \times 10^{-34} \text{ J.s}) \, (3 \times 10^8 \text{ m.s}^{-1}) \, (1 \text{ eV})}{(500 \times 10^3 \text{ eV}) \, (1.6 \times 10^{-19} \text{ J})}$$

$$\approx 2.49 \times 10^{-12} \text{ m}. \tag{4}$$

7. *Calculation of the wavelength λ_2 associated with the scattered photon*:

The wavelength λ_2 is given by:

$$\Delta\lambda = \lambda_2 - \lambda_1 \quad \Rightarrow \quad \lambda_2 = \Delta\lambda + \lambda_1$$
$$= 4.85 \times 10^{-12} \text{ m} + 2.49 \times 10^{-12} \text{ m} \approx 7.34 \times 10^{-12} \text{ m}. \tag{5}$$

8. *Energy E_2 of the diffused photon*:

Energy E_2 is given by:

$$E_2 = \frac{hc}{\lambda_2} = \frac{(6.63 \times 10^{-34} \text{ J.s}) \, (3 \times 10^8 \text{ m.s}^{-1}) \, (1 \text{ eV})}{(7.34 \times 10^{-12} \text{ m}) \, (1.6 \times 10^{-19} \text{ J})} \approx 1.69 \times 10^5 \text{ eV}$$

$$= 169 \text{ keV}. \tag{6}$$

Exercise 30

One photon of the wavelength λ_1 collides elastically with one free electron of mass m_e initially at rest in the reference frame of the supposed Galilean laboratory. The photon is scattered at an angle θ compared to its initial direction and its wavelength becomes λ_2 while the electron moves in a direction making an angle φ with the direction of the incident photon.

1. (a) *What are the conserved quantities during the collision?*
   (b) *Define the momenta and energies of the photon and the electron before and after the collision. What are the relationships that link these quantities?*
   (c) *Establish the relationship which links λ_1, λ_2, and θ. What are the limits of $\lambda_2 - \lambda_1$? Calculate numerically $h/m_e c$.*
   (d) *Express the ratio of the energy difference ΔE_e of the recoil electron to the energy of the incident photon $h\nu_1$ as a function of λ_2, $h/m_e c$, and θ.*
   (e) *Deduce the condition on the energy of the incident photon to obtain a noticeable Compton's scattering.*
   (f) *Knowing that $\Delta E_e = 300$ keV and $\lambda_2, = 7.5$ pm, with what types of photons should the target be irradiated in hopes of obtaining a noticeable Compton's scattering?*

2. *Show that it is not possible for the photon to be totally absorbed by the electron, thus transferring all its energy to the electron in the form of kinetic energy.*
3. *One photon interacts with one electron at rest. We observe the scattering of a photon of the same wavelength as that of the incident photon. Give a plausible explanation of this phenomenon. We give:*
   - *Planck's constant: $h = 6.63 \times 10^{-34}$ J.s*
   - *Mass of the electron: $m_e = 9 \times 10^{-31}$ kg*
   - *Speed of light: $c = 3 \times 10^{8}$ ms^{-1}*
   - *1 eV = 1.6×10^{-19} J*
   - *1 keV = 10^{3} eV*

Solution

1. (a) *Quantities conserved during the collision*:

Let us consider the collision between one photon coming from the left and one electron at rest. The system considered in this study is the photon+electron set. There is no energy exchange with the outside. So the energy and the momentum are conserved during the collision.

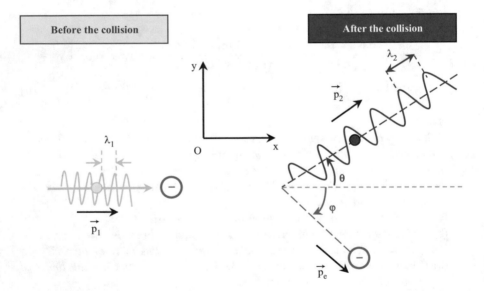

(b) *Definition of the momenta and energies of the photon and the electron before and after the collision*:

The momenta and energies of the photon and the electron before and after the collision are summarized in the following table:

	Before the collision		After the collision	
	Incident photon	Electron at rest	Scattered photon	Recoil electron
Energy	$h\nu_1$	$m_e c^2$	$h\nu_2$	$E_e = \sqrt{P_e^2 c^2 + m_e^2 c^4}$
Momentum	$\dfrac{h\nu_1}{c}\begin{pmatrix}1\\0\end{pmatrix}$	0	$\dfrac{h\nu_2}{c}\begin{pmatrix}\cos\theta\\\sin\theta\end{pmatrix}$	$P_e\begin{pmatrix}\cos\varphi\\-\sin\varphi\end{pmatrix}$

– *Relationships that link these magnitudes*:

A. *Conservation of energy*:

$$E_1^P(\text{incident photon}) + E_1^e(\text{electron at rest}) = E_2^P(\text{scattered photon}) \\ + E_2^e(\text{recoil electron}). \tag{1}$$

Namely:

$$h\nu_1 + m_e c^2 = h\nu_2 + E_e. \tag{2}$$

By dividing (2) by c, that equation turns into:

$$\frac{h\nu_1}{c} + m_e c = \frac{h\nu_2}{c} + \frac{E_e}{c}. \tag{3}$$

B. *Conservation of momentum*:

$$P_1^P(\text{incident photon}) + P_1^e(\text{electron at rest}) = P_2^P(\text{scattered photon}) \\ + P_2^e(\text{recoil electron}). \tag{4}$$

– *According to the x-axis, we get*:

$$\frac{h\nu_1}{c} + 0 = \frac{h\nu_2}{c}\cos\theta + P_e\cos\varphi. \tag{5}$$

– *According to the y-axis, we obtain*:

$$0 + 0 = \frac{h\nu_2}{c}\sin\theta - P_e\sin\varphi. \tag{6}$$

(c) *Establishment of the relationship that links λ_1, λ_2, and θ:*

Let us eliminate P_e, E_e, and φ by starting from the following equation given in the above table:

$$E_e = \sqrt{P_e^2 c^2 + m_e^2 c^4} \quad \Rightarrow \quad E_e^2 = P_e^2 c^2 + m_e^2 c^4 \quad \Rightarrow \quad P_e^2$$

$$= \left(\frac{E_e}{c}\right)^2 - m_e^2 c^2. \tag{7}$$

From (3), we get:

$$\frac{h\nu_1}{c} + m_e c = \frac{h\nu_2}{c} + \frac{E_e}{c} \quad \Rightarrow \quad \left(\frac{E_e}{c}\right)^2 = \left(\frac{h\nu_1}{c} + m_e c - \frac{h\nu_2}{c}\right)^2. \tag{8}$$

From (5), we find:

$$\frac{h\nu_1}{c} + 0 = \frac{h\nu_2}{c} \cos\theta + P_e \cos\varphi \quad \Rightarrow \quad P_e \cos\varphi = \frac{h\nu_1}{c} - \frac{h\nu_2}{c} \cos\theta. \tag{9}$$

From (6), we obtain:

$$0 + 0 = \frac{h\nu_2}{c} \sin\theta - P_e \sin\varphi \quad \Rightarrow \quad P_e \sin\varphi = \frac{h\nu_2}{c} \sin\theta. \tag{10}$$

By squaring (9) and (10) and adding to both sides of (9) the corresponding sides of (10), we get: .

$$P_e^2 \left(\sin^2\varphi + \cos^2\varphi\right) = P_e^2 = \left(\frac{h\nu_1}{c} - \frac{h\nu_2}{c} \cos\theta\right)^2 + \left(\frac{h\nu_2}{c} \sin\theta\right)^2. \tag{11}$$

By incorporating (11) and (8) in (7), we find:

$$\left(\frac{h\nu_1}{c} - \frac{h\nu_2}{c} \cos\theta\right)^2 + \left(\frac{h\nu_2}{c} \sin\theta\right)^2 = \left(\frac{h\nu_1}{c} + m_e c - \frac{h\nu_2}{c}\right)^2 - m_e^2 c^2. \tag{12}$$

After expanding and simplification, we obtain:

$$\frac{1}{\nu_2} - \frac{1}{\nu_1} = \frac{h}{m_e c^2} (1 - \cos\theta). \tag{13}$$

Knowing that frequencies and wavelengths are inversely proportional, we deduce Compton's equation, and (14) transforms into:

$$\frac{1}{\dfrac{c}{\lambda_2}} - \frac{1}{\dfrac{c}{\lambda_1}} = \frac{\lambda_2}{c} - \frac{\lambda_1}{c} = \frac{h}{m_e c^2}\,(1 - \cos\theta) \quad \text{that is:} \quad \lambda_2 - \lambda_1$$

$$= \frac{h}{m_e c}\,(1 - \cos\theta). \tag{14}$$

– *Limits of $\Delta\lambda = \lambda_2 - \lambda_1$:*

According to (14), we can say that $\Delta\lambda$ is maximal if $\cos\theta = -1$, namely:

$$\Delta\lambda_{\text{Max}} = \frac{2h}{m_e c}. \tag{15}$$

According to (14), we can say that $\Delta\lambda$ is minimal if $\cos\theta = -1$, namely:

$$\Delta\lambda_{\text{Max}} = 0. \tag{16}$$

The wavelength shift $\Delta\lambda$ is therefore between these two limit values:

$$0 < \lambda_2 - \lambda_1 < \frac{2h}{m_e c}. \tag{17}$$

– Numerical calculation of $h/m_e c$:

$$\frac{h}{m_e c} = \frac{6.63 \times 10^{-34}\ \text{J.s}}{\left(9.1 \times 10^{-31}\ \text{kg}\right)\left(3 \times 10^8\ \text{m.s}^{-1}\right)} = 0.243 \times 10^{-11}\ \text{m} = 2.43\ \text{pm.} \tag{18}$$

(d) *Ratio of the energy difference ΔE_e of the recoil electron to the energy of the incident photon $h\nu_1$ as a function of λ_2, $h/m_e c$, and θ:*

Starting from (2), we obtain:

$$h\nu_1 + m_e c^2 = h\nu_2 + E_e \quad \Rightarrow \quad \Delta E_e = E_e - m_e c^2 = h\,(\nu_1 - \nu_2). \tag{19}$$

Let us divide (19) by the energy of the incident photon $h\nu_1$, and it becomes:

$$\frac{\Delta E_e}{h\nu_1} = \frac{h\,(\nu_2 - \nu_1)}{h\nu_1} = \frac{\nu_2 - \nu_1}{\nu_1}. \tag{20}$$

From (14), we get:

$$\frac{1}{\nu_2} - \frac{1}{\nu_1} = \frac{\nu_1 - \nu_2}{\nu_1 \nu_2} = \frac{h}{m_e c^2}\,(1 - \cos\theta) \quad \Rightarrow \quad \frac{\nu_1 - \nu_2}{\nu_1}$$

$$= \frac{h_2}{m_e c^2}\,(1 - \cos\theta). \tag{21}$$

By incorporating (21) in (20), we obtain the sought equation:

$$\frac{\Delta E_e}{h\nu_1} = \frac{h\nu_2}{m_e c^2} \, (1 - \cos\theta). \tag{22}$$

(e) *Condition on the energy of the incident photon to observe Compton's scattering*:

It is given by:

$$\frac{\Delta E_e}{h\nu_1} = \frac{h\nu_2}{m_e c^2} \, (1 - \cos) < \frac{2h\nu_2}{m_e c^2} = \frac{\dfrac{2h}{m_e c^2}}{\dfrac{1}{2}} = \frac{\dfrac{2h}{m_e c^2}}{\dfrac{c}{\lambda_2}} = \frac{\dfrac{2h}{m_e c}}{\lambda_2} \approx \frac{5}{\lambda_2}, \tag{23}$$

that is, in the end:

$$\frac{\Delta E_e}{h\nu_1} < \frac{5}{\lambda_2} \quad \Rightarrow \quad h\nu_1 > \frac{5\Delta E_e}{\lambda_2}. \tag{24}$$

(f) *Photon types in hope of obtaining Compton's scattering*:

From (24), we arrive at:

$$\nu_1 > \frac{5\Delta E_e}{\lambda_2 h} = \frac{5\,(300 \times 10^3 \text{ eV})\,(1.6 \times 10^{-19} \text{ J})}{(7.5 \times 10^{-12} \text{ m})\,(6.63 \times 10^{-34} \text{ J.s})} = 4.83 \times 10^{31} \text{ Hz}. \tag{25}$$

To obtain a noticeable Compton's scattering, we must irradiate the target with γ-rays ($\nu > 3 \times 10^{16}$ Hz).

2. *Total absorption of the incident energy by the electron*:

Let us stand in a reference frame where the electron is at rest before the photon-electron interaction. Let us write the energies and momenta involved before and after the collision.

	Before the collision		After the collision	
	Incident photon	Electron at rest	Scattered photon	Recoil electron
Energy	$h\nu_1$	$m_e c^2$	0	$\sqrt{P_e^2 c^2 + m_e^2 c^4}$
Momentum	$\dfrac{h\nu_1}{c}$	0	0	P_e

The considered system is the photon+electron set. There is no exchange of energy with the outside at the moment of the impact.

A. *Conservation of energy*:

It is given by the following equation:

$$h\nu_1 + m_e c^2 = \sqrt{P_e^2 c^2 + m_e^2 c^4}. \tag{26}$$

B. *Conservation of momentum according to the x-axis only*:

It is given by the following equation:

$$\frac{h\nu_1}{c} + 0 = P_{\text{e}}. \tag{27}$$

Let us square (26), and it becomes:

$$\left(h\nu_1 + m_{\text{e}}c^2\right)^2 = (h\nu_1)^2 + m_{\text{e}}^2 c^4 + 2m_{\text{e}}c^2 h\nu_1 = P_{\text{e}}^2 c^2 + m_{\text{e}}^2 c^4. \tag{28}$$

After simplification, we obtain:

$$(h\nu_1)^2 + 2m_{\text{e}}c^2 h\nu_1 = P_{\text{e}}^2 c^2. \tag{29}$$

Let us square (27), and it becomes:

$$\left(\frac{h\nu_1}{c}\right)^2 = P_{\text{e}}^2. \tag{30}$$

By doing the ratio (30)/(29), we arrive at:

$$\frac{2m_{\text{e}}c^2}{h\nu_1} = 0. \tag{31}$$

This result is aberrant because m_{e} and/or c cannot be zero. In reality, it is the hypothesis of the exercise that is false. Indeed, during the collision of one photon with one electron at rest, the photon cedes only a part of its energy to the recoil electron and scatters with a lesser energy.

3. *Scattering of the photon of which the wavelength has not varied*:

According to (16), the wavelengths of the incident photon λ_1 and the scattered photon λ_2 are equal if $\Delta\lambda_{\text{Max}}$ is zero, namely, $\cos\theta = 1 \Rightarrow \theta = 0°$. The scattered photon leaves on the same straight line and in the same direction as the incident photon ($\theta = 0$). The two photons have the same wavelength and the same energy ($\lambda_1 = \lambda_2$ and $E_1 = E_2$), which means that the electron does not carry any energy ($E_{\text{C}} = 0$). This collision is tangential.

Exercise 31

A 100 keV photon interacts with one free electron by Compton's scattering. The mass of the electron is $m_e = 0.511$ MeV/c^2.
1. *Calculate the maximal and minimal values that can take the energy of the scattered photon.*
2. *Calculate the kinetic energy carried by the Compton electron in each case.*

Solution
– *Maximal value of the energy of the scattered photon*:

By starting from Eq. (13) demonstrated in Exercise 29, it becomes:

$$\frac{1}{\nu_2} - \frac{1}{\nu_1} = \frac{h}{m_e c^2}\,(1 - \cos\theta) \quad \Rightarrow \quad \frac{1}{h\nu_2} - \frac{1}{h\nu_1} = \frac{1}{m_e c^2}\,(1 - \cos\theta) \qquad (1)$$

where $h\nu_1$ and $h\nu_2$ are respectively the energies of the incident photon and the scattered photon, from which we deduce the expression of the energy of the scattered photon:

$$\frac{1}{h\nu_2} = \frac{1}{m_e c^2}\,(1 - \cos\theta) + \frac{1}{h\nu_1}. \qquad (2)$$

The energy of the scattered photon is maximal if $\cos\theta = 1 \Rightarrow \theta = 0°$: the photon continues its way with energy equal to:

$$\frac{1}{(h\nu_2)_{\text{Max}}} = 0 + \frac{1}{h\nu_1} \quad \Rightarrow \quad (h\nu_2)_{\text{Max}} = h\nu_1 = 100\ \text{keV}. \qquad (3)$$

– *Minimal value of the energy of the scattered photon*:

The energy of the scattered photon is minimal if $\cos\theta = -1 \Rightarrow \theta = 180°$: the photon turns back with energy equal to:

$$\frac{1}{(h\nu_2)_{\text{min}}} = \frac{2}{m_e c^2} + \frac{1}{h\nu_1} = \frac{2}{m_e c^2} + \frac{1}{h\nu_1} = \frac{2}{511\ \text{keV}} + \frac{1}{100\ \text{keV}}$$
$$= 0.0139\ \text{keV}^{-1}, \qquad (4)$$

namely:

$$(h\nu_2)_{\text{min}} = \frac{1}{0.0139\ \text{eV}^{-1}} = 71.9\ \text{keV}. \qquad (5)$$

1. *Kinetic energy carried by the Compton electron in each case*:

The kinetic energy carried by the recoil electron is given by:

$$E_c = h\nu_1 - h\nu_2. \tag{6}$$

- *Backscatter $\theta = 180°$:*

The energy carried by the electron is maximal because the scattered photon carries a minimal energy:

$$(E_c)_{\text{Max}} = h\nu_1 - (h\nu_2)_{\text{min}} = 100 \text{ keV} - 71.9 \text{ keV} = 28.1 \text{ eV}. \tag{7}$$

- *Tangential collision $\theta = 0°$:*

The energy carried by the electron is minimal because the scattered photon carries a maximal energy:

$$(E_c)_{\text{mim}} = h\nu_1 - (h\nu_2)_{\text{Max}} = 100 \text{ keV} - 100 \text{ keV} = 0 \text{ eV}. \tag{8}$$

Exercise 32

One photon of the wavelength λ, of the energy E_1, and the momentum P_1 arriving from the left along the x-axis strikes one electron of speed v, energy E_2, and momentum $P_2 = \gamma m_e v$ coming from the right following an oblique direction which makes with the x-axis an angle θ_2. After the collision, the photon scatters in a direction defined by the angle θ_1' with respect to the x-axis with a wavelength λ', energy E_1', and a momentum P_1', and the recoil electron deviates from its initial direction by an angle θ_2' with respect to the x-axis with a speed v', a momentum $P_2' = \gamma' m_e v'$, and energy E_1'. m_e represents the mass of the electron at rest. We give:

$$\gamma = \frac{1}{\sqrt{1 - \dfrac{v^2}{c^2}}} \quad \text{and} \quad \gamma' = \frac{1}{\sqrt{1 - \dfrac{v'^2}{c^2}}}.$$

(i) (a) *What are the conserved quantities during the collision?*
   (b) *Define the momenta and energies of the photon and the electron before and after the collision:*
   (c) *What are the relationships that link the incidental and scattering quantities?*
(ii) *Let us now consider the particular case of an elastic collision between one photon of wavelength λ, energy E_1, and momentum P_1 and one electron at rest of mass m_e.*
   (a) *What are the conserved quantities during the collision?*
   (b) *Define the momenta and energies of the photon and the electron before and after the collision:*
   (c) *What are the relationships that link the incidental and scattering quantities?*

(iii) *In this third and last part of the exercise, we study the inverse Compton's scattering. This is the scattering of electrons onto photons, transferring them a large part of their energy. Its interest is important in astrophysics in situations where high-energy electrons scatter on low-energy photons. To simplify, we consider a frontal collision between the electron and the photon.*
  (a) *Calculate the ratio between the wavelengths of the photon before and after the collision.*
  (b) *What is the ratio of their energies?*

Solution

1. (a) *Quantities conserved during the collision*:

The system considered in this study is the photon+electron set. There is no energy exchange with the outside. Therefore, the energy and the momentum are conserved during this relativistic elastic collision.

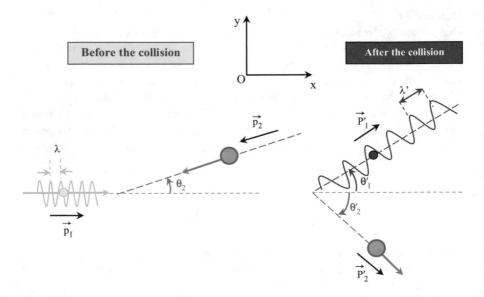

(b) *Definition of the momenta and energies of the photon and the electron before and after the collision*:

According to Eq. (4.53) of the course, the momentum of the photon P_i has the following expression:

$$E_i = P_i c = h\nu_i = h\frac{c}{\lambda_i} \;\Rightarrow\; P_i = \frac{E_i}{c} = \frac{h}{\lambda_i}. \qquad (1)$$

According to Eq. (4.56) of the course, the energy of the electron E_e is linked to its momentum P_e by:

$$P_e^2 c^2 = E_e^2 - m_e^2 \, c^4 \quad \Rightarrow \quad E_e^2 = P_e^2 c^2 + m_e^2 \, c^4. \tag{2}$$

We have also seen in the course that in special relativity, Einstein's equation states that at any mass m_e corresponds an energy $m_e c^2$ called *rest mass energy* and that every particle in motion includes both the rest mass energy and the kinetic energy E_c, namely, according to (4.50) of the course:

$$E_e = mc^2 = E_c + m_e c^2 \tag{3}$$

where $E_e = mc^2$ is the energy of the particle and m its relativistic mass given by:

$$m = \frac{m_e}{\sqrt{1 - \dfrac{v^2}{c^2}}} = \gamma m_e. \tag{4}$$

Moreover, any material particle in motion (one electron, for example) has a momentum given by Eq. (4.52) of the course:

$$P_e = mv = \gamma m_e v. \tag{5}$$

Under of what has just been said and on the basis of the figure, the momenta and energies of the photon and the electron before and after the collision can be summarized as follows:

	Before the collision		After the collision	
	Incident photon	Mobile electron	Scattered photon	Recoil electron
Energy	$E_1 = \dfrac{hc}{\lambda}$	$E_2 = \gamma m_e c^2$	$E_1' = P_1' c$	$E_2' = \sqrt{P_2'^2 c^2 + m_e^2 c^4}$
Momentum	$\dfrac{h}{\lambda}\begin{pmatrix} 1 \\ 0 \end{pmatrix}$	$\gamma m_e v \begin{pmatrix} -\cos\theta_2 \\ -\sin\theta_2 \end{pmatrix}$	$P_1' \begin{pmatrix} \cos\theta_1' \\ \sin\theta_1' \end{pmatrix}$	$P_2' \begin{pmatrix} \cos\theta_2' \\ -\sin\theta_2' \end{pmatrix}$

(c) *The relationships that link these quantities*:

The photon+electron system being isolated, let us write there is conservation of energy and momentum during the collision:

A. *Conservation of energy*:

$$\begin{aligned} E_1 \text{(incident photon)} + E_2 \text{(mobile electron)} &= E_1' \text{(scattered photon)} \\ &+ E_2' \text{(recoil electron)}. \end{aligned} \tag{6}$$

According to the table, we get:

$$E = \frac{hc}{\lambda} + \gamma m_e c^2 = P_1' c + \sqrt{{P_2'}^2 c^2 + m_e^2 c^4}. \tag{7}$$

B. *Conservation of momentum*:

$$\begin{aligned} E_1(\text{incident photon}) + P_2(\text{mobile electron}) = \\ P_1'(\text{scattered photon}) + E_2'(\text{recoil electron}). \end{aligned} \tag{8}$$

– *According to the x-axis, we get*:

$$P_x = \frac{h}{\lambda} - \gamma m_e v \cos\theta_2 = P_1' \cos\theta_1' + P_2' \cos\theta_2'. \tag{9}$$

– *According to the y-axis, we find*:

$$P_y = 0 - \gamma m_e v \sin_2 = P_1' \sin\theta_1' - P_2' \sin\theta_2'. \tag{10}$$

From (7), we have:

$$E = P_1' c + \sqrt{{P_2'}^2 c^2 + m_e^2 c^4} \;\Rightarrow\; \left(E - P_1' c\right)^2 = {P_2'}^2 c^2 + m_e^2 c^4. \tag{11}$$

By adding the squares of (9) and (10) and by using the remarkable identity $\sin^2\theta_2' + \cos^2\theta_2' = 1$, we deduce:

$${P_2'}^2 = \left(P_x - P_1' \cos\theta_1'\right)^2 + \left(P_y - P_1' \sin\theta_1'\right)^2. \tag{12}$$

Let us divide (11) by c^2, and it becomes:

$${P_2'}^2 + m_e^2 c^2 = \left(\frac{E}{c} - P_1'\right)^2. \tag{13}$$

Let us incorporate (12) into (13), and we obtain:

$$\begin{aligned} \left(P_x - P_1' \cos\theta_1'\right)^2 + \left(P_y - P_1' \sin\theta_1'\right)^2 + m_e^2 c^2 &= \left(\frac{E}{c} - P_1'\right)^2 \\ &= \frac{E^2}{c^2} + {P_1'}^2 - \frac{2P_1' E}{c}. \end{aligned} \tag{14}$$

Let us expand (14), and we find:

$$P_x^2 + P_1'^2 \cos^2\theta_1' - 2P_xP_1' \cos\theta_1' + P_y^2 + P_1'^2 \sin^2\theta_1' - 2P_yP_1' \sin\theta_1'$$
$$= \frac{E^2}{c^2} + P_1'^2 - \frac{2P_1'E}{c} - m_e^2c^2. \tag{15}$$

Let us simplify (15), and it transforms into:

$$P_x^2 + P_y^2 - \frac{E^2}{c^2} = 2P_xP_1' \cos\theta_1' + 2P_yP_1' \sin\theta_1' - \frac{2P_1'E}{c} - m_e^2c^2. \tag{16}$$

After factorization of P_1', we deduce from (16):

$$P_1' = \frac{\dfrac{E^2}{c^2} - P_x^2 - P_y^2 - m_0^2c^2}{\dfrac{2E}{c} - 2P_x \cos\theta_1' - 2P_y \sin\theta_1'}. \tag{17}$$

Let us express the numerator of (17) otherwise. To do this, let us replace E, P_x, and P_y by their expressions (7), (9), and (10), and we obtain:

$$\frac{E^2}{c^2} - P_x^2 - P_y^2 - m_e^2c^2 = \left(\frac{h}{\lambda} + \gamma m_e c\right)^2 - \left(\frac{h}{\lambda} - \gamma m_e v \cos\theta_2\right)^2 - (-\gamma m_e v \sin\theta_2)^2 - m_e^2c^2$$
$$= \left(\frac{h}{\lambda}\right)^2 + \gamma^2 m_e^2 c^2 + \frac{2hm_ec}{\lambda} - \left(\frac{h}{\lambda}\right)^2 + (\gamma m_e v)^2 \cos^2\theta_2 + \frac{2h\gamma m_e v \cos\theta_2}{\lambda}$$
$$+ (\gamma m_e v)^2 \sin^2\theta_2 - m_e^2c^2 = \gamma^2 m_e^2 c^2 + \frac{2h\gamma m_e c}{\lambda} + (\gamma m_e v)^2 \left(\sin^2\theta_2 + \cos^2\theta_2\right)$$
$$+ \frac{2h\gamma m_e v \cos\theta_2}{\lambda} - m_e^2c^2 = \frac{2h\gamma m_e}{\lambda}\left(c + v\cos\theta_2\right) + \gamma^2 m_e^2 c^2 - m_e^2c^2 + \gamma^2 m_e^2 v^2. \tag{18}$$

As according to (4), we have:

$$\frac{m_e^2}{1 - \dfrac{v^2}{c^2}} = \gamma^2 m_e^2 \quad \Rightarrow \quad \gamma^2 m_e^2 c^2 - m_e^2c^2 + \gamma^2 m_e^2 v^2 = 0 \tag{19}$$

and we then obtain the momentum of the diffused photon P_1':

$$P_1' = \frac{\dfrac{hm_e}{\lambda}\left(c + v\cos\theta_2\right)}{\dfrac{E}{c} - P_x \cos\theta_1' - P_y \sin\theta_1'}. \tag{20}$$

The wavelength λ' of the scattered photon is:

$$\lambda' = \frac{h}{P_1'} = \frac{h\left(\dfrac{E}{c} - P_x\,cos\,\theta_1' - P_y\,sin\,\theta_1'\right)}{\dfrac{h\gamma m_e}{\lambda}\,(c + v\,cos\,\theta_2)}. \tag{21}$$

Moreover, from (9), we have:

$$P_x = P_1'\,cos\,\theta_1' + P_2'\,cos\,\theta_2' \;\Rightarrow\; P_2'\,cos\,\theta_2' = P_x - P_1'\,cos\,\theta_1'. \tag{22}$$

From (10), we have:

$$P_y = P_1'\,sin\,\theta_1' - P_2'\,sin\,\theta_2' \;\Rightarrow\; P_2'\,sin\,\theta_2' = P_y - P_1'\,sin\,\theta_1'. \tag{23}$$

By doing the ratio (23)/(22), we obtain:

$$tan\,\theta_2' = \frac{sin\,\theta_2'}{cos\,\theta_2'} = \frac{P_y - P_1'\,sin\,\theta_1'}{P_x - P_1'\,cos\,\theta_1'}. \tag{24}$$

By adding the squares of (22) and (23) and by using the remarkable identity $sin^2\theta_2' + cos^2\theta_2' = 1$, we arrive at the expression of P_2':

$$P_2' = \sqrt{\left(P_x - P_1'\,cos\,\theta_1'\right)^2 + \left(P_y - P_1'\,sin\,\theta_1'\right)^2}. \tag{25}$$

By starting from the table data, we end up with the expression of the energy of the scattered photon:

$$E_1' = P_1'c = \frac{hc}{\lambda'}. \tag{26}$$

From (7), we deduce the expression of the energy of the scattered electron E_2':

$$\frac{hc}{\lambda} + \gamma m_e c^2 = P_1'c + E_2' \;\Rightarrow\; E_2' = \frac{hc}{\lambda} + \gamma m_e c^2 - P_1'c$$
$$= \frac{hc}{\lambda} + \gamma m_e c^2 - \frac{hc}{\lambda'}. \tag{27}$$

The momentum of the scattered electron is given by:

$$P_2' = \gamma' m_e v' = \frac{m_e v'}{\sqrt{1 - \dfrac{v'^2}{c^2}}} \;\Rightarrow\; P_2'^{\,2} = \frac{m_e^2\,v'^2 c^2}{c^2 - v'^2}. \tag{28}$$

After rearrangement of (28), we end up with the expression of the speed v' of the scattered electron:

$$v' = \frac{P_2' c}{\sqrt{P_2'^2 + m_e^2 c^2}}. \tag{29}$$

2. (a) *Quantities conserved during the collision*:

The considered system in this study is the photon+electron set. There is no energy exchange with the outside. Therefore, the energy and the momentum are conserved during this relativistic elastic collision.

(b) *Definition of the momenta and energies of the photon and electron before and after the collision*:

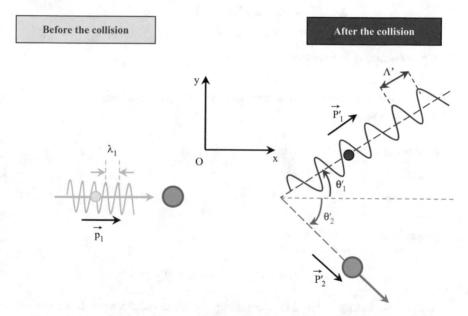

Since the electron is at rest, its speed is zero ($v = 0$) and consequently $\gamma = 1$. The momenta and the energies of the photon and the electron before and after the collision can be summarized as follows:

	Before the collision		After the collision	
	Incident photon	Electron at rest	Scattered photon	Recoil electron
Energy	$E_1 = \dfrac{hc}{\lambda}$	$E_2 = m_e c^2$	$E_1' = P_1' c$	$E_2' = \sqrt{P_2'^2 c^2 + m_e^2 c^4}$
Momentum	$\dfrac{h}{\lambda}\begin{pmatrix} 1 \\ 0 \end{pmatrix}$	0	$P_1'\begin{pmatrix} \cos\theta_1' \\ \sin\theta_1' \end{pmatrix}$	$P_2'\begin{pmatrix} \cos\theta_2' \\ -\sin\theta_2' \end{pmatrix}$

(c) *The relationships that link these quantities*:

The photon+electron system being isolated, let us write that there is conservation of energy and momentum during the collision.

A. *Conservation of energy*:

E_1(incident photon) $+ E_2$(electron at rest) $= E'_1$(scattered photon) $+$ E'_2(recoil electron). (30)

According to the table, we obtain:

$$E = \frac{hc}{\lambda} + m_e c^2 = P'_1 c + \sqrt{P'^2_2 c^2 + m_e^2 c^4}. \tag{31}$$

B. *Conservation of momentum*:

E_1(incident photon) $+ P_2$(electron at rest) $= P'_1$(scattered photon) $+$ E'_2(recoil electron). (32)

– *According to the x-axis, we find*:

$$P_x = \frac{h}{\lambda} + 0 = P'_1 \cos\theta'_1 + P'_2 \cos\theta'_2 = \frac{h}{\lambda}. \tag{33}$$

– *According to the y-axis, we get*:

$$P_y = 0 + 0 = P'_1 \sin\theta'_1 - P'_2 \sin\theta'_2 = 0. \tag{34}$$

Taking as starting point (21) and taking into account the new data ($\theta_2 = 0$, $\gamma = 1$, $P_y = 0$), we can write:

$$\lambda' = \frac{\lambda\left(\frac{E}{c} - P_x \cos\theta'_1\right)}{m_e c}. \tag{35}$$

By injecting (31) and (33) into (35), it turns into:

$$\lambda' = \frac{\lambda\left[\left(\frac{hc}{\lambda c} + m_e c^2\right) - \frac{h}{\lambda} \cos\theta'_1\right]}{m_e c} = \lambda\left(\frac{h}{\lambda m_e c} + 1 - \frac{h}{\lambda m_e c} \cos\theta'_1\right). \tag{36}$$

After rearrangement of (36), it becomes:

$$\lambda' = \lambda + \frac{h}{m_e c}(1 - cos\,\theta'_1). \tag{37}$$

Equation (37) shows that the ratio $\lambda_C = h/m_e c$ has the dimension of a wavelength. It is called *Compton wavelength of the electron* and has a value of 0.002426 nm (or 2.426×10^{-12} m). Let us look for the expression of the scattering angle of the electron θ'_2:

From (33), we have:

$$P_x = P'_1\,cos\,\theta'_1 + P'_2\,cos\,\theta'_2 \quad \Rightarrow \quad P'_2\,cos\,\theta'_2 = P_x - P'_1\,cos\,\theta'_1. \tag{38}$$

From (34), we have:

$$P_y = P'_1\,sin\,\theta'_1 - P'_2\,sin\,\theta'_2 = 0 \quad \Rightarrow \quad P'_2\,sin\,\theta'_2 = P'_1\,sin\,\theta'_1 - P_y. \tag{39}$$

Doing the ratio (33)/(34), we obtain:

$$\frac{1}{tan\,\theta'_2} = \frac{cos\,\theta'_2}{sin\,\theta'_2} = \frac{P_x - P'_1\,cos\,\theta'_1}{P'_1\,sin\,\theta'_1 - P_y} = \frac{\frac{h}{\lambda} - \frac{h}{\lambda'}cos\,\theta'_1}{\frac{h}{\lambda'}\,sin\,\theta'_1 - 0}. \tag{40}$$

In (40), we have replaced P_x by h/λ according to (33) and P'_1 by h/λ' according to (1), P_y is zero according to (34), and this equation transforms into:

$$\frac{1}{tan\,\theta'_2} = \frac{\frac{1}{\lambda} - \frac{1}{\lambda'}cos\,\theta'_1}{\frac{1}{\lambda'}\,sin\,\theta'_1} = \frac{\frac{\lambda' - \lambda\,cos\,\theta'_1}{\lambda\lambda'}}{\frac{1}{\lambda'}\,sin\,\theta'_1} = \frac{\lambda' - \lambda\,cos\,\theta'_1}{\lambda\,sin\,\theta'_1}. \tag{41}$$

Let us inject (37) into (41), and we get:

$$\frac{1}{tan\,\theta'_2} = \frac{\left[\lambda + \frac{h}{m_e c}(1 - cos\,\theta'_1)\right] - \lambda\,cos\,\theta'_1}{\lambda\,sin\,\theta'_1} = \frac{\lambda + \frac{h}{m_e c}(1 - cos\,\theta'_1) - \lambda\,cos\,\theta'_1}{\lambda\,sin\,\theta'_1}$$

$$= \frac{\lambda(1 - \lambda\,cos\,\theta'_1) + \frac{h}{m_e c}(1 - cos\,\theta'_1)}{\lambda\,sin\,\theta'_1} = \frac{(1 - \lambda\,cos\,\theta'_1)\left(\lambda + \frac{h}{m_e c}\right)}{\lambda\,sin\,\theta'_1} = \frac{(1 - cos\,\theta'_1)\left(1 + \frac{h}{\lambda m_e c}\right)}{sin\,\theta'_1}$$

$$= \frac{(1 - cos\,\theta'_1)\left(1 + \frac{h}{m_e c}\right)}{sin\,\theta'_1} \tag{42}$$

with $\nu = h/\lambda'$, the frequency of the incident photon. Using the following trigonometric relationships:

$$1 - cos\,\theta'_1 = 2\,sin^2\frac{\theta'_1}{2} \quad and \quad sin\,\theta'_1 = 2\,sin\frac{\theta'_1}{2}\,cos\frac{\theta'_1}{2} \tag{43}$$

we finally arrive at:

$$\frac{1}{tan\,\theta'_2} = \frac{2\,sin^2\dfrac{\theta'_1}{2}\left(1+\dfrac{h\nu}{m_e c}\right)}{2\,sin\dfrac{\theta'_1}{2}\,cos\dfrac{\theta'_1}{2}} = \left(1+\frac{h\nu}{m_e c}\right)\frac{sin\dfrac{\theta'_1}{2}}{cos\dfrac{\theta'_1}{2}} = \left(1+\frac{E_1}{E_2}\right)tan\frac{\theta'_1}{2}. \tag{44}$$

According to the table, we have $E_1 = hc/\lambda = h\nu$ (the incident photon energy) and $E_2 = m_e c^2$ (the energy of the electron at rest before the collision). Now let us look for the expression of the momentum P'_2 of the recoil electron. To do this, let us add the squares of (38) and (39):

$$P'_2{}^2 cos^2\theta'_2 + P'_2{}^2\,sin^2\theta'_2 = \left(P_x - P'_1\,cos\,\theta'_1\right)^2 + \left(P'_1\,sin\,\theta'_1 - P_y\right)^2. \tag{45}$$

As according to (33), we have $P_x = h/\lambda'$; according to (34), we have $P_y = 0$; according to (1), we have $P'_1 = h/\lambda'$; and as $sin^2\theta'_2 + cos^2\theta'_2 = 1$, we finally get:

$$P'_2{}^2 = \left(\frac{h}{\lambda} - \frac{h}{\lambda'}\,cos\,\theta'_1\right)^2 + \left(\frac{h}{\lambda'}\,sin\,\theta'_1\right)^2 \quad \Rightarrow \quad P'_2$$
$$= h\sqrt{\left(\frac{1}{\lambda} - \frac{1}{\lambda'}\,cos\,\theta'_1\right)^2 + \left(\frac{1}{\lambda'}\,sin\,\theta'_1\right)^2}. \tag{46}$$

By injecting of (37), the scattering photon E'_1 can be written as:

$$E'_1 = h\nu' = \frac{hc}{\lambda'} = \frac{hc}{\lambda + \dfrac{h}{m_e c}(1 - cos\,\theta'_1)} = \frac{hc}{\lambda\left[1+\dfrac{h}{\lambda m_e c}(1 - cos\,\theta'_1)\right]}. \tag{47}$$

In the end, we obtain:

$$E'_1 = \frac{hc}{\lambda\left[1+\dfrac{h}{\lambda m_e c}(1-cos\theta'_1)\right]} = \frac{\dfrac{hc}{\lambda}}{\left[1+\dfrac{h}{\lambda m_e c}(1-cos\theta'_1)\right]} = \frac{\dfrac{hc}{\lambda}}{\left[1+\dfrac{\dfrac{hc}{\lambda}}{m_e c^2}(1-cos\theta'_1)\right]}$$

$$= \frac{h\nu}{\left[1+\dfrac{h}{m_e c^2}(1-cos\theta'_1)\right]} = \frac{E_1}{\left[1+\dfrac{E_1}{E_2}(1-cos\theta'_1)\right]}. \tag{48}$$

Let us look for the energy E'_2 of the scattered electron taking as starting point (31):

$$\frac{hc}{\lambda} + m_e c^2 = P_1' c + \sqrt{P_2'^2 c^2 + m_e^2 c^4} \quad \Rightarrow \quad E_1 + m_e c^2 = E_1' + E_2'. \tag{49}$$

By injecting (48) into (49), we find:

$$E_2' = m_e c^2 + E_1 - E_1' = m_e c^2 + E_1 - \frac{E_1}{\left[1 + \frac{E_1}{E_2}\left(1 - cos\,\theta_1'\right)\right]}$$

$$= m_e c^2 + \frac{E_1\left[1 + \frac{E_1}{E_2}\left(1 - cos\,\theta_1'\right)\right] - E_1}{\left[1 + \frac{E_1}{E_2}\left(1 - cos\,\theta_1'\right)\right]} = m_e c^2 + \frac{E_1\left[1 + \frac{E_1}{E_2}\left(1 - cos\,\theta_1'\right) - 1\right]}{\left[1 + \frac{E_1}{E_2}\left(1 - cos\,\theta_1'\right)\right]}$$

$$= m_e c^2 + \frac{E_1\left[\frac{E_1}{E_2}\left(1 - cos\,\theta_1'\right)\right]}{\left[1 + \frac{E_1}{E_2}\left(1 - cos\,\theta_1'\right)\right]}. \tag{50}$$

3. (a) *Calculation of the ratio between the wavelengths of the photon before and after the collision*:

On the basis of the exercise hypotheses, we have:

- $\theta_2 = 0°$ (the electron and the incident photon are on the same straight line).
- $\theta_1' = 180°$, after the collision the photon turns back carrying with it a higher energy than it had before the collision.

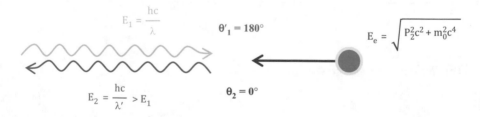

By virtue of what has just been said and on the basis of the figure, the momenta and the energies of the photon and the electron before and after the collision are:

(i) *Energies*:

$$E = \frac{hc}{\lambda} + \sqrt{P_2^2 c^2 + m_e^2 c^4}. \tag{51}$$

(ii) *Momenta*:

– *According to the x-axis, we obtain*:

$$P_x = \frac{h}{\lambda} - P_2.$$ (52)

– *According to the y-axis, we find*:

$$P_y = 0 - 0 = 0.$$ (53)

Taking as a starting point Eq. (21), we obtain:

$$\lambda' = \frac{h\left(\frac{E}{c} - P_x \cos\theta_1' - P_y \sin\theta_1'\right)}{\frac{h\gamma m_e}{\lambda}(c + v\cos\theta_2)} = \frac{\lambda\left(\frac{E}{c} - P_x \cos\theta_1' - P_y \sin\theta_1'\right)}{m_e(c + v\cos\theta_2)}.$$ (54)

By taking into account the hypotheses mentioned above and by incorporating (52) and (53) into (54), we find the ratio between the wavelengths of the photon before and after the collision:

$$\frac{\lambda'}{\lambda} = \frac{\left(\frac{E}{c} - P_x \cos 180° - (0)\sin 180°\right)}{m_e(c + v\cos 0°)} = \frac{\frac{E}{c} + P_x}{m_e(c + v)} = \frac{\frac{hc}{\lambda} + \sqrt{P_2^2 c^2 + m_0^2 c^4} + \frac{h}{\lambda} - P_2}{m_e(c + v)}$$

$$= \frac{\frac{h}{\lambda}(c + 1)\sqrt{P_2^2 c^2 + m_e^2 c^4} - P_2}{\gamma m_e(c + v)} = \xi.$$ (55)

(b) *Ratio of their energies*:

It is given by:

$$\frac{\lambda'}{\lambda} = \frac{\frac{hc}{E_2}}{\frac{hc}{E_1}} = \frac{E_1}{E_2} = \xi \quad \Rightarrow \quad \frac{E_2}{E_1} = \frac{1}{\xi} = \frac{\gamma m_e(c + v)}{\frac{h}{\lambda}(c + 1)\sqrt{P_2^2 c^2 + m_e^2 c^4} - P_2}.$$ (56)

Chapter 3
Black Body

Abstract This chapter presents the radiation theory of the black body. It is a purely fictitious ideal body which allows the obtaining of the basic laws of thermal radiation. In classical thermodynamics, its equivalent is the ideal gas, which is first studied before the studying of real gas. The object which comes closest to the black body model is the inside of an oven. The study of the black body stumbled on the explanation of the ultraviolet catastrophe and led to the quantum revolution. This point is discussed in this chapter.

The chapter begins with a precise course, followed by exercises in order of increasing difficulty. These exercises are solved in a very detailed way; they allow students to assimilate the course and help them prepare for exams. It is intended for first-year university students. It can also benefit bachelor's and doctoral students because of the deliberately simplified presentation of the treated concepts.

3.1 Introduction

The mechanisms of heat transfer naturally depend on temperature. Three modes of transfer exist: conduction, convection, and radiation. In the conduction and convection, this dependence mainly involves differences in temperature; on the other hand, in radiation, the importance of exchanges is strongly related to the temperature level. According to this level, the nature of the radiation will be different. For the human body, for example, two types of radiation coexist: it receives the solar radiation (visible) and emits in the infrared (invisible). At the moment, the solar radiation nature and the mechanisms which govern them are not fully established. Two theories exist, corpuscular and wave theories: they are complementary and allow explaining the different phenomena observed to date.

It is known that radiation propagates in the vacuum at a constant speed $c \approx 3 \times 10^8$ m/s. The nature of the radiation and its frequency depend on the broadcasting source. The wavelength of a radiation λ is defined by the ratio of the light speed c to its frequency ν. All bodies emit radiations; however, we can only feel the radiation of which the wavelengths are between 0.1 and 100 μm. In this so-called

S. Khene, *Topics and Solved Exercises at the Boundary of Classical and Modern Physics*, Undergraduate Lecture Notes in Physics, https://doi.org/10.1007/978-3-030-87742-2_3

231

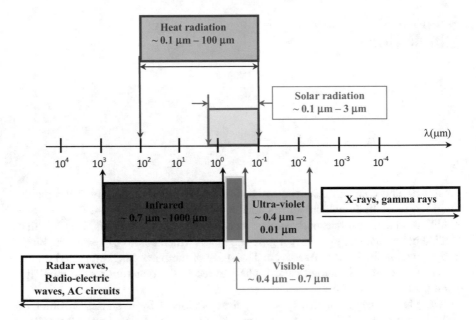

Fig. 3.1 Domains of the electromagnetic spectrum

thermal domain, radiation may warm a solid, liquid, gaseous, inert, or living body. Inside this very restricted domain, the human eye is only able to perceive a very small domain called *visible domain*, ranging from 0.38 to 0.76 μm (Fig. 3.1).

3.2   Reminders

Before defining the black body and addressing the laws that describe its behavior, it is useful for the understanding of this course to recall some basic photometric quantities. Let us consider for that purpose a body carried at a temperature T. This body emits some energy from its surface in the form of thermal radiation, namely, in the form of electromagnetic waves; this energy is emitted at all wavelengths and in all directions. If this body is not alone in space, it also receives some energy in the form of radiation from other bodies of the surrounding environment (Fig. 3.2).

3.2.1   Photometric Quantities

(i) A photometric quantity is said to be *monochromatic* if its wavelengths are within a very narrow range λ and $\lambda + d\lambda$. In practice, radiation is characterized by a single wavelength. This quantity is noted in the following by the index λ.

Fig. 3.2 Emitting (in blue) and receiving (in black) surfaces of radiation

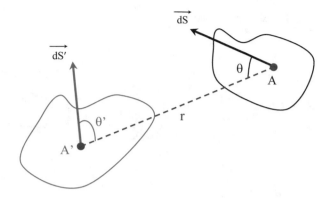

(ii) A photometric quantity is said to be *total (or polychromatic)* if it involves the whole spectrum of wavelengths.

(iii) A photometric quantity is said to be *directional* if the radiation depends on the direction of the emission. This quantity is noted in the following by the index Ox indicating the emission axis.

(iv) A photometric quantity is called *hemispherical* if it contains wavelengths belonging to the half-space of a sphere.

3.2.2 Solid Angle

The solid angle is the three-dimensional analogue of the plane angle. The latter is defined as the ratio of the length of the arc on the radius. Although the plane angle is the ratio of two lengths, its unit is the radian (rad). The projection of a solid angle on a sphere gives an area on a surface. This angle is expressed in the international system in steradians (symbol sr). It is therefore a dimensionless unit. A solid angle of 1 steradian delimits on a sphere of radius 1, from the center of this sphere, a surface of area 1. The solid angle which intercepts the entire sphere is therefore 4π steradians since the area of a sphere of radius R is $4\pi R^2$.

(i) *Elementary solid angle of an inclined surface* (Fig. 3.3):

It is given by the following equation:

$$d\Omega = \frac{\overrightarrow{dS}.\vec{u}}{r^2}.$$ (3.1)

(ii) *Total solid angle*:

The total solid angle Ω is calculated by the integration of expression (3.1) over the entire surface (S) intercepted by this angle. Namely:

Fig. 3.3 Elementary solid
angle

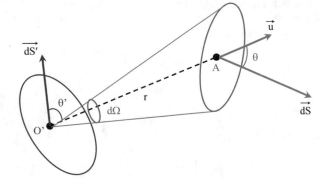

Fig. 3.4 Projection of an
object on a sphere

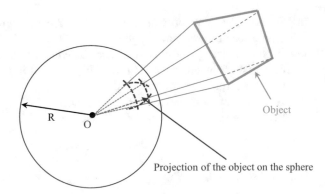

$$\Omega = \iint d\Omega = \iint_{(S)} \frac{\overrightarrow{dS} \cdot \overrightarrow{u}}{r^2}. \tag{3.2}$$

To calculate the solid angle under which we see any object from a point O, we
project this object on a sphere of the radius R and centered at this point (Fig. 3.4).

If this surface (S) is the surface of a spherical portion belonging to this sphere, the
solid angle Ω takes the following form:

$$\Omega = \frac{1}{R^2} \iint_{(S)} dS = \frac{S}{R^2}, \quad \Omega \text{ in steradians (sr).} \tag{3.3}$$

The surface (S) and the radius R are expressed in m^2 and m, respectively. The
solid angle under which we see, from the center of the sphere of the center O and the
radius R, a spherical cap of surface (S) of which the apparent diameter extends over
an angle of 2θ is written as:

Fig. 3.5 Solid angle of a
cone of revolution, θ being
the half-angle at the top

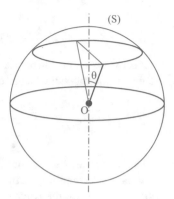

$$\Omega = \frac{S}{R^2} = 2\pi(1 - \cos\theta). \tag{3.4}$$

So, the complete space is seen under a solid angle of 4π sr. The human eye sweeps around 2π sr that is half of the space or a half-sphere or a hemisphere. It can be seen that the solid angle appears to be the measure of the area of the spherical cap cut by the cone on the sphere of the center O and of the unit radius (Fig. 3.5).

3.2.3   Radiated Flux

The radiated flux is the luminous power radiated by a source in all space. It is expressed in watts (W). This flux, noted in the following Φ, is emitted for different wavelengths. It is given by:

$$\Phi = \int_0^\infty \Phi_\lambda \, d\lambda \tag{3.5}$$

where Φ_λ is the monochromatic flux. It is generally expressed in W/μm (if λ is in μm).

3.2.4   Energy Intensity

Energy intensity characterizes the emission of a body in a given direction Ox around an elementary solid angle $d\Omega$:

$$I_{Ox} = \frac{d_{Ox}}{d}.$$ (3.6)

It is expressed in watts/steradians (W/sr). Since the radiated flux is polychromatic, we define the monochromatic energy intensity in (W.sr^{-1}.µm^{-1}) by:

$$I_{Ox} = \int_0^\infty I_{Ox,\lambda} \, d\lambda.$$ (3.7)

The vector set $\overrightarrow{I_{Ox}}$ defines a surface called *emission indicator*. For an isotropic punctual source emitting in the same way in all directions, its emission indicator is a sphere centered on the source.

3.2.5 Emittance

The emittance is the total flux emitted by a surface (S) per unit area, namely:

$$M = \frac{d\Phi}{dS}.$$ (3.8)

It is expressed in watts/square meters (W/m^2). This emittance can also be defined as a function of the wavelength. It is then about the monochromatic emittance M_λ, namely:

$$M_\lambda = \frac{dM}{d\lambda} \quad \Rightarrow \quad M = \int_0^\infty M_\lambda \, d\lambda.$$ (3.9)

It is generally expressed in W.m^{-2}.µm^{-1} (if λ is in µm).

3.2.6 Luminance

Let us suppose that a person looks at an elementary surface dS surrounding a point O of this surface under an angle θ with respect to the normal $\vec{n}$ characterizing dS (Fig. 3.6).

Let us consider an elementary solid angle $d\Omega$ of the summit O that is incident on the pupil of the observer's eye. The luminous flux $d\Phi$ that penetrates into the eye is naturally proportional to the surface dS and to the solid angle $d\Omega$. But, it is also proportional to the cosine of the angle θ because under an angle $\theta = \pi/2$ rad, the surface dS cannot be perceived anymore. There is therefore a factor of proportionality noted L_{Ox}, called *luminance* which verifies the following equation:

Fig. 3.6 Luminous flux emitted by dS and perceived by the human eye

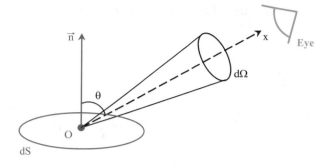

$$d\Phi = L_{Ox}\, dS\, d\Omega \cos\theta \;\;\Rightarrow\;\; L_{Ox} = \frac{d\Phi}{dS\, d\Omega \cos\theta} = \frac{1}{dS\;\cos\theta} \qquad (3.10)$$

where I is the energy intensity; the luminance L_{Ox} is thus expressed in $W.m^{-2}.sr^{-1}$. The monochromatic luminance, in $W.m^{-3}.sr^{-1}.m^{-1}$, is given by:

$$L_{\lambda,Ox} = \frac{L_{Ox}}{d\lambda} = \frac{d\Phi}{dS\, d\Omega\, d\lambda \cos\theta}. \qquad (3.11)$$

3.2.7 Illuminance

If a surface emits a radiation, it also receives a radiation. The illumination E corresponds to the total flux received by a surface dS:

$$E = \frac{d\Phi}{dS}. \qquad (3.12)$$

It is expressed as the emittance in W/m^2. The emittance and the illumination should not be confused because the illumination is a quantity relative to the reception of the radiation, whereas the emittance is a physical quantity that characterizes the emission of the radiation.

3.3 Lambert's Law

We say that a source verifies the Lambert law if its luminance does not depend on the direction of the emission. Most emissive bodies verify this property. We then obtain a simple relationship between the luminance L and the emittance M:

$$M_\lambda = \pi L_\lambda \;\; or \;\; M = \pi L. \qquad (3.13)$$

Demonstration:

By definition of the emittance in W/m^2, we have (see (3.8)):

$$M = \frac{d\Phi}{dS} \qquad (3.14)$$

with:

$$d\Phi = \int_{1/2 \text{ space}} d\Phi' \qquad (3.15)$$

where $d\Phi$ is the flux emitted in the whole half-space and $d\Phi'$ the flux emitted inside the solid angle. According to (3.10), we obtain:

$$d\Phi' = L \, dS \cos\theta \, d\Omega. \qquad (3.16)$$

By injecting (3.16) in (3.15), we get:

$$d\Phi = \int_{1/2 \text{ space}} L \, dS \cos\theta \, d\Omega = L \, dS \int_{1/2 \text{ space}} \cos\theta \, d\Omega \qquad (3.17)$$

because L and dS are constant, dS being the emitting surface (that of the source).

Let us consider a half-sphere of the unit radius, centered at O. The solid angle $d\Omega$ cuts out a surface $d\sigma$ on the surface of this half-sphere. Let $d\sigma'$ be the projection of $d\sigma$ on the basic plane (Fig. 3.7). Knowing that:

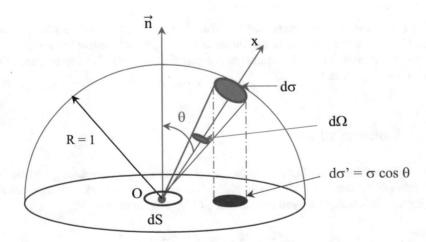

Fig. 3.7 Half-sphere of the unit radius, centered at O

$$d\Omega = \frac{d\sigma}{1^2} = d\sigma \qquad (3.18)$$

we obtain for the case of the portion of the sphere intercepted by the solid angle $d\Omega$:

$$d\Phi = L\,dS \int_{1/2 \text{ space}} \cos\theta\,d\sigma = L\,dS \int_{\text{basic circle}} d\sigma' = L\,dS\,\pi. \qquad (3.19)$$

By starting from (3.14), we finally arrive at:

$$M = \frac{d\Phi}{dS} = \frac{L\,dS\,\pi}{dS} = \pi L. \qquad (3.20)$$

3.4 Black Body

The name of the black body was introduced for the first time by the physicist Gustav Kirchhoff in 1862. The resulting studies allowed Max Planck to discover the quantization of electromagnetic interactions that was one of the foundations of quantum physics. The black body is an ideal body for which the absorption coefficient α is equal to 1. In other words, the absorbed flux is equal to the incident flux regardless of the value of λ. It is therefore a perfect transmitter and a perfect absorber. It is also used as a reference and nicknamed *standard of the radiation*. The wavelengths of the received radiation and those of the emitted radiation are not necessarily the same. Indeed, a human body illuminated by the sun (so in the visible) re-emits at room temperature in the infrared. The black body is a purely fictitious body which allows obtaining the basic laws of thermal radiation. In classical thermodynamics, its equivalent is the perfect gas that is studied first before studying real gases. The real object that comes closest to this model is the interior of an oven. To be able to study the radiation in this cavity, one of its faces is pierced with a small hole letting out a tiny fraction of the internal radiation. It is moreover an oven that was used by Wilhelm Wien to determine the electromagnetic emission laws as a function of temperature. A black body can be colored. Indeed, although the sun is very bright, its spectrum is superimposed on that of a black body of temperature 5777 K.

In what follows, all physical quantities relating to the black body will be provided with the exponent 0 $\left(M^0, M_\lambda^0, L_{\lambda,\text{Ox}}^0 \ldots\right)$.

3.5 Planck's Law

From the quantum theory, Max Planck established a relationship between the black body monochromatic emittance M_λ^0, its temperature, and the wavelength λ:

$$\frac{dM^0}{d\lambda} = M_\lambda^0(T) = \frac{C_1 \lambda^{-5}}{e^{\frac{C_2}{\lambda T}} - 1} \tag{3.21}$$

with λ, the wavelength (in m or μm) and T, the temperature in K. The coefficients C_1 and C_2 have the following expressions:

$$C_1 = 2hc^2\pi \quad \text{and} \quad C_2 = h\frac{c}{k_B} \tag{3.22}$$

where c is the speed of light in vacuum ($c = 2.998 \ 10^8$ m/s), h Planck's constant ($h = 6.63 \times 10^{-34}$ J.S), and k_B Boltzmann's constant ($k_B = 1.38 \times 10^{-23}$ J/K). The constants C_1 and C_2 have the following values:

$$C_1 = 3.74 \times 10^{-16} \text{ W.m}^2 \quad \text{and} \quad C_2 = 1.44 \times 10^{-2} \text{ m.K.} \tag{3.23}$$

The Planck law is the basic law for all which concerns the radiation emission. Figure 3.8 gives the plot of the monochromatic emittance $M_\lambda^0(T)$ as a function of the wavelength λ for different values of the temperature T. This qualitative graph reveals that:

- The x-axis is the asymptote of all curves when the wavelength tends to infinity.
- These curves go through a maximum which is all the larger the higher the temperature.
- The wavelength corresponding to this maximum which is all the smaller the higher the temperature of the body.
- For a given wavelength, the monochromatic emittance is all the more high as the temperature is high.

3.5.1   Laws Derived from the Planck Law

Although the Planck law is easily programmable, some derived forms greatly facilitate calculations. So:

(i) By starting from the Planck law, we note that for the short wavelengths λ that characterize the visible domain (see Fig. 3.8), $exp(C_2/\lambda T)$ becomes very large compared to 1. By neglecting 1 in the denominator, we obtain the following approximation:

$$M_\lambda^0(T) = \frac{C_1 \lambda^{-5}}{e^{\frac{C_2}{\lambda T}} - 1} \approx \frac{C_1 \lambda^{-5}}{e^{\frac{C_2}{\lambda T}}} \approx C_1 \lambda^{-5} e^{-\frac{C_2}{\lambda T}}. \tag{3.24}$$

By returning to Fig. 3.8, we see that Eq. (3.24) is only valid in the yellow band of the visible range.

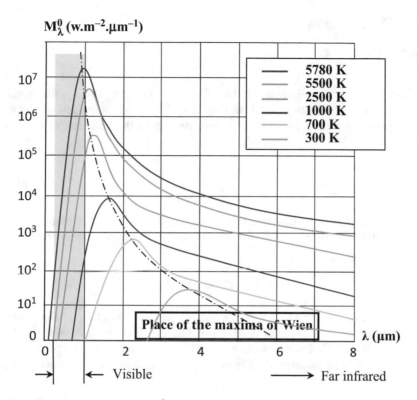

Fig. 3.8 Monochromatic emittance $M_\lambda^0(T)$ of the black body as a function of the wavelength λ for different values of temperature

(ii) For the long wavelengths λ which are situated in the far infrared (see Fig. 3.8), $exp(C_2/\lambda T)$ becomes very small compared to 1 allowing making the following approximation:

$$e^\varepsilon \approx 1 + \varepsilon \quad \text{for} \quad \varepsilon \text{ very small} \qquad (3.25)$$

namely:

$$M_\lambda^0(T) = \frac{C_1 \lambda^{-5}}{e^{\frac{C_2}{\lambda T}} - 1} \approx \frac{C_1 \lambda^{-5}}{1 + \frac{C_2}{T} - 1} \approx \frac{C_1 T}{C_2 \lambda^4}. \qquad (3.26)$$

Equation (3.26) can be applied in the far-infrared domain.

3.6   Wien's Laws

3.6.1   *Wien's First Law*

The first law of Wien tells us about the value of the wavelength λ_{max}, for which the monochromatic emittance is maximal (Fig. 3.9). To do this, it suffices to maximize the function by setting its derivative with respect to the wavelength λ equal to zero, i.e.:

$$\frac{M_\lambda^0(T)}{\lambda} = 0. \tag{3.27}$$

We obtain after calculations a relationship between the temperature T and the wavelength of the maximum of the emittance λ_{max} (the demonstration will be given later):

$$\lambda_{max} T = C^{te} = 2880 \; \mu m.K. \tag{3.28}$$

The spectral emittance thus presents a maximum for a value λ_{max} which obeys a simple law called *the first law of Wien or the Wien displacement law*, with λ_{max} in μm and T in kelvins. This law expresses the fact that for a black body, the product of the temperature and the wavelength of the peak of the curve $M_\lambda^0 = f(T)$ is always equal to a constant.

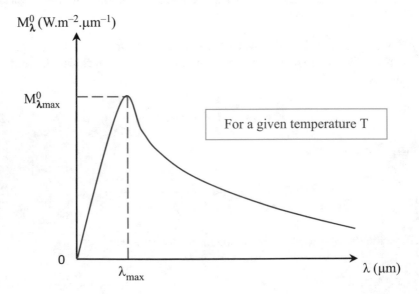

Fig. 3.9 For a given temperature T and a wavelength λ_{max}, the monochromatic emittance M_λ^0 reaches its maximum

Table 3.1 Some values of λ_{max} and T of which the product is 2880 µm.K

Temperature (K)	Maximum wavelength (µm)
300	9.60 (infrared)
600	4.80
900	3.20
1500	1.92
2500	1.15
4000	0.72
5760	0.50 (solar radiation)

The first law of Wien allows knowing the temperature of a body assimilated to a black body by the only position of the maximum of the spectral emittance and to understand the radiation-related phenomena, such as the greenhouse effect. Table 3.1 shows that a body at room temperature (300 K) mainly emits the infrared radiation. It is precisely the radiation that is present around us, but that we cannot perceive because our eye is sensitive only to the radiation of which the wavelength is between 0.4 and 0.7 µm (visible). By comparison, the sun of which the surface temperature is of the order of 5790 K sends us the maximum of radiation for a wavelength of 0.5 µm (yellow) in the visible range. This is exactly the wavelength for which our eye has a maximal efficiency. In addition, this law indicates that the total emitted amount of energy increases with the black body temperature: the hotter the body, the more it emits the electromagnetic energy. We also notice that the maximum of energy is emitted for a wavelength which decreases with the temperature. This means that when the temperature of a body rises, it emits first in the infrared. Then, when it begins to emit in the visible, this body is first reddening, then turns to yellow, then to white. When the temperature becomes very intense, as in some stars, the emitted light is mainly in short wavelengths, and the object seems to us blue: a blue star is therefore hotter than a red star!.

3.6.1.1 Demonstration 1 of the First Wien Law

In the Planck formula (see (3.21)), we have:

$$\frac{dM^0}{d\lambda} = M_\lambda^0(T) = \frac{C_1\lambda^{-5}}{\frac{C_2}{e^{\lambda T}} - 1}.$$

(3.29)

Let us put:

$$U = \frac{C_2}{\lambda T},$$

(3.30)

namely:

$$\lambda = \frac{C_2}{UT} \;\Rightarrow\; \lambda^5 = \frac{C_2^5}{U^5 \, T^5}. \tag{3.31}$$

By injecting (3.31) in (3.29), we obtain:

$$M_\lambda^0(T) = \frac{C_1}{\lambda^5 \left(e^{\frac{C_2}{\lambda T}} - 1 \right)} = \frac{C_1}{\dfrac{C_2^5}{U^5 \, T^5} \left(e^U - 1 \right)} = \frac{C_1 T^5}{C_2^5} \, \frac{U^5}{e^U - 1}. \tag{3.32}$$

Let us derive with respect to U, and cancel the obtained expression because we are at a maximum of the curve; we obtain:

$$\frac{dM_\lambda^0(T)}{dU} = \left(\frac{C_1 T^5}{C_2^5} \right) \left[\frac{5U^4(e^U - 1) - e^U \, U^5}{(e^U - 1)^2} \right] = 0. \tag{3.33}$$

As:

$$\frac{C_1 T^5}{C_2^5} > 0 \;\; \text{and} \;\; U \in \,]0; \infty[\tag{3.34}$$

and by setting the numerator of the function U in brackets equal to zero, we obtain:

$$5U^4 \left(e^U - 1 \right) - e^U \, U^5 = 0 \;\Rightarrow\; 5 - 5e^U = U. \tag{3.35}$$

The numerical solution by means of a calculator or by the point-by-point plot gives $U = 4.9651$, namely:

$$\frac{C_2}{\lambda_{\max} T} = 4.9651 \;\; \text{and therefore,} \;\; \lambda_{\max} T \approx 2880 \;\mu m.K. \tag{3.36}$$

3.6.1.2 Demonstration 2 of the First Wien Law

In the visible domain $(0.4 - 0.7 \;\mu m)$, λ is very small, so $exp(C_2/\lambda T)$ is very large compared to 1 (see the denominator of Eq. (3.29)). This allows us to make the same approximation as in Eq. (3.24), namely:

$$M_\lambda^0(T) \approx C_1 \lambda^{-5} e^{-\frac{C_2}{\lambda T}}. \tag{3.37}$$

By derivation with respect to λ and by setting of the derived expression equal to zero, we find:

$$\frac{dM_\lambda^0(T)}{d\lambda} \approx \frac{d}{d\lambda}\left(C_1\lambda^{-5}e^{-\frac{C_2}{\lambda T}}\right) = \frac{C_1}{e^{\frac{C_2}{\lambda T}}}\left(\frac{-5\lambda + C_2}{T\lambda^7}\right) = 0. \tag{3.38}$$

As:

$$\frac{C_1}{e^{\frac{C_2}{\lambda T}}} \neq 0 \tag{3.39}$$

it remains to set the numerator of the second factor equal to zero:

$$-5\lambda + C_2 = 0 \quad \Rightarrow \quad \lambda_{max}T = \frac{C_2}{5}. \tag{3.40}$$

But (see (3.22)):

$$\frac{C_2}{5} = h\frac{c}{5k_B} \tag{3.41}$$

we finally get:

$$\lambda_{max}T = \frac{hc}{5k_B} = \frac{(6.626 \times 10^{-34} \text{ J.s}) \left(2.998 \times 10^8 \text{ m.s}^{-1}\right)}{5 \left(1.38 \times 10^{-23} \text{ J.K}^{-1}\right)} \tag{3.42}$$
$$= 2.879 \times 10^{-3} \text{ m.K} \approx 2880 \text{ µm.K}.$$

3.6.2   Wien's Second Law

The second Wien law gives the relationship between the maximal monochromatic emittance $M_{\lambda max}^0$ and the body temperature T, namely:

$$M_{\lambda max}^0(T) = BT^5 \tag{3.43}$$

with T in kelvins and $B = 1.28 \times 10^{-5}$ W.m^{-3}.K^{-5} if λ_{max} is in meters. This law reflects the importance of the temperature with respect to the emittance. Thus, the ratio of the maximal emittance of the sun ($T \approx 6000$ K) and a body at room temperature (300 K) is 3×10^6.

3.6.2.1   Demonstration

The demonstration of Eq. (3.43) is easy. Indeed, the vertical coordinate of the extremum $M_{\lambda max}^0$ is determined by writing that $\lambda = \lambda_{max}$ in the Planck law, namely:

$$M^0_{\lambda_{\max}}(T) = \frac{C_1 \lambda_{\max}^{-5}}{e^{\frac{C_2}{\lambda_{\max}T}} - 1}.$$ (3.44)

Let us divide the two members of the previous equality by T^5, and it becomes:

$$\frac{M^0_{\lambda_{\max}}(T)}{T^5} = \frac{C_1 \lambda_{\max}^{-5} T^{-5}}{e^{\frac{C_2}{\lambda_{\max}T}} - 1}.$$ (3.45)

Knowing that $\lambda_{\max}T = C^{te} = 2880$ μm. K (the first Wien law), we finally obtain:

$$M^0_{\lambda_{\max}}(T) = BT^5.$$ (3.46)

3.7   The Stefan-Boltzmann Law

The Planck law describes the radiation rate of a black body in the vacuum as a function of its temperature T and the wavelength λ of the observed radiation, while the Stefan-Boltzmann law describes the emitted heat rate of the radiation from a black body. This law was discovered in an independent way, firstly by means of the experiments of Joseph Stefan in 1879 and then theoretically by Ludwig Boltzmann in 1884. It can be obtained by summing all the monochromatic emittances for all the wavelengths:

$$M^0 = \int_0^\infty M^0_\lambda(T) \, d\lambda.$$ (3.47)

We obtain after integration:

$$M^0 = \sigma_0 T^4$$ (3.48)

where T is in K and $\sigma_0 = 5.68 \times 10^{-8}$ W.m^{-2}.K^{-4}. σ_0 is called *the Stefan-Boltzmann constant*. This law is often written in the following form:

$$M^0 = \varepsilon \, \sigma_0 T^4$$ (3.49)

where ε is the emissivity ($\varepsilon = 1$ for a black body and $\varepsilon < 1$ for a real body which will be defined later). The sun is assimilated to a black body of temperature 5800 K and of emittance equal to 6.4×10^7 W.m^{-2}.

3.7.1 Demonstration

Let us calculate the total emittance M^0 of the black body by integrating the mono-chromatic emittance $M_\lambda^0(T)$:

$$M^0 = \int_0^\infty M_\lambda^0(T)d\lambda = \int \frac{C_1\lambda^{-5}}{e^{\frac{C_2}{\lambda T}} - 1}\, d\lambda. \tag{3.50}$$

Let us put:

$$X = \frac{C_2}{\lambda T}. \tag{3.51}$$

This gives:

$$\lambda = \frac{C_2}{X T} \quad \Rightarrow \quad d\lambda = -\frac{C_2}{X^2 T}\, dX. \tag{3.52}$$

The integral (3.50) becomes:

$$M^0 = -\int_{+\infty}^0 C_1 \left(\frac{XT}{C_2}\right)^5 \frac{1}{e^X - 1} \frac{C_2}{X^2 T}\, dX = C_1 \left(\frac{T}{C_2}\right)^4 \int_0^{+\infty} \frac{X^3}{e^X - 1} dX. \tag{3.53}$$

We obtain (the calculation of this integral will be given in the form of an exercise):

$$\int_0^{+\infty} \frac{X^3}{e^X - 1}\, dX = \frac{\pi}{15}. \tag{3.54}$$

By injecting (3.54) into (3.53), we find:

$$M^0 = C_1 \left(\frac{T}{C_2}\right)^4 \frac{\pi}{15}. \tag{3.55}$$

Knowing that (see (3.22)):

$$C_1 = 2hc^2\pi \quad \text{and} \quad C_2 = h\frac{c}{k_B} \tag{3.56}$$

we finally arrive at:

$$M^0 = \frac{2\pi^5 k_B^4}{15c^2 h^3} T^4 = \sigma_0 T^4. \tag{3.57}$$

3.8   Useful Interval of Radiation

If the Planck law allows to easily calculate the monochromatic emittance of a body for a given temperature T and a wavelength λ, it is often useful to know the energy flux emitted by a body at a temperature T in a wavelength range $d\lambda = \lambda_1 - \lambda_2$. This flux is generally expressed with respect to the total flux that is given by the total emittance fraction $F_{\lambda_1 - \lambda_2}$, defined by:

$$F_{\lambda_1 - \lambda_2} = \frac{\int_{\lambda_1}^{\lambda_2} M_\lambda^0(T) d\lambda}{M^0} = \frac{1}{M^0} \left(\int_{\lambda_1}^{0} M_\lambda^0(T)\, d\lambda + \int_{0}^{\lambda_2} M_\lambda^0(T)\, d\lambda \right)$$

$$= \frac{1}{M^0} \left(\int_{0}^{\lambda_2} M_\lambda^0(T)\, d\lambda - \int_{0}^{\lambda_1} M_\lambda^0(T)\, d\lambda \right) = F_{0 - \lambda_2} - F_{0 - \lambda_1}. \tag{3.58}$$

This fraction of the total emittance allows defining the useful range of radiation as being the spectral range between $0.5\ \lambda_{\max}$ and $5\ \lambda_{\max}$ ($\lambda_{\max}$ is the wavelength of the maximum of the monochromatic emittance). In this interval, the black body emits 96% of its radiation. $F_{0 - \lambda_1}$ represents the fraction of the emittance corresponding to the wavelengths lower than λ_1 for a given temperature T. It therefore depends on two parameters λ and T. Although this relationship is programmable, there are tables that give its value for different temperatures. In this context, it would be useful to have a table providing, for each temperature, the value of the following integral:

$$F_{0 - \lambda} = \frac{\int_{0}^{\lambda_2} M_\lambda^0(T) d\lambda}{M^0} = \frac{\int_{0}^{\lambda_2} M_\lambda^0(T) d\lambda}{\sigma_0\, T^4} = \frac{1}{\sigma_0} \int_{0}^{\lambda} \frac{C_1}{\lambda^5 T^4 \left(e^{\frac{C_2}{\lambda T}} - 1 \right)}. \tag{3.59}$$

The disadvantage is that it would be necessary to have a table with two inputs (T and λ). To remedy this, we use the product λT as a unique variable, and we put:

$$F_{0 - (\lambda T)} = \frac{1}{\sigma_0} \int_{0}^{\lambda T} \frac{C_1 d(\lambda T)}{(\lambda T)^5 \left(e^{\frac{C_2}{\lambda T}} - 1 \right)} \tag{3.60}$$

and we thus obtain a table with a unique entry on the variable (λT). We then calculate the energy emitted by a black body at the temperature T between λ_1 and λ_2 by:

$$F_{\lambda_1 - \lambda_2} = F_{0 - (\lambda_2 T)} - F_{0 - (\lambda_1 T)}. \tag{3.61}$$

An example of the calculation will be given in the form of an exercise.

3.9   Ultraviolet Catastrophe

The problem of the black body is a problem that was at the origin of the quantum revolution due to the ultraviolet catastrophe that it generated. Indeed, physicists cared about radiation laws of the black body from the beginning of the twentieth century and calculated, under the terms of classical physics laws, the energy density radiated in each wavelength range:

$$E(\lambda, T) \, d\lambda = \frac{8\pi k_B \, T}{\lambda^4} \, d\lambda \qquad (3.62)$$

where $E(\lambda,T) \, d\lambda$ was the energy emitted in the interval λ and $\lambda + d\lambda$ from a black body at the temperature T, and k_B was Boltzmann's constant. The theoretical and experimental curves coincided reasonably well for the short frequencies, but for the high frequencies, there was no agreement between them. The experimental values were much lower than the theoretical ones which should tend towards infinity. It is what we called at the time *ultraviolet catastrophe*. It was necessary to introduce quantum physics to explain this difference and to find the way to match the theory and the experiment. This is Planck who obtained for the first time a good agreement between the theory and the experiment, assuming that the electromagnetic energy, instead of being continuous as in the classical theory, was in fact discontinuous and therefore could only take discrete values, multiple of hc/λ, where c was the speed of light in vacuum and h was Planck's constant. What was only a calculation artifice allowed finding a formula that perfectly corresponded to the experiment: this was the Planck law. This law, as we could expect it, gave the classical formula if h tended to zero, namely, if we considered the electromagnetic energy as continuous. Thus, taking into account the discrete nature of energy is essential in order to find the right expression of the energy density emitted by a black body, but if the quantization of the energy is decisive, and knowing that the spectra of the emission of different materials are very varied, why does the radiation law of the black body depend only on the temperature of the material and not on its nature? In fact, we must remember that there are two kinds of discrete characters: the discrete character of energy levels of atoms, which is at the origin of the discrete character of the atomic emission spectra, and which depends on the considered material, and that of the radiation, which consists of interpreting the radiation as consisting of photons, each containing a quantum of energy hc/λ, and which does not depend on the nature of the considered material. The Planck law describes the radiation at thermal equilibrium, independently of materials. Thus, in the radiation law of the black body, the quantization of energy corresponds to *energy packets*, multiples of hc/λ which is none other than the energy of one photon. It is thus the radiation that is quantized here and not the quantization of energy levels of the black body.

3.10   Real Body

3.10.1   Introduction

The black body is the standard of radiation; the evaluation of the emission of a real body is thus compared to that of a black body placed under the same conditions of temperatures and wavelengths. This comparison is established using global hemispheric and monochromatic coefficients called *emissivities*. The emissivity of real bodies depends on the nature of the surface (dielectric, conductor, etc.), the state of surfaces (defect of flatness, roughness, etc.), and the chemical state of these surfaces (oxidation, grease, paint, etc.).

3.10.2   Definition of Different Emissivities

3.10.2.1   Monochromatic Hemispheric Emissivity

The monochromatic hemispheric emissivity is given by:

$$\varepsilon_\lambda = \frac{M_\lambda}{M_\lambda^0}. \tag{3.63}$$

3.10.2.2   Global Hemispheric Emissivity

The global hemispheric emissivity is given by:

$$\varepsilon_\lambda = \frac{M}{M^0} = \frac{\int_0^\infty M_\lambda d\lambda}{\sigma_0 T^4} = \frac{\int_0^\infty M_\lambda d\lambda}{\sigma_0 T^4} = \frac{\int_0^\infty \varepsilon_\lambda M_\lambda^0 d\lambda}{\sigma_0 T^4}. \tag{3.64}$$

This emissivity can also depend on the direction. It is then called *directional* ε_{Ox}.

3.10.3   Gray Body

If for the real body the emissivity is constant whatever are the wavelength and the direction of the emission, this body is called *gray* (Fig. 3.10). When it is not possible to consider the body as gray for the whole spectrum, this approximation can be reduced to wide bands of the considered spectrum. It will then be necessary to use the total emittance fraction to quantify the emission in various emission bands.

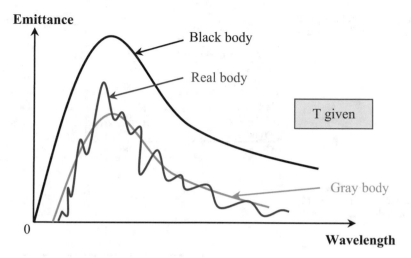

Fig. 3.10 Spectra of black, gray, and real bodies

3.10.4   *Usual Real Body Emissivities*

The energetic properties (luminance, emittance) of a real body are deduced from those of the black body having the same temperature by a simple multiplication by emissivity. The thermal study allows highlighting two classes of the radiative behavior.

A. *Electrically conductive materials*: The monochromatic emissivity of these bodies decreases as the wavelength increases.
B. *Dielectric materials*: The monochromatic emissivity increases with the wavelength in the near infrared. These bodies follow the Lambert law.

   In practice, we look for the values of these emissivities in tables. These give the global or hemispheric emissivity rarely the monochromatic or directional emissivity. Moreover, these parameters vary according to the surface state of materials and the temperature (Table 3.2).

3.10.5   *Absorption of Real Bodies*

When a radiation reaches the surface of a body, one part is reflected, a second is transmitted, and a third is absorbed by the mass of the receiver (Fig. 3.11). Only this last part corresponds to the energy contribution of the body.
   The different coefficients that characterize this physical phenomenon are:

– Reflection coefficient ρ, defined by:

Table 3.2 Global emissivity of some materials for several temperatures

Temperature	20 °C	250 °C	500 °C
Polished stainless steel	0.15	0.18	0.22
Altered stainless steel	0.85	0.85	0.85
Polished iron	0.06	0.09	0.14
Oxidized iron	0.80	0.80	0.80
Carbon	0.95		
Water	0.96		
Wood	0.90		
Glass	0.90		

Fig. 3.11 Incident E, reflected ρE, transmitted τE, and absorbed αE fluxes

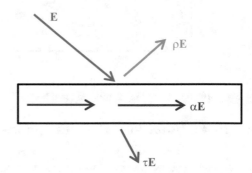

$$\rho = \frac{\Phi_r}{E} = \frac{\text{Reflected flux}}{\text{Illumination}}. \tag{3.65}$$

– Absorption coefficient α, defined by:

$$\alpha = \frac{\Phi_a}{E} = \frac{\text{Absorbed flux}}{\text{Illumination}}. \tag{3.66}$$

– Transmission coefficient τ, defined by:

$$\tau = \frac{\Phi_r}{E} = \frac{\text{Transmitted flux}}{\text{Illumination}}. \tag{3.67}$$

These three coefficients are linked by the following equation:

$$\alpha + \rho + \tau = 1. \tag{3.68}$$

These coefficients strongly depend on the wavelength. This is the case of the glass that lets pass the short wavelengths and absorbs the long ones. This phenomenon is at the basis of the explanation of the greenhouse effect. These coefficients also depend on the direction of the radiation.

3.10.6 Kirchhoff's Law

There is a relationship between emissivities ε and absorptivities α. This law is called *the Kirchhoff law*. It indicates that for each wavelength and each direction of space, monochromatic emissivities and absorptivities are equal:

$$\varepsilon_{Ox,\lambda} = \alpha_{Ox,\lambda}. \tag{3.69}$$

In the case of an illumination and an emission obeying the Lambert law, the Kirchhoff law can be extended to hemispherical monochromatic quantities:

$$\varepsilon_\lambda = \alpha_\lambda. \tag{3.70}$$

In the particular case of gray bodies, this law is reduced to:

$$\varepsilon = \alpha. \tag{3.71}$$

It is often used not for the entire spectrum, but for useful spectral bands.

3.10.6.1 Demonstration

We consider a small body placed in a closed enclosure and perfectly isolated from the outside. This body is in thermal equilibrium with the enclosure at the same temperature T. The radiative flux emitted by an elementary surface of the body in the direction of the enclosure is strictly equal to the radiative flux absorbed by this elementary surface coming from the enclosure. In other words, the walls of the enclosure behave like black bodies.

3.11 Radiative Exchanges

3.11.1 Radiative Exchanges Between Black Surfaces

3.11.1.1 View Factors

Let us consider two elementary surfaces dS_1 and dS_2, respectively, belonging to surfaces (S_1) and (S_2) and let us write the expression of the total flux emitted by dS_1 and arriving on dS_2 (Fig. 3.12):

$$d\Phi_{12} = \frac{L_1^0 dS_1 \cos\theta_1 dS_2 \cos\theta_2}{d^2}. \tag{3.72}$$

The elementary surface dS_1 obeys the Lambert law, namely:

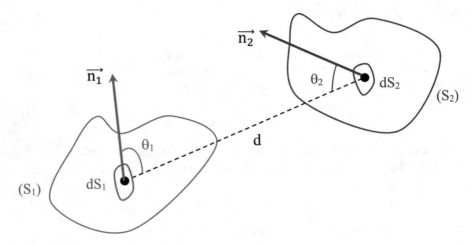

Fig. 3.12 Radiative exchanges between black surfaces

$$L_1^0 = \frac{M_1^0}{\pi}. \tag{3.73}$$

To know the total flow emitted by (S_1) and arriving on (S_2), we integrate on (S_2) then on (S_1):

$$\Phi_{12} = \int_{(S_1)} \int_{(S_2)} \frac{M_1^0 dS_1 \cos\theta_1 \, dS_2 \cos\theta_2}{d^2}. \tag{3.74}$$

Knowing that the total emittance in $W.m^{-2}$ is by definition (see (3.8)):

$$M_1^0 = \frac{\Phi_1}{S_1} \quad \Rightarrow \quad \Phi_1 = M_1^0 S_1 \tag{3.75}$$

we finally obtain the proportion of the total flux emitted by the surface (S_1) and arriving at (S_2):

$$F_{12} = \frac{\Phi_{12}}{\Phi_1} = \frac{\Phi_{12}}{M_1^0 S_1} = \frac{1}{S_1} \int_{(S_1)} \int_{(S_2)} \frac{dS_1 \cos\theta_1 \, dS_2 \cos\theta_2}{d^2}. \tag{3.76}$$

From (3.76), we notice that the proportion of the total flux emitted by (S_1) and arriving at (S_2) appears as a purely geometrical quantity. It is called *view factor* or *configuration factor* or *form factor* or *shape factor* of the surface (S_2) towards the surface (S_1). By the same reasoning, it is possible to determine F_{21}, the proportion of the flux emitted by (S_2) and arriving at the surface (S_1), namely:

$$F_{12} = \frac{1}{S_2} \int_{(S_2)} \int_{(S_1)} \frac{dS_2 \cos\theta_2 \, dS_1 \cos\theta_1}{d^2}.$$ (3.77)

3.11.1.2   Relationships Between View Factors

(i) *Reciprocity relationship*:

The symmetry of expressions F_{12} and F_{21} leads to the reciprocity of view factors:

$$S_1 \, F_{12} = S_2 \, F_{21}.$$ (3.78)

(ii) *Total influence*:

When all the radiation coming from the surface (1) reaches the surface (2), we say there is a total influence:

$$F_{12} = 1.$$ (3.79)

This property can be generalized by considering a closed volume consisting of n isothermal surfaces. The flux emitted by the surface (i) reaches whatever its trajectory in a surface (j) with $i \leq j \leq n$. We obtain:

$$\sum_{j=1}^{j=n} F_{ij} = 1.$$ (3.80)

(iii) *Calculation of view factors*:

For general configurations, the estimation of view factors is very complex. However, some methods simplify the calculation such as the Ondracek method, the Monte Carlo method, and the use of formulas and abacuses. Moreover, the use of the reciprocal and total influence relationships allows a simplification of these calculations in certain configurations.

3.11.2   *Concept of the Net Flux*

All surfaces emit and receive the radiation which is totally absorbed in the case of the black body. If Φ_i is the emitted flux and $\sum_{j=1}^{n} \Phi_{ji}$ the received flux coming from other surfaces, the net flux relative to the surface (S_i) corresponds to the difference of the emitted flux minus the absorbed flux by this surface, namely:

$$\Phi^i_{net} = \Phi_i - \sum_{j=1}^{n} \Phi_{ji} = \Phi_i - \sum_{j=1}^{n} F_{ji}\Phi_j \qquad (3.81)$$

because according to (3.73), we had for the surfaces 1 and 2:

$$F_{12} = \frac{\Phi_{12}}{\Phi_1} \quad \Rightarrow \quad \Phi_{12} = F_{12}\Phi_1 \qquad (3.82)$$

which we generalize for any surfaces i and j as follows:

$$\Phi_{ji} = F_{ji}\,\Phi_i \qquad (3.83)$$

thus justifying the writing of the third member of equality (3.81). Since these surfaces are black, these fluxes can be expressed as a function of emittances as follows:

$$\Phi^i_{net} = M^0_i S_i - \sum_{j=1}^{n} F_{ji}\, M^0_j S_j \qquad (3.84)$$

because according to (3.8), we had for the surface 1:

$$M^0_1 = \frac{\Phi_1}{S_1} \quad \Rightarrow \quad \Phi_1 = M^0_1\, S_1 \qquad (3.85)$$

which we can generalize for any surface i as follows:

$$\Phi_i = M^0_i\, S_i \qquad (3.86)$$

thus justifying the writing of the second member of equality (3.84). The relationships of the reciprocity and the total influence allow us to write:

$$S_j\, F_{ji} = S_i\, F_{ij} \quad \text{and} \quad \sum_{j=1}^{n} F_{ij} = 1. \qquad (3.87)$$

By injecting (3.87) into (3.84), we find:

$$\begin{aligned} \Phi^i_{net} &= M^0_i S_i \sum_{j=1}^{n} F_{ij} - \sum_{j=1}^{n} F_{ij}\, M^0_j S_i \\ &= \sum_{j=1}^{n} M^0_i S_i F_{ij} - \sum_{j=1}^{n} F_{ij}\, M^0_j S_i. \end{aligned} \qquad (3.88)$$

In the end, we arrive at:

$$\Phi^i_{net} = \sum_{j=1}^{n} S_i F_{ij}\left(M^0_i - M^0_j \right). \qquad (3.89)$$

Equation (3.89) is very important for two reasons. The first is that according to the sign of Φ^i_{net}, it is possible to know if the black surface loses energy by radiation

Table 3.3 Analogue representation of thermal and electrical exchanges

Radiation	Electricity
Net flux exchanged by radiation between two black surfaces (S_1) and (S_2) $$\Phi = S_1 F_{12} \left(M_1^0 - M_2^0 \right)$$	Intensity of the electric current which is established between two nodes of potentials V_1 and V_2 connected by the resistance R_{12} $$I = \frac{1}{R_{12}} \left(V_1 - V_2 \right)$$
Difference between the emittances of two black surfaces (S_1) and (S_2) $$M_1^0 - M_2^0$$	Potential difference between two nodes (1) and (2) of the circuit $$V_1 - V_2$$
View factor – surface $$S_1 F_{12}$$	Conductance between two nodes (1) and (2) at potentials V_1 and V_2 connected by the resistance R_{12} $$K = \frac{1}{R_{12}}$$

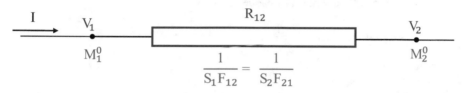

Fig. 3.13 Analogue schematization between two black surfaces

$\left(\Phi_{net}^i > 0 \right)$ or if it gains it $\left(\Phi_{net}^i < 0 \right)$ or if the energy losses are compensated by the gains $\left(\Phi_{net}^i = 0 \right)$. In the latter case, the surface is said to be *adiabatic* with respect to the external environment. The second reason is that this expression recalls the expression of the electric current which flows in an electrical circuit:

$$I_{ij} = \frac{V_i - V_j}{R_{ij}} \tag{3.90}$$

where I_{ij} is the intensity of the electric current, $V_i - V_j$ the potential difference between two nodes (i) and (j) of the circuit, and R_{ij} the electrical resistance. This observation leads us to propose an analogue representation of the thermal and electrical exchanges that can be summarized as follows (Table 3.3, Fig. 3.13).

3.11.3 Radiative Exchanges Between Gray Opaque Surfaces

The case of radiative exchanges between gray opaque surfaces is more complex than that of black surfaces because gray opaque surfaces reflect the radiation, but do not transmit it. To take this phenomenon into account, a new notion is introduced, namely, *radiosity*, noted in the following J. The radiosity of a gray opaque surface is defined as the sum of the emitted flux and the reflected flux per unit area, i.e., the flux that leaves the surface (Fig. 3.14). It is expressed in W/m^2.

$$J = \varepsilon M^0 + \rho E \qquad (3.91)$$

where J is the radiosity, εM^0 the flux emitted by the gray opaque surface with ε the emissivity and M^0 the emittance, and ρE the flux reflected by the gray opaque surface with E the illumination and ρ the coefficient of the reflection. The net flux of a gray opaque surface (S_i) is written as:

$$\Phi^i_{net} = \varepsilon M^0_i S_i - \varepsilon E \, S_i. \qquad (3.92)$$

Here $\varepsilon M^0 S_i$ is the emitted flux and $\alpha E \, S_i$ the absorbed flux, both expressed in W (we recall that for opaque surfaces, there is no flux transmission ($\tau = 0$)). The introduction of the radiosity quantity is of great interest to gray surfaces where its coefficients have values independent of the nature of the emitted and incident radiation. We can then write for opaque surfaces:

$$\alpha + \rho + \tau = 1 \;\; \Rightarrow \;\; \rho = 1 - \alpha = 1 - \varepsilon \;\; \Rightarrow \;\; \varepsilon = \alpha. \qquad (3.93)$$

It becomes:

$$\Phi^i_{net} = \varepsilon S_i \left(M^0_i - E \right). \qquad (3.94)$$

According to (3.91) and (3.93), we obtain:

$$J_i = \varepsilon M^0_i + \rho E \;\; \Rightarrow \;\; E = \frac{J_i - \varepsilon M^0_i}{\rho} = \frac{J_i - \varepsilon M^0_i}{1 - \varepsilon}. \qquad (3.95)$$

By injecting (3.93) into (3.92), we get the expression of the net flux as a function of the radiosity:

$$\Phi^i_{net} = \varepsilon S_i \left(M^0 - \frac{J_i - \varepsilon M^0_i}{1 - \varepsilon} \right) = \frac{\varepsilon}{1 - \varepsilon} \left(M^0_i - J_i \right) S_i. \qquad (3.96)$$

The exchange between opaque gray surfaces (S_i) and (S_j) can also be expressed according to their respective radiosities J_i and J_j. Indeed, the surface (S_i) sends a radiative flux throughout the $S_i J_i$ space. Only the fraction $F_{ij} S_i J_i$ reaches the surface

Fig. 3.14 Illuminance E, emitted flux εM^0, reflected flux ρE, and absorbed flux αE

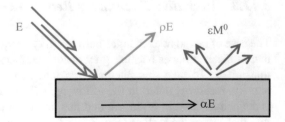

(S_j). Reciprocally, from the surface (S_j), it arrives at (S_i) the fraction $F_{ij}S_jJ_j$. The net flux exchanged between the surfaces (S_i) and (S_j) is then:

$$\Phi_{net}^{ij} = S_iF_{ij}J_i - S_iF_{ij}J_j = S_iF_{ij}(J_i - J_j). \tag{3.97}$$

This equation is similar to Eq. (3.89) which describes the radiative exchanges between black surfaces if we replace the radiosity J by the emittance M^0.

3.12   Solved Exercises

Exercise 1
By using polar coordinates, calculate the solid angle under which we see a crown.

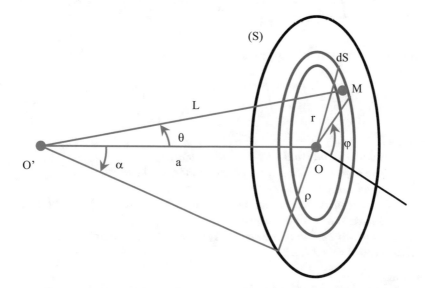

Solution
Let us consider a crown of the radius ρ, of the center O, seen by an observer placed on O'. Let us look for the angle delimited by the cone of the summit O' and the disk of the surface (S). Let $L = O'M$ and $a = O'O$ (see figure). The solid angle under which we see the surface (S) from the point O' is given by:

$$\Omega = \iint_{(S)} \frac{dS\cos\theta}{L^2} = \int_\varphi \int_r \frac{r\,d\varphi\,dr\cos\theta}{L^2}. \tag{1}$$

By looking at the figure, we see that all the surface elements dS are equidistant from O' and their perpendiculars make the same angle θ with the radius vector, namely:

$$r = a\tan\theta \quad \text{and} \quad a = L\cos\theta. \tag{2}$$

We differentiate dr:

$$dr = a\frac{d\theta}{cos^2\theta}. \tag{3}$$

By injecting (3) and (2) into (1), we find:

$$\Omega = \int_\varphi \int_\theta \frac{(a\tan\theta)\,d\varphi \left(a\dfrac{d\theta}{cos^2\theta}\right)\cos\theta}{\left(\dfrac{a}{\cos\theta}\right)^2} = \int_\varphi \int_\theta \sin\theta\,d\varphi\,d\theta. \tag{4}$$

By integrating φ from 0 to 2π and θ from 0 to α, we finally arrive at:

$$\Omega = \int_{\varphi=0}^{\varphi=2\pi} \int_{\theta=0}^{\varphi=\alpha} \sin\theta\,d\theta\,d\varphi = 2\pi(1 - \cos\alpha). \tag{5}$$

Exercise 2
1. *What is a black body and why is it called black?*
2. *Quote five objects which have a behavior close to the black body.*
3. *Can a black body be colored?*
4. *How can scientists claim to know the temperature of a star without setting foot there to make a direct measurement?*
5. *What is the wavelength of the emissivity peak of the human body at $37\,°C$ and in what spectral range is it?*

Solution
1. In physics, a black body designates an ideal object of which the emitted electro-magnetic wave spectrum depends only on its temperature T. Indeed, at thermal equilibrium, the emission and the absorption are balanced and the radiation really emitted depends only on the temperature (thermal radiation). It is called *black* because it absorbs all the received radiation without any reflection or transmission.
2. An oven, a star, a light bulb, the pupil of the human eye, and a cavity with a tiny hole have a behavior close to that of the black body.
3. Yes, a black body can be colored. Indeed, the Sun, for example, is considered a black body. At low spectral resolution, the Sun spectrum is superimposed on that of a black body of temperature $T = 5777$ K.
4. The answer comes from the analysis of the distribution of the luminous intensity of the spectrum of the star. From this analysis, we find the value of the wavelength

of the maximal light intensity λ_{max}, and from there, we lead to the temperature of the star through the first law of Wien:

$$T(K) = \frac{2880\,(\mu m.K)}{\lambda_{max}(\mu m)}. \tag{1}$$

5. We consider the human body as a black body and we apply the first law of Wien:

$$\lambda_{max}(m) = \frac{2880\,(\mu m.K)}{T\,(K)} = \frac{2880 \times 10^{-6}(m)}{(37 + 273)(K)} = 9.29 \times 10^{-6}\,m = 9.29\,\mu m. \tag{2}$$

This wavelength is in the infrared range of the electromagnetic spectrum.

Exercise 3

1. *Why do sunspots appear black with regard to the surrounding atmosphere?*
2. *Is a red star warmer than a blue star?*

Solution

1. The sunspots seem black relative to the surrounding atmosphere because they are cooler than the surrounding atmosphere. Indeed, the hotter a body, the more it radiates and the more it loses energy for the benefit of the external environment.
2. No, a star radiates like a black body; an equilibrium temperature leading to the red color will always be lower than that of a black body of blue temperature.

Exercise 4

Calculate in two ways the monochromatic emittance of a tungsten filament of temperature $T = 1403$ K. Conclude. We give:
- *Constant of Wien $B = 1.28 \times 10^{-5}$ $W.m^{-3}.K^{-5}$*
- *Speed of light $c = 2.998 \times 10^{8}$ $m.s^{-1}$*
- *Planck's constant 6.63×10^{-34} J.s*
- *Boltzmann's constant $k_B = 1.38 \times 10^{-23}$ J/K*

Solution

1. *First method*:

   The monochromatic emittance is given by the second law of Wien, namely:

   $$M^0_{\lambda\,max}(T) = B\,T^5 = \left(1.28 \times 10^{-5}\,W.m^{-3}.K^{-5}\right)\left(1403^5\,K^5\right)$$
   $$= 6.96 \times 10^{10}\,W.m^{-3} \tag{1}$$

if λ_{max} is expressed in meters.

2. *Second method*:

Let us first look for the value of the maximal wavelength by using the first law of Wien:

$$\lambda_{max}(\mu m) = \frac{2880\ (\mu m.K)}{T\ (K)} = 2.053\ \mu m = 2.053 \times 10^{-6}\ m. \tag{2}$$

The monochromatic emittance of the tungsten filament can be directly calculated from the Planck law:

$$\frac{dM^0}{d\lambda} = M_\lambda^0\ T = \frac{C_1 \lambda^{-5}}{e^{\frac{C_2}{\lambda T}} - 1} \tag{3}$$

with λ, the wavelength in meters and T, the temperature in kelvins. The coefficients C_1 and C_2 have the following expressions:

$$C_1 = 2hc^2\pi \quad \text{and} \quad C_2 = h\frac{c}{k_B}. \tag{4}$$

Here c is the speed of light in vacuum ($c = 2.998 \times 10^8$ m/s), h the Planck constant ($h = 6.63 \times 10^{-34}$ J.s), and k_B the Boltzmann constant ($k_B = 1.38 \times 10^{-23}$ J/K). The constants C_1 and C_2 have the following values:

$$C_1 = 3.74 \times 10^{-16}\ W.m^2 \quad \text{and} \quad C_2 = 1.44 \times 10^{-2}\ m.K. \tag{5}$$

The monochromatic emittance of the tungsten filament is:

$$M_\lambda^0(T) = \frac{\left(3.74 \times 10^{-16}\ W.m^2\right)\left(2.053 \times 10^{-6}\ m\right)^{-5}}{e^{\frac{(1.44 \times 10^{-2})}{(2.053 \times 10^{-6})(1403)}} - 1} = 6.96 \times 10^{10}\ W.m^{-3}. \tag{6}$$

The Planck law describes the variation of the monochromatic emittance over the whole spectral domain and allows calculating M_λ^0 for any wavelength λ including for λ_{max}, whereas the Wien laws are only valid for λ_{max}.

Exercise 5
To heat a room, a cylindrical radiator of diameter $D = 2$ cm and of length $L = 0.5$ m is used. This radiator radiates like a black body and emits through its side surface a flux of 1 kW. We neglect the exchanges by convection and conduction.
1. *Calculate in Celsius the temperature θ of the radiator.*
2. *Determine the wavelength λ_{max} for which the spectral energy density emitted from the radiator is maximal. In what spectral domain does it emit?*

3. *What should be the temperature of the radiator so that this wavelength is 2 μm? What would be the released flux? We give $\sigma_0 = 5.68 \times 10^{-8}$ W.m^{-2}.K^{-4} (constant of Stefan-Boltzmann).*

Solution

Data:

- Diameter of the cylindrical radiator: $D = 2$ cm $= 2 \times 10^{-2}$ m
- Length of the cylindrical radiator: $L = 0.5$ m $= 5 \times 10^{-1}$ m
- Flux radiated from the lateral surface of the radiator: $\Phi^0 = 1$ kW $= 1000$ W
- Side surface of the cylindrical radiator:

$$S = 2\pi \frac{D}{2} L = (2)(3.14)\left(\frac{2 \times 10^{-2} \text{ m}}{2}\right)(5 \times 10^{-1} \text{ m}) = 3.14 \times 10^{-2} \text{ m}^2. \quad (1)$$

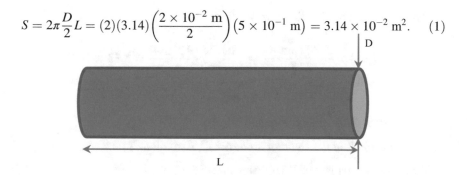

1. *Calculation of the temperature θ (in Celsius) of the radiator:*

Let T be the temperature of the radiator in kelvins, θ its temperature in degrees Celsius, and Tr the temperature of the heated room. There is a simple relationship between T and θ:

$$T(\text{K}) = \theta(°\text{C}) + 273. \quad (2)$$

If we neglect the heat absorbed by the radiator, which is very likely because $T \gg$ Tr, the only flux taken into account is that emitted from the radiator in the form of thermal radiation. According to the Stefan-Boltzmann law, we have:

$$M^0 = \frac{\Phi^0}{S} = \varepsilon \sigma_0 T^4 \quad \Rightarrow \quad \Phi^0 = \varepsilon \sigma_0 S T^4 \quad (3)$$

where ε is the emissivity ($\varepsilon = 1$ for a black body). From (3), we calculate the value of the radiator temperature in kelvins:

$$T = \left(\frac{\Phi_0}{\sigma_0 S}\right)^{\frac{1}{4}} = \left[\frac{1000 \ \text{W}}{\left(5.68 \times 10^{-8} \ \text{W.m}^{-2}.\text{K}^{-4}\right)\left(3.14 \times 10^{-2} \ \text{m}^2\right)}\right]^{\frac{1}{4}}$$

$$= 865.5 \ \text{K} \tag{4}$$

that is in degrees Celsius:

$$\theta(^\circ\text{C}) = T(\text{K}) - 273 = 865.5 - 273 = 592.5 \ ^\circ\text{C}. \tag{5}$$

2. *Calculation of the maximal wavelength*:

   According to the Wien first law, we find:

$$\lambda_{\max}(\mu\text{m}) = \frac{2880 \ (\mu\text{m.K})}{T(\text{K})} = \frac{2880 \times 10^{-6} \ \text{m.K}}{865.5 \ \text{K}} = 3.33 \times 10^{-6} \ \text{m}$$

$$= 3.33 \ \mu\text{m}. \tag{6}$$

   The radiator emits in the infrared.

3. *Temperature T' of the radiator so that λ_{max} is equal to 2 µm*:

   By using the first law of Wien, we have:

$$T'(\text{K}) = \frac{2880 \ (\mu\text{m.K})}{\lambda_{\max}(\mu\text{m})} = \frac{2880 \ (\mu\text{m.K})}{2 \ \mu\text{m}} = 1440 \ \text{K}. \tag{7}$$

   Released flux:

   By using the Stefan-Boltzmann law, we find:

$$\frac{\Phi'^0}{\Phi^0} = \frac{\sigma_0 \ S \ T'^4}{\sigma_0 \ S \ T^4} \ \Rightarrow \ \Phi'^0 = \Phi^0 \left(\frac{T'}{T}\right)^4 = (1000 \ \text{W})\left(\frac{1440 \ \text{K}}{865.5 \ \text{K}}\right)^4 = 7.66 \ \text{W}. \tag{8}$$

Exercise 6
A source (S) of surface $S = 0.01 \ m^2$ behaves like a black body. Its emittance is $M^0 = 3140 \ W.m^{-2}$. It emits towards three screens of the same surface $S' = 0.01 \ m^2$, at the distance $d = 1 \ m$ from the source, arranged as shown in the figure.
1. *Calculate the temperature of the source (S).*
2. *What are the energy fluxes that reach each of the screens? We give $\sigma_0 = 5.68 \times 10^{-8} \ W.m^{-2}. \ K^{-4}$ (the Stefan-Boltzmann constant).*

Solution

1. *Source temperature (S):*

   According to the Stefan-Boltzmann law, we have:

$$M^0 = \varepsilon \sigma_0 T^4 \quad \Rightarrow \quad T = \left(\frac{M^0}{\sigma_0}\right)^{\frac{1}{4}} = \left(\frac{3140 \text{ W.m}^{-2}}{5.68 \times 10^{-8} \text{ W.m}^{-2}.\text{K}^{-4}}\right)^{\frac{1}{4}} = 484.93 \text{ K}$$

$$\approx 485 \text{ K}. \tag{1}$$

   Here, ε is the emissivity ($\varepsilon = 1$ for a black body).

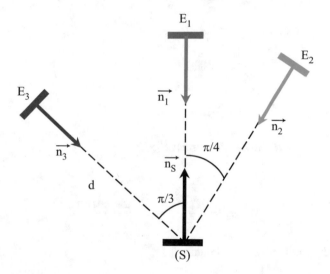

2. *Energy fluxes that fall on each of the three screens E_1, E_2, and E_3:*

   Let us consider the two elementary surfaces dS_1 and dS_2, respectively, belonging to surfaces (S_1) and (S_2) and let us write the expression of the total flux emitted by dS_1 and arriving at dS_2; it is given by:

$$d\Phi_{12} = \frac{L_1^0 \, dS_1 \cos\theta_1 \, dS_2 \cos\theta_2}{d^2}. \tag{2}$$

   The elementary surface dS_1 obeys the Lambert law, namely:

$$L_1^0 = \frac{M_1^0}{\pi}. \tag{3}$$

   By injecting (2) into (3), we obtain:

$$d\Phi_{12} = \frac{M_1^0 \, dS_1 \cos\theta_1 dS_2 \cos\theta_2}{\pi d^2}. \tag{4}$$

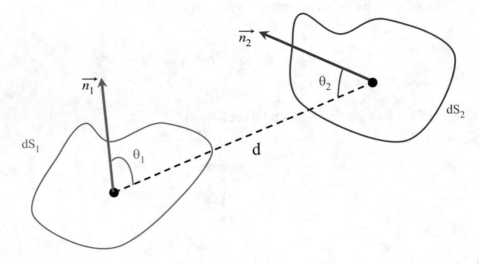

2.1. *Energy flux that reaches the screen E_1:*

It is given by:

$$d\Phi_{SE_1} = \frac{M_1^0 dS_S \cos\theta_S \, dS_{E_1} \cos\theta_{E_1}}{\pi d^2} \tag{5}$$

with according to the figure and data:

$$\cos\theta_S = \left(\overrightarrow{n_S}.\overrightarrow{d}\right) = \cos 0 = 1, \quad \cos\theta_{E_1} = \left(\overrightarrow{n_1}.\overrightarrow{d}\right) = \cos 0 = 1,$$
$$dS_S = dS_{E_1} = S = 10^{-2} \text{ m}^2. \tag{6}$$

We finally get:

$$d\Phi_{SE_1} = \frac{(3140 \text{ W.m}^{-2)})(10^{-2} \text{ m}^2)(1)(10^{-2} \text{ m}^2)(1)}{(3.14)(1^2 \text{ m}^2)} = 0.1 \text{ W}. \tag{7}$$

2.2. *Energy flux that reaches the screen E_2:*

It is given by:

$$d\Phi_{SE_2} = \frac{M_1^0 dS_S \cos_S dS_{E_2} \cos\theta_{E_2}}{\pi d^2}. \tag{8}$$

According to the figure and data, we have:

$$\cos\theta_S = \left(\overrightarrow{n_S}.\overrightarrow{d}\right) = \cos\left(\frac{\pi}{4}\right) = 0.71, \quad \cos\theta_{E_2} = \left(\overrightarrow{n_2}.\overrightarrow{d}\right) = \cos 0 = 1,$$
$$dS_S = dS_{E_2} = S = 10^{-2} \ \text{m}^2. \tag{9}$$

We finally get:

$$d\Phi_{SE_1} = \frac{\left(3140 \ \text{W.m}^{-2)}\right)\left(10^{-2} \ \text{m}^2\right)(0.71)\left(10^{-2} \ \text{m}^2\right)(1)}{(3.14)\left(1^2 \ \text{m}^2\right)} = 0.07 \ \text{W}. \tag{10}$$

2.3. Energy flux that reaches the screen E₃:

It is given by:

$$d\Phi_{SE_3} = \frac{M_1^0 dS_S \cos\theta_S \, dS_{E_3} \cos\theta_{E_3}}{\pi d^2}. \tag{11}$$

According to the figure and data, we have:

$$\cos\theta_S = \left(\overrightarrow{n_S}.\overrightarrow{d}\right) = \cos\left(\frac{\pi}{3}\right) = 0.5, \quad \cos\theta_{E_3} = \left(\overrightarrow{n_2}.\overrightarrow{d}\right) = \cos 0 = 1,$$
$$dS_S = dS_{E_3} = S = 10^{-2} \text{m}^2. \tag{12}$$

We finally get:

$$d\Phi_{SE_1} = \frac{\left(3140 \ \text{W.m}^{-2)}\right)\left(10^{-2} \ \text{m}^2\right)(0.5)\left(10^{-2} \ \text{m}^2\right)(1)}{(3.14)\left(1^2 \ \text{m}^2\right)} = 0.05 \ \text{W}. \tag{13}$$

Exercise 7

We want to calibrate a thermal radiation receiver. For that purpose, we have a furnace equipped with a circular opening of $D_F = 20$ mm in diameter, of which the luminance is $L_F = 3.72 \times 10^5$ W.m^{-2}. The sensitive part of the receiver has an area of $S_R = 1.6 \times 10^{-5}$ m². The receiver is placed parallel to the opening.

1. At what distance is the value of its illumination $E_R = 1000$ W.m^{-2}?

2. The receiver is inclined by 20° with respect to the furnace. What is its illumination?

3. For a given inclination, what is the shape of the curve giving the illumination as a function of the furnace-receiver distance r?

268   3 Black Body

Solution

1. *Furnace-receiver distance for which the illumination is $E_R = 1000$ W.m^{-2}:*

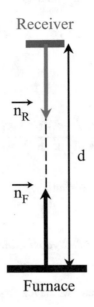

Receiver

Furnace

Data:

– Diameter D_F of the circular opening of the oven: $D_F = 20$ mm $= 20 \times 10^{-3}$ m.
– Surface S_F of the circular opening of the furnace:

$$S = \pi \left(\frac{D_F}{2}\right)^2 = (3.14) \left(\frac{20 \times 10^{-3} \text{ m}}{2}\right)^2 = 3.14 \times 10^{-4} \text{ m}^2.$$

– Sensitive area of the receiver $S_R = 1.6 \times 10^{-5}$ m^2.

$$\cos\theta_F = \left(\overrightarrow{n_F}.\overrightarrow{d}\right) = \cos 0 = 1, \quad \cos\theta_R = \left(\overrightarrow{n_R}.\overrightarrow{d}\right) = \cos 0 = 1.$$

The expression of the flux emitted by the surface of the furnace S_F and arriving at the surface of the receiver S_R is given by:

$$\Phi_{FR} = \frac{L^0 \, S_F \cos\theta_F \, S_R \cos\theta_R}{d^2}. \tag{1}$$

The resulting illumination on the surface S_R of the receiver is given by:

$$E = \frac{\Phi_{FR}}{S_R}. \tag{2}$$

It is expressed as the emittance in W/m^2. The emittance and the illumination should not be confused because the illumination is a quantity relative to the reception of the radiation, whereas the emittance is a physical quantity that characterizes the emission of the radiation. As the furnace and the receiver behave like black bodies, all the flux emitted by the furnace is received by the receiver (perfect emitters and absorbers). That is, by injecting (1) into (2):

$$E = \frac{L^0 S_F \cos \theta_F \, S_R \cos \theta_R}{S_R d^2} = \frac{L^0 S_F}{d^2} \;\Rightarrow\; d = \sqrt{\frac{L^0 S_F}{E}}$$

$$= \sqrt{\frac{\left(3.72 \times 10^5 \text{ W.m}^{-2}\right)\left(3.14 \times 10^{-4} \text{ m}^2\right)}{1000 \text{ W.m}^{-2}}} = 0.342 \text{ m}. \tag{3}$$

2. *Illumination of the inclined receiver of 20° relative to the furnace*:

Since the distance d between the furnace and the receiver is kept constant, the illumination $E(\theta_R, d)$ has the following expression:

$$E(\theta_R, d) = \frac{L^0 S_F \cos \theta_F \, S_R \cos \theta_R}{S_R d^2} = \frac{L^0 S_F \cos \theta_R}{d^2} \tag{4}$$

with $\cos \theta_F = 1$. On the basis of the ratio between (4) and (3), we obtain:

$$\frac{E(\theta_R, d)}{E} = \frac{\dfrac{L^0 S_F \cos \theta_R}{d^2}}{\dfrac{L^0 S_F}{d^2}} = \cos \theta_R \;\Rightarrow\; E(\theta_R, d) = E \cos \theta_R$$

$$= \left(1000 \text{ W.m}^{-2}\right) \cos 20° = 940 \text{ W.m}^{-2}. \tag{5}$$

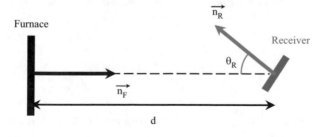

3. *Function $E(\theta_R, r)$ for fixed θ_R:*

$$\frac{E(\theta_R, r)}{E(\theta_R, d)} = \frac{\dfrac{L^0 S_F \cos \theta_R}{r^2}}{\dfrac{L^0 S_F \cos \theta_R}{d^2}} = \frac{d^2}{r^2}, \tag{6}$$

that is:

$$E(\theta_R, r) = \frac{d^2 E(\theta_R, d)}{r^2} = \frac{\text{Cte}}{r^2}. \tag{7}$$

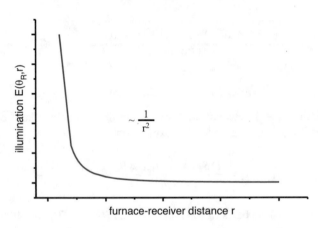

furnace-receiver distance r

Exercise 8

The Sun emits, in all directions of space, electromagnetic waves, a part of which arrives on Earth. We suppose that the Sun behaves as a black body.

1. *Write the heat flux $d\Phi'$ emitted by an element dS_S of the surface of the Sun at the temperature T_S towards an element dS_R of the surface of Earth.*

2. *By taking the Sun-Earth distance $R_{S-T} \approx r$, deduce that the heat flux emitted from the whole surface of the Sun towards an element dS_R of the surface of Earth is equal to:*

$$d\Phi = \sigma_0\, T_S^4\, \frac{R_S^2}{R_{S-T}^2}\, dR_T\, \cos \theta_T.$$

3. *How the quantity $E = \dfrac{d\Phi}{dS_T}$ is called? The maximal value of this quantity at very high altitude is $E_{Max} = 1390\ \text{W.m}^{-2}$. Show that we can calculate the temperature T_S of the Sun if we know the apparent radius of the Sun defined by $\alpha \approx \dfrac{R_S}{R_{S-T}}$. Is this quantity easily measurable? Calculate T_S. We give $\alpha = 4.64 \times 10^{-3}$ and*

$\sigma_0 = 5.68 \times 10^{-8}$ *W.m^{-2}.K^{-4} (Boltzmann's constant), $R_S = 6.96 \times 10^5$ km, and*
$R_{S-T} = 150 \times 10^6$ *km.*

4. *In what domain is situated the wavelength λ_{max} of the Sun that corresponds to a maximum of its spectral emittance and what is the useful range of the solar radiation?*

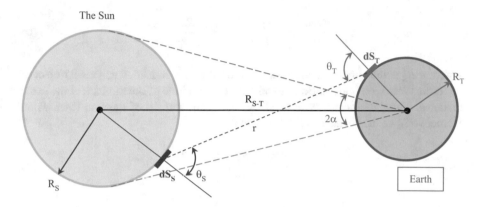

Solution

1. *Expression of the heat flux $d\Phi'$:*

   It is given by:

   $$d\Phi' = \frac{M_S^0 \, dS_S \cos\theta_E \, dS_T \cos\theta_T}{\pi r^2} = \frac{\sigma_0 T_S^4 \, dS_S \cos\theta_S \, dS_T \cos\theta_T}{\pi r^2} \tag{1}$$

   knowing that the Sun behaves like a black body and obeys the laws of Lambert and Stefan-Boltzmann:

   $$M_S^0 = \pi L_S^0 \;\Rightarrow\; L_S^0 = \frac{M_S^0}{\pi} = \frac{\sigma_0 T_S^4}{\pi}. \tag{2}$$

2. *Heat flux emitted by the entire surface of the Sun:*

   The solid angle $d\Omega$, under which from dS_T we see dS_S, is given by:

   $$d\Omega = \frac{dS_S \cos\theta_S}{r^2}. \tag{3}$$

   By assuming the following approximation:

   $$r \approx R_{S-T} \tag{4}$$

the solid angle $d\Omega$ becomes close to the following expression:

$$d\Omega \approx \frac{dS_S \cos \theta_S}{R_{S-T}^2}.$$ (5)

By integrating (5) on all the elements dS_S that constitute the outer surface of the Sun, we obtain:

$$\Omega = \iint_{(S_S)} d\Omega = \frac{\pi R_S^2}{R_{S-T}^2}$$ (6)

where πR_S^2 is the circular section occupied by the Sun. Indeed, a spherical object of the radius R is seen under the same solid angle as a circle of the radius R at the same distance. Thus, the heat flux emitted by the entire surface of the Sun towards an element dS_R of the surface of Earth is equal to:

$$d\Phi = \frac{\sigma_0 T_S^4 R_S^2 dR_T \cos \theta_T}{R_{S-T}^2}.$$ (7)

3.1. *Physical meaning of the quantity φ*:

The quantity E is given by:

$$E = \frac{d\Phi}{dS_T} = \frac{\sigma_0 T_S^4 \cos \theta_T}{R_{S-T}^2}.$$ (8)

Equation (8) represents by definition the illumination of Earth by the Sun.

3.2. *Measurement of the apparent radius of the Sun*:

The ratio $\alpha \approx \dfrac{R_S}{R_{S-T}}$ represents the apparent radius of the Sun; it is directly measured with a telescope.

3.3. *Calculation of the temperature T_S at the surface of the Sun*:

As the radius of the Sun is small in front of the distance which separates Earth and this star, we can consider that the apparent radius of the Sun α is approximately equal to:

$$\alpha \approx \frac{R_S}{R_{S-T}} = \frac{6.96 \times 10^5 \text{ km}}{150 \times 10^6 \text{ km}} = 4.64 \times 10^{-3}.$$ (9)

Knowing by hypothesis that:

$$E_{max} = \sigma_0 \, T_S^4 \, \frac{R_S^2}{R_{S-T}^2} = \sigma_0 \, T_S^4 \, \alpha^2 = 1390 \ \text{W.m}^{-2} \tag{10}$$

because $\cos \theta_T = 1$ in (8) if the illumination is maximal, the temperature T_S of the Sun is then given by:

$$T_S = \left(\frac{\varphi_{max}}{\sigma_0 \, \alpha^2} \right)^{\frac{1}{4}} = \left[\frac{1390 \ \text{W.m}^{-2}}{\left(5.68 \times 10^{-8} \ \text{W.m}^{-2}.\text{K}^{-4} \right) \left(4.64 \times 10^{-3} \right)} \right]^{\frac{1}{4}}$$

$$= 5800 \ \text{K}. \tag{11}$$

4.1. *Calculation of the maximal wavelength*:

According to the Wien first law, it becomes:

$$\lambda_{max}(\mu m) = \frac{2880 \ (\mu m.K)}{T(K)} = \frac{2880 \times 10^{-6} \ \text{m}}{5800 \ \text{K}} \approx 0.50 \times 10^{-6} \ \text{m}$$

$$= 0.50 \ \mu m. \tag{12}$$

The Sun emits in the visible.

4.2. *Useful domain*:

The useful emission domain of the Sun is between $0.5 \ \lambda_{max} = 0.25 \ \mu m$ and $5 \ \lambda_{max} = 2.5 \ \mu m$.

Exercise 9

A spherical satellite behaves like a black body.
1. *In the vacuum, at an altitude where the solar energy flux has an emittance M' equal to 1400 W.m^{-2}, determine the temperature T_S of the satellite.*
2. *Back on Earth, the satellite receives a solar illumination $E'' = 700$ W.m^{-2}. The ambient temperature at ground level is $T_a = 17\,°C$, what is the new temperature T_S of the satellite, if we neglect any exchange of energy by conduction and convection?*

Solution
1. *Temperature T_S of the satellite*:

The energy flux Φ' received by the satellite from the Sun is equal to:

$$M' = \frac{\Phi'}{S'} \quad \Rightarrow \quad \Phi' = M'S' = M'\pi r^2 \tag{1}$$

where $S' = \pi r^2$ is the circular section of the spherical satellite exposed to the Sun rays. Indeed, a spherical object of the radius r is seen under the same solid angle as a circle of the radius r at the same distance. The satellite in turn emits an energy flux through its entire spherical cap $S = 4\pi r^2$. According to the Stefan-Boltzmann law, we have:

$$M = \frac{\Phi}{S} = \varepsilon \sigma_0 T_S^4 \quad \Rightarrow \quad \Phi = \varepsilon \sigma_0 S T_S^4 = 4\pi r^2 \sigma_0 T_S^4 \tag{2}$$

where M is the emittance of the satellite, ε the emissivity ($\varepsilon = 1$ for a black body), and σ_0 the Stefan-Boltzmann constant. At thermal equilibrium, we have:

$$\Phi = \Phi' \quad \Rightarrow \quad 4\pi r^2 \sigma_0 T_S^4 = M' \pi r^2. \tag{3}$$

In the end, we obtain:

$$T_S = \left(\frac{M'}{4\,\sigma_0}\right)^{\frac{1}{4}} = \left[\frac{1400 \text{ W.m}^{-2}}{4\,(5.68 \times 10^{-8} \text{ W.m}^{-2}.\text{K}^{-4})}\right]^{\frac{1}{4}} = 280 \text{ K} = 7\,°\text{C}. \tag{4}$$

2. *New satellite temperature T_S*:

E'' being the solar illumination that radiates the satellite, the energy flux Φ'' received by the satellite from the Sun is equal to:

$$E'' = \frac{\Phi''}{S''} \quad \Rightarrow \quad \Phi'' = E'' S'' = E'' \pi r^2 \tag{5}$$

where $S'' = \pi r^2$ is the circular section of the spherical satellite exposed to the Sun rays. Indeed, a spherical object of the radius r is seen under the same solid angle as a circle of the radius r at the same distance. The energy flux received by the spherical cap $S' = 4\pi r^2$ of the satellite from the ambient air (which surrounds it) is:

$$\Phi' = \varepsilon \sigma_0 S' T_a^4 = 4\pi r^2 \sigma_0 T_a^4. \tag{6}$$

The satellite in turn emits a flux through its spherical cap $S = S' = 4\pi r^2$. According to the Stefan-Boltzmann law, we have:

$$M = \frac{\Phi}{S} = \varepsilon \sigma_0 T_S^4 \quad \Rightarrow \quad \Phi = \varepsilon \sigma_0 S T_S^4 = 4\pi r^2 \sigma_0 T_S^4 \tag{7}$$

where M is the emittance of the satellite, ε the emissivity ($\varepsilon = 1$ for a black body), and σ_0 the Stefan-Boltzmann constant. At thermal equilibrium, we have:

$$\Phi_{\text{emitted}} = \Phi'_{\text{received}} + \Phi'_{\text{received}} \quad \Rightarrow \quad 4\pi r^2 \sigma_0 T_S^4 = 4\pi\, r^2 \sigma_0 T_a^4 + E'' \pi\, r^2. \tag{8}$$

Knowing that T_a (K) $= 17\,°\text{C} + 273 \text{ K} = 290 \text{ K}$, it becomes in the end:

$$T_S = \left(T_a^4 + \frac{E''}{4\sigma_0}\right)^{\frac{1}{4}} = \left[290^4\,\mathrm{K}^4 + \frac{700\ \mathrm{W.m}^{-2}}{4\left(5.68 \times 10^{-8}\ \mathrm{W.m}^{-2}.\mathrm{K}^{-4}\right)}\right]^{\frac{1}{4}} = 317\ \mathrm{K}$$

$$= 44\ {}^\circ\mathrm{C}. \tag{9}$$

Exercise 10

The demonstration in the course of the Stefan-Boltzmann law reveals in (3.47) the following integral:

$$\int_0^\infty \frac{X^3}{e^X - 1}\,dX.$$

Show that it is equal to $\pi/15$.

Solution

Knowing the development of the sum of the following geometric series:

$$\sum_{n=0}^{\infty} X^n = \frac{X}{1 - X} = \frac{1}{\dfrac{1}{X} - 1} \tag{1}$$

we obtain:

$$\frac{1}{e^X - 1} = \frac{1}{\dfrac{1}{e^{-X}} - 1} = \sum_{n=0}^{\infty}\left(e^{-X}\right)^n = \sum_{n=0}^{\infty} e^{-nX} \tag{2}$$

and consequently:

$$\int_0^\infty \frac{X^3}{e^X - 1}\,dX = \int_0^\infty \sum_{n=0}^{\infty}\left(X^3 e^{-nX}\right)dX = \sum_{n=0}^{\infty}\int_0^\infty \left(X^3 e^{-nX}\right)dX. \tag{3}$$

We integrate three times by parts Eq. (3), and we find:

$$\sum_{n=0}^{\infty}\int_0^\infty \frac{3X^2 e^{-nX}}{X}\,dX \quad \text{then} \quad \sum_{n=0}^{\infty}\int_0^\infty \frac{-6X e^{-nX}}{X^2}\,dX \quad \text{then} \quad \sum_{n=0}^{\infty}\int_0^\infty \frac{-6X e^{-nX}}{X^2}\,dX. \tag{4}$$

This gives:

$$\sum_{n=0}^{\infty} \frac{6}{X^4} = 6\sum_{n=0}^{\infty} \frac{1}{X^4}, \tag{5}$$

that is:

$$\sum_{n=0}^{\infty} \frac{1}{X^4} = \frac{\pi^4}{90} \Rightarrow \int_0^{\infty} \frac{X^3}{e^X - 1} \, dX = 6 \sum_{n=0}^{\infty} \frac{1}{X^4} = \frac{6^4}{90} = \frac{\pi^4}{15}. \quad (6)$$

Exercise 11

Let u be the spectral and the volume density of the black body radiation at the temperature T. This physical quantity, which represents the electromagnetic energy per unit volume, in the frequency band between v and $v + dv$ has the following form: $d\varepsilon = u(v,T) \, dv$.

1. What is the dimensional equation of $u(v,T)$?
2. In classical thermodynamics, $u(v,T)$ has the following form: $u_{cl}(v,T) = A \, c^x \, v^y$ $(k_B T)^z$ where c is the speed of light, A an unknown numeric constant, and k_B Boltzmann's constant. $k_B T$ is an energy that depends on temperature. Calculate the exponents x, y, and z. What is this formula called?
3. By estimating the total energy per unit of volume ε, show that the previous expression is physically illogical. Does the paradox come from high or low frequencies?
4. It is quantum physics, which resolves the problem. Indeed, the energy density is in fact of the form:

$$u_q(v, T) = u_{cl}(v, T) f\left(\frac{h}{k_B T}\right)$$

where $f\left(\frac{h}{k_B T}\right)$ is a function to be found. By considering the fact that it allowed to overcome the problem appeared in 3, deduce the limits of the function $f\left(\frac{h}{k_B T}\right)$ at origin 0 and at infinity.

5. For high frequencies, Wien proposed the following approximate experimental formula:

$$u_w(v, T) \approx v^3 \exp\left(-\frac{av}{T}\right)$$

where a was a constant. Show that that implies for $v \to \infty$, $f \approx \exp\left(-\frac{av}{T}\right)$.

6. Planck later found a certain formula f, by trying to reconcile the Rayleigh-Jeans formula with that of Wien. It has the following form:

$$f = \frac{\frac{hv}{k_B T}}{e^{\frac{h}{k_B T}} - 1}.$$

Show that we find these limits in high and low frequencies.

7. Trace qualitatively the three behaviors.

Solution

1. *Dimensional equation of u(ν,T):*

$$d\varepsilon = u(\nu, T)d\nu \quad \Rightarrow \quad u(\nu, T) = \frac{d\varepsilon}{d\nu}. \tag{1}$$

The quantity $d\varepsilon$ is by hypothesis energy per unit volume, so it has the dimension of energy per volume. The quantity $d\nu$ is, by hypothesis, a frequency. The frequency is the inverse of the period and vice versa. We finally get:

$$[u(\nu, T)] = \frac{[d]}{[V][d]} = \frac{ML^2T^{-2}}{L^3T^{-1}} = ML^{-1}T^{-1}. \tag{2}$$

2. *Determination of the exponents x, y, and z:*

A is dimensionless, so the dimensional equation of u_{cl} is:

$$[u_{cl}(\nu, T)] = [c]^x [\nu]^y [\text{energy}]^z = \left(LT^{-1}\right)^x \left(T^{-1}\right)^y \left(ML^2T^{-2}\right)^z$$
$$= L^{x+2z} T^{-x-y-2z} M^z. \tag{3}$$

By comparing with (2), we obtain the following system of equations:

$$z = 1$$
$$x + 2z = -1 \quad \Rightarrow \quad x = -2z - 1 = -3 \tag{4}$$
$$-x - y - 2z = -1 \quad \Rightarrow \quad y = -x - 2z + 1 = 2.$$

We finally get:

$$u_{cl}(\nu, T) = A \frac{\nu^2}{c^3} k_B T. \tag{5}$$

Formula (5) is that of Rayleigh-Jeans named after the two scientists John William Strutt Rayleigh and James Jeans who proposed it in order to express the distribution of the energy spectral luminance of the thermal radiation of the black body as a function of temperature in the area of low frequencies.

3. *Evaluation of the total energy per unit volume:*

$$\varepsilon_{cl} = \int_0^\infty u_{cl}(\nu, T) \, d\nu = \int_0^\infty A \frac{\nu^2}{c^3} k_B T \, d\nu = A \frac{k_B T}{c^3} \int_0^\infty \nu^2 \, d\nu = A \frac{k_B T}{c^3} \left[\frac{\nu^3}{3} \right]_0^\infty = \infty. \tag{6}$$

As shown in Eq. (6), this integral diverges at high frequencies, which is aberrant from the point of view of physics.

4. *Limits of the function f at origin 0 and at infinity*:

$$u_{\mathrm{q}}(\nu, T) = u_{\mathrm{cl}}(\nu, T) f\left(\frac{h\nu}{k_{\mathrm{B}}T}\right) = A\frac{\nu^2}{c^3}k_{\mathrm{B}}T f\left(\frac{h\nu}{k_{\mathrm{B}}T}\right). \tag{7}$$

When ν tends to 0, we have to find again the classical expression u_{cl}, so:

$$\lim_{\nu\to 0} f\left(\frac{h\nu}{k_{\mathrm{B}}T}\right) = 1. \tag{8}$$

When ν goes to infinity, the total energy must remain finite. The function $f\left(\frac{h\nu}{k_{\mathrm{B}}T}\right)$ must make up the divergence in ν^2 of u_{cl}, that is:

$$\lim_{\nu\to\infty} \nu^2 f\left(\frac{h\nu}{k_{\mathrm{B}}T}\right) = 0. \tag{9}$$

5. *Limit of f when ν tends to infinity*:

The Wien formula is written as:

$$u_w(\nu, T) \approx \nu^3 \exp\left(-\frac{a\nu}{T}\right). \tag{10}$$

We compare (9) with the Wien formula in (10) which, by hypothesis, is valid for high frequencies. It follows that:

$$\lim_{\nu\to\infty} f\left(\frac{h\nu}{k_{\mathrm{B}}T}\right) \;\to\; \nu\exp\left(-\frac{a\nu}{T}\right). \tag{11}$$

6. *Limit of f at low and high frequencies*:

The Planck formula is:

$$f = \frac{\dfrac{h\nu}{k_{\mathrm{B}}T}}{e^{\frac{h}{k_{\mathrm{B}}T}} - 1}. \tag{12}$$

− At low frequencies $h\nu \ll k_{\mathrm{B}}T$:

$$e^{\frac{h\nu}{k_{\mathrm{B}}T}} \approx 1 + \frac{h\nu}{k_{\mathrm{B}}T}. \tag{13}$$

The function f tends to 1, and we find again $u_p = u_{cl}$, that is, the behavior predicted by the Rayleigh-Jeans formula.

- At high frequencies $h\nu \gg k_B T$:

$$e^{\frac{h\nu}{k_B T}} - 1 \approx e^{\frac{h\nu}{k_B T}} \tag{14}$$

that is:

$$f = \frac{h\nu}{k_B T} e^{-\frac{h\nu}{k_B T}} \approx \nu \, e^{-\frac{h\nu}{k_B T}}. \tag{15}$$

We find again the behavior planned by Wien with $a = h/k_B$.

7. *Plot of the three behaviors*:

The Rayleigh-Jeans law (in green) is in agreement with the experimental results at low frequencies, but strongly in disagreement with these results at high frequencies. This inconsistency between the observations and the predictions of classical physics is commonly called as *ultraviolet catastrophe*. The Wien law (in red) is correctly defined for high radiation frequencies. The law of Planck (in black) gives the good behavior of the emission at all the frequencies; it constitutes a fundamental aspect of the development of quantum mechanics at the beginning of the twentieth century.

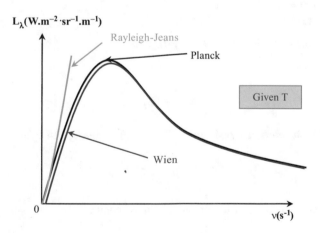

Exercise 12

An electrical current of intensity $I = 140$ A flows in a cylindrical copper tube, of which the length L and the thickness e are 75 cm and 0.125 mm, respectively, such that $e = R_e - R_i$ where R_e and R_i represent the outer and inner radii of the tube, respectively. To simplify the problem, we assume that $R_e + R_i \approx D = 1.5$ cm with D, the outside diameter of the cylinder. The resistivity ρ of copper is 1.7 $\mu m.cm$.

1. *What is the electric power dissipated by Joule effect?*
2. *This power is radiated from the surface of the tube at temperature $T = 683$ K.
   Knowing that copper behaves like a real body, calculate its emissivity ε.*

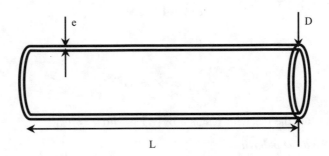

Solution

Data:

- $I = 140$ A
- $L = 75$ cm $= 0.75$ m
- $e = 0.125$ mm $= 1.25 \times 10^{-4}$ m
- $D = 1.5$ cm $= 1.5 \times 10^{-2}$ m
- $\rho = 1.7$ $\mu\Omega.$cm $= 1.7 \times 10^{-8}$ $\Omega.$m

1. *Electric power dissipated by Joule effect*:

   It is given by:

   $$P = RI^2 \tag{1}$$

   where R is the electrical resistance of the copper tube given by:

   $$R = \rho \frac{L}{S} \tag{2}$$

   where S is the surface of the cross section of copper. That is:

   $$S = \pi R_e^2 - \pi R_e^2 = \pi \left(R_e^2 - R_e^2 \right) = \pi (R_e - R_e)(R_e + R_e) = \pi e D. \tag{3}$$

   By injecting (3) and (2) into (1), we obtain:

$$P = RI^2 = \rho\frac{L}{S}I^2 = \rho\frac{L}{S}I^2 = \rho\frac{L}{\pi\,eD}I^2$$

$$= \left(1.7\times10^{-8}\ \Omega.m\right)\frac{0.75\ m}{(3.14)\left(1.5\times10^{-2}\ m\right)\left(1.25\times10^{-4}m\right)}(140\ A)^2 \quad (4)$$

$$= 42.34\ W.$$

2. Emissivity ε of copper:

The copper tube emits an energy flux Φ by its external lateral surface S_L of the radius R_e such that $S_L = 2\pi R_e L = \pi DL$. According to the Stefan-Boltzmann law, we have:

$$M = \frac{\Phi}{S} = \varepsilon\sigma_0 T_S^4 \;\Rightarrow\; \varepsilon = \frac{\Phi}{\sigma_0\,S_L\,T_S^4} = \frac{\Phi}{\sigma_0(\pi DL)T_S^4}$$

$$= \frac{42.34\ W}{\left(5.68\times10^{-8}\ W.m^{-2}.K^{-4}\right)(3.14)\left(1.5\times10^{-2}m\right)(0.75\ m)\left(683^4\ K^4\right)} \approx 0.1$$

$$(5)$$

M being the emittance of the tube in W/m^2 and $\sigma_0 = 5.68\ 10^{-8}\ W.m^{-2}.K^{-4}$, the Stefan-Boltzmann constant.

Exercise 13

A filament lamp with a power $P = 75\ W$ operates at a voltage of $U = 220\ V$. This lamp behaves like a real body. It is formed of a cylindrical filament of tungsten which radiates by its lateral surface. It is surrounded by a sphere of glass in which we create a vacuum. To obtain a white light, the temperature of the filament must reach a temperature $T_F = 2500\ K$. Calculate the diameter D and the length L of the filament knowing that the emissivity of the lamp is $\varepsilon = 0.3$, and the resistivity of the filament is $\rho = 88\ \mu\Omega.cm$. We give $\sigma_0 = 5.68\times10^{-8}\ W.m^{-2}.K^{-4}$, the Stefan-Boltzmann constant.

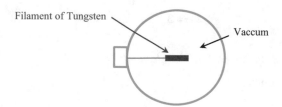

Filament of Tungsten

Vaccum

Solution

1. Calculation of the diameter D of the filament:

Knowing that the electric power P of the filament is equal to the radiated energy flux Φ, the emittance of the lamp is given by the Stefan-Boltzmann law:

$$M = \frac{\Phi}{S_L} = \varepsilon\,\sigma_0\,T_F^4 \;\Rightarrow\; \Phi = P = M S_L = \varepsilon\,\sigma_0\,T_F^4\,S_L = \varepsilon\,\sigma_0\,T_F^4\,(\pi D L)$$
$$= (0.3)(5.68 \times 10^{-8}\ \text{W.m}^{-2}.\text{K}^{-4})(2500^4\ \text{K}^4)(3.14)\big[(D \times L)\,\text{m}^2\big] \qquad (1)$$
$$= 75\ \text{W}$$

where S_L represents the lateral surface of the filament through which the white light radiates and T_F its temperature. That is:

$$D \times L = \frac{75\ \text{W}}{(0.3)\ (5.68 \times 10^{-8}\ \text{W.m}^{-2}.\text{K}^{-4})(2500^4\ \text{K}^4)(3.14)}$$
$$= 3.6 \times 10^{-5}\ \text{m}^2. \qquad (2)$$

Moreover, the electrical resistance of the cylindrical filament is given by:

$$R = \frac{U^2}{P} = \rho\frac{L}{S} = \rho\frac{L}{\pi\dfrac{D^2}{4}} = \frac{4\rho L}{\pi D^2} \;\Rightarrow\; \frac{L}{D^2} = \frac{\pi U^2}{4\rho P}$$

$$= \frac{(3.14)(220^2\ \text{V}^2)}{4\ (88 \times 10^{-8}\ \Omega.\text{m})\ (75\ \text{W})} = 5.76 \times 10^8\ \text{m}^{-1}. \qquad (3)$$

where S is the circular section of the filament. By doing the ratio (2)/(3), we find:

$$\frac{(2)}{(3)} = \frac{D \times L}{\dfrac{L}{D^2}} = D^3 = \frac{3.6 \times 10^{-5}\ \text{m}^2}{5.76 \times 10^8\ \text{m}^{-1}} = 0.625 \times 10^{-13}\ \text{m}^3 \;\Rightarrow$$
$$D = \sqrt[3]{0.625 \times 10^{-13}}\ \text{m}^3 \qquad (4)$$

that is:

$$D = 3.97 \times 10^{-5}\ \text{m}. \qquad (5)$$

2. *Calculation of the length L of the filament*:

From (2), we obtain:

$$D \times L = 3.6 \times 10^{-5}\ \text{m}^2 \;\Rightarrow\; L = \frac{3.6 \times 10^{-5}\ \text{m}^2}{3.97 \times 10^{-5}\ \text{m}} = 0.91\ \text{m}. \qquad (6)$$

Exercise 14

By assuming that the Sun is a black body at 5800 K, what is the fraction of energy that it emits in the visible range (0.4 – 0.8 μm)? We give the following table:

$F_{0-\lambda T}$	λT (μm.K)
0.082	2000
0.126	2320
0.281	3000
0.586	4640
0.925	10 000

Solution

Let's calculate the product λT:

$$\lambda_1 T = (0.4 \text{ μm})(5800 \text{ K}) = 2320 \text{ μm.K and } \lambda_1 T = (0.8 \text{ μm})(5800 \text{ K})$$
$$= 4640 \text{ μm.K.} \tag{1}$$

Let's calculate the emitted energy between λ_1 and λ_2:

The values of $F_{0-\lambda_{1T}} = F_{0-2320}$ μm.K and $F_{0-\lambda_{2T}} = F_{0-24\ 640}$ μm.K are given in the table, namely:

$$F_{\lambda_1-\lambda_2} = F_{0-\lambda_2 T} - F_{0-\lambda_{1T}} = 0.586 - 0.126 = 0.46. \tag{2}$$

Therefore, 46% of the solar energy is radiated in the visible.

Exercise 15

Let be the two concentric spheres (S_1) and (S_2), of the radii R_1 and R_2 such that $R_1 < R_2$. The flux emitted by the sphere (S_1) is totally absorbed by the sphere (S_2). Calculate as a function of (S_i) the view factors F_{ij} (i = 1 or 2, j = 1 or 2).

Solution

1. *Calculation of F_{12}:*

When all the radiation coming from surface (1) reaches surface (2), there is a total influence, namely:

$$F_{12} = 1. \tag{1}$$

2. *Calculation of F_{21}:*

The symmetry of the expressions of F_{12} and F_{21} leads to the reciprocity of form factors:

$$S_1 F_{12} = S_2 F_{21} \quad \Rightarrow \quad F_{21} = \frac{S_1}{S_2} F_{12} = \frac{S_1}{S_2} \times 1 = \frac{S_1}{S_2}. \tag{2}$$

3. Calculation of F_{22}:

The generalization of the reciprocity relationship (2) at a closed volume consisting of n isothermal surfaces allows writing that the flux emitted by surface (i) reaches, whatever its trajectory, a surface (j) with $i \leq j \leq n$. For the flux emitted by the sphere (2) ($i = 2$), it becomes:

$$\sum_{j=1}^{j=n} F_{ij} = 1 \quad \Rightarrow \quad \sum_{j=1}^{j=2} F_{ij} = F_{21} + F_{22} = 1 \quad \Rightarrow \quad F_{22} = 1 - \frac{S_1}{S_2}$$

$$= \frac{S_2 - S_1}{S_2}. \tag{3}$$

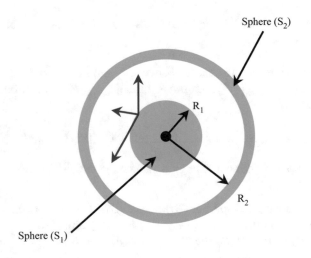

Sphere (S_2)

R_1

R_2

Sphere (S_1)

4. Calculation of F_{11}:

Using the generalization relationship mentioned in (3) and taking into account that this time it is sphere (1) that emits the flux ($i = 1$), it becomes:

$$\sum_{j=1}^{j=n} F_{ij} = 1 \quad \Rightarrow \quad \sum_{j=1}^{j=2} F_{ij} = F_{11} + F_{12} = 1 \quad \Rightarrow \quad F_{11} = 1 - 1 = 0. \tag{4}$$

Chapter 4
Thermodynamics

Abstract This chapter studies macroscopic thermodynamics, which is the science of energy balance and prediction of the evolution of systems during the modification of their external or internal parameters. It studies energy transformations, matter, and equilibrium states of systems. It presents a universal character and applies to all physics and chemistry areas. The Gibbs paradox is one of the topics treated in this chapter.

The chapter begins with a precise course, followed by exercises in order of increasing difficulty. These exercises are solved in a very detailed way; they allow students to assimilate the course and help them prepare for exams. It is intended for first-year university students. It can also benefit bachelor's and doctoral students because of the deliberately simplified presentation of the treated concepts.

4.1 Introduction

Thermodynamics is the science of energy balance and prediction of the evolution of systems during the modification of their external or internal parameters. It studies the transformations of energies, matter, and equilibrium states of systems. It presents a universal character and applies to all the physics and chemistry areas. There are two ways to apprehend it, macroscopic thermodynamics and statistical thermodynamics.

4.1.1 Macroscopic Thermodynamics

It is based on two fundamental principles which we add three other principles:

- First principle of thermodynamics. This principle deals with the conservation of energy and introduces the internal energy function U.
- Second principle of thermodynamics. This principle concerns the evolution of systems and introduces the entropy function S.

© The Author(s), under exclusive license to Springer Nature Switzerland AG 2021
S. Khene, *Topics and Solved Exercises at the Boundary of Classical and Modern Physics*, Undergraduate Lecture Notes in Physics,
https://doi.org/10.1007/978-3-030-87742-2_4

- Third principle of thermodynamics or the Nernst principle which states that the entropy of a pure body is zero at $T = 0$ K.
- Fourth principle or zero principle of thermodynamics, which specifies that the temperature is an identifiable intensive quantity.
- Fifth principle of thermodynamics that derives from the Onsager theory which emphasizes a linear relationship between volumetric currents and thermodynamic forces.

This approach is purely phenomenological. It has the advantage of involving very few variables. It is well adapted to the understanding of various thermodynamic phenomena at the macroscopic scale.

4.1.2 Statistical Thermodynamics

This approach takes into account each real elementary particle contained in the studied system. It implements many variables and generates relatively complex calculations. It has the advantage of better understanding the matter behavior at the atomic scale.

Macroscopic thermodynamics and statistical thermodynamics are not contradictory. Indeed, it is possible, by starting from statistical thermodynamics with suitable hypotheses and a statistical treatment of some variables, to end up with macroscopic quantities described by macroscopic thermodynamic.

4.2 Fundamental Notions

It is very important to understand the physical meaning of each term used in thermodynamics in order to better apprehend its laws.

4.2.1 State of a System

Thermodynamics studies the matter and energy exchanges between a material environment called *system* and its surrounding called *external environment* or just *outside*. A macroscopic system is a portion of the space delimited by a real or an imaginary surface containing the studied material. It consists of a large number of atoms or molecules. This system is characterized by the average values of variables that are used to describe it and that are accessible to measurements. The outside is all that does not belong to the system. By convention, everything that the system receives from the outside is positively counted and all that it yields to the outside

Table 4.1 Different types of thermodynamic systems

System	Matter exchange	Energy exchange	Example
Isolated	No	No	Calorimeter
Closed	No	Yes	Electric batteries
Open	Yes	Yes	Human being

is negatively counted. These considerations allow one to define different types of systems met in macroscopic thermodynamics (Table 4.1).

The state of a macroscopic thermodynamic system is characterized by a minimum number of measurable physical quantities called *independent state variables*. These variables allow to reconstitute experimentally and unambiguously the state of the system with a set of perfectly defined properties. The most used variables are those of Gibbs that we can subdivide into two categories:

- Physical parameters such as the system temperature T, its pressure P, its volume V, etc.
- Composition parameters such as the mass m of the system, the number of moles n, etc.

The number of independent variables to describe a system depends on its nature and the studied problem. The choice of these parameters is free. Once these parameters are chosen, the values of other parameters are calculated from particular relationships called *state equations*.

4.2.1.1 Temperature

The intuitive approach we have about the notion of temperature is based on our physiological sensations. Temperature is not measurable because a measurable physical quantity can be compared to another known quantity of the same kind called *standard*. This is not the case with temperature, which is essentially an identifiable quantity. Its determination is based on the notion of thermal equilibrium between the measuring instrument and the studied system. It is based on the dilation and the pressure of solids, liquids, or gaseous bodies or any other physical property that varies with temperature. This general principle is applied in very different ways depending on temperature ranges and the nature of materials to be studied. The greater the kinetic energy of the atoms which make up the system, the higher its temperature. When these atoms are perfectly motionless, the body has no internal vibration: this state defines the zero temperature. Conversely, the temperature scale is open to infinity. A more accurate temperature definition will be given later (see Sect. 4.2.2.13). The unit of temperature in the international system is the kelvin, of symbol K. By convention, unit names are common names; they are written in lowercase such as the joule, ampere, volt, watt, pascal, henry, etc. There are other measurement systems prior to that and still in use such as the Celsius, Centigrade, and Fahrenheit scales.

Table 4.2 Some examples of temperatures

T (°C)	−273.15 °C	0 °C	20 °C
T (K)	0 K	273.15 K	293.15

Table 4.3 Some examples of temperatures

T (°C)	0 °C	20 °C	100 °C
T (°F)	32 °F	68 °F	212 °F

- *Celsius temperature scale*:

Its unit is the degree Celsius (°C) and not the celsius (C). We say, for example, that the current temperature in Annaba is 32 degrees Celsius (32 °C) and not 32 celsiuses (32 C). This scale of temperature is built on two remarkable points, the melting of the ice at 0 °C and the boiling of the water under the atmospheric pressure at 100 °C.

- *Kelvin temperature scale*:

This scale is defined from the temperature of the triple point of water (coexistence of water in liquid, solid, and gaseous forms), thermodynamic fixed point, at 273.16 kelvins (= 0.01 °C). Its unit is therefore the kelvin (K). This is the absolute temperature scale with the same increment as the Celsius temperature scale. Its values are either zero or positive (Table 4.2). The absolute zero is situated at −273.15 °C. It corresponds to a total absence of microscopic agitation. This scale is deduced from the Celsius scale by the following affine function:

$$T(\text{K}) = T(°\text{C}) + 273.15. \tag{4.1}$$

- *Fahrenheit temperature scale*:

Its unit is the degree Fahrenheit (°F) and not the fahrenheit (F). We say, for example, that the current temperature in New York is 74 degrees Fahrenheit (74 °F) and not 74 fahrenheits (74 F). This scale is used in Anglo-Saxon countries. It assigns a range of 180 °F between the solidification temperature of water and its boiling point. It fixes the solidification point of water at +32 °F and its boiling point at +212 °F (Table 4.3). This scale is deduced from the Celsius scale by the following affine function:

$$T(°\text{F}) = T(°\text{C}) \times 1.8 + 32. \tag{4.2}$$

- *Centigrade scale*:

Contrary to the Celsius temperature scale which is based on its unique reference point, the 0 °C, this scale of measurement is such that the points 0 and 100 are fixed. It is called *centigrade* because the two reference points are 100° apart. This scale is the ancestor of the Celsius scale. They are slightly different.

4.2.1.2 Pressure

The pressure P, exerted by a fluid (gas or liquid) at a point M of a real or an imaginary surface (S) that encompasses it, is defined by the ratio of the elementary force $\overrightarrow{dF}$ applied in the normal direction to this surface on the surface element dS, surrounding this point (Fig. 4.1):

$$\overrightarrow{dF} = P \, dS \, \vec{n}_{\text{ext}} \tag{4.3}$$

In the international system (SI), the unit of pressure, which is a force per unit area, is the pascal (Pa) where 1 pascal amounts to 1 N.m^{-2}. It is also possible to express the pressure in atmospheres, mm of mercury, and mm of water, of which the correspondence is:

$$1 \text{ atm} = 760 \text{ mm Hg} = 10.33 \text{ m } H_2O = 101\,325 \text{ Pa.} \tag{4.4}$$

In the C.G.S. system, the pressure is expressed in baryes (ba) with $1 \text{ ba} = 1 \text{ dyn}/1 \text{ cm}^2$ and $1 \text{ Pa} = 10 \text{ ba}$. Although they are outside the international system, some units continue to be widely used because they are better adapted to the dimensions of the studied phenomena. Among these accessory units, we have the bar with $1 \text{ bar} = 10^5 \text{ Pa}$ and the torr with $1 \text{ Torr} = 133.4 \text{ Pa}$. The pressure exerted by a gas on a surface is due to the innumerable collisions of molecules of fluid with the wall. It is given by:

$$P = \frac{1}{3} m n_V \left\langle \vec{v}^2 \right\rangle \tag{4.5}$$

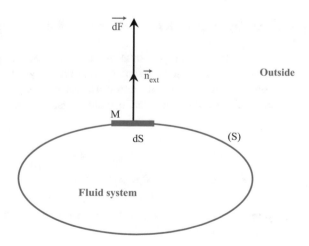

Fig. 4.1 Pressure notion

where m is the mass of one molecule, $n_V = N/V$ the number of molecules per unit volume, and $\left\langle \vec{v}^2 \right\rangle$ the mean squared speed. For a liquid, this pressure originates from the force of gravity:

$$\vec{\nabla} P = \rho\left(\vec{r}\right) \vec{g}\left(\vec{r}\right) \tag{4.6}$$

where $\vec{g}\left(\vec{r}\right)$ is the acceleration due to gravity and $\rho\left(\vec{r}\right)$ the density at the point $\vec{r}$ of fluid.

4.2.1.3 Intensive and Extensive Variables

In the physical sciences, an extensive variable is a physical quantity that is proportional to the size of the system. The volume, mass, energy, and number of moles of a system are examples of extensive variables. Conversely, an intensive variable is a quantity that does not depend on the quantity of matter present in the studied system. The temperature, pressure, volume, surface density, and chemical affinity are examples of intensive variables. In general, intensive and extensive quantities are bound. In this context, let us mention the internal energy of a gas, its temperature, momentum, and speed. As far as possible, it is better to characterize a system by intensive variables.

4.2.1.4 Homogeneous and Heterogeneous Systems

A homogeneous system is formed of one phase. A heterogeneous system is formed of several phases.

4.2.1.5 Open, Closed, Isolated, and Insulated Systems

An open system exchanges some matter and some energy with the outside environment, whereas a closed system exchanges only some energy with the external environment. An isolated system does not exchange either matter or energy with the external environment. An insulated system does not exchange heat with the outside.

4.2.2 Evolution of a System

A system evolves from one equilibrium state to another as a result of energy exchanges with the outside world.

4.2.2.1   State of Rest

The state of rest of a system satisfies two conditions:

- There is no exchange of energy between the system and the outside, and between the different parts of the system.
- The system is stable against external disturbances.

4.2.2.2   State of Equilibrium

The equilibrium state of a system satisfies three conditions:

- The temperature T is the same at every point of the system (thermal equilibrium).
- The system pressure does not change over time (mechanical equilibrium).
- There is no change in the chemical composition of the system over time (chemical equilibrium).

4.2.2.3   Transformation of a System

A transformation is an evolution of the system from an initial state to a final state under the action of an external disturbance (Fig. 4.2).

4.2.2.4   Cyclic Transformation

A transformation is called *cyclic* if the final state is identical to the initial state; otherwise it is called *open* or simply *transformation*.

Fig. 4.2 Schematization of thermodynamic transformation

4.2.2.5 Irreversible Transformation

An irreversible transformation is induced by a sudden modification of the external environment. The system evolves rapidly through badly defined intermediate states. Only the initial state and the final state are well-defined equilibrium states (Fig. 4.3a). Once the final state of equilibrium is reached, if the outside is brought back to its previous state, the system returns to the initial state without going through the same path as the first one.

4.2.2.6 Quasi-static Transformation

The quasi-static transformation is a progressive modification of the external environment which leaves the system time to equilibrate at each stage. During the return, the system passes through the same intermediate equilibrium states, but in the opposite direction.

4.2.2.7 Reversible Transformation

The reversible transformation is formed of a succession of equilibrium states infinitely close to each other. This is the limiting case of the quasi-static transformation (Fig. 4.3b). It satisfies three conditions:

- The system evolves by a succession of equilibrium states infinitely close to each other.
- The values of system parameters are known at all times.

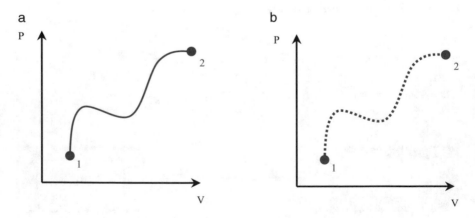

Fig. 4.3 Transformation. (**a**) Irreversible: the system quickly evolves via badly defined intermediate states. (**b**) Reversible: the system evolves by a succession of equilibrium states infinitely close to each other

- The possibility for a system to spontaneously and accurately regain its state immediately prior to an infinitesimal modification of external conditions. It is the same for the surrounding environment. This is an evolution of the system for which no entropy is produced. Entropy is a measurable physical quantity that is most often associated with a state of disorder, randomness, or uncertainty. It characterizes the degree of disorganization or unpredictability of the information content of a system. It will be formally defined in Sect. 4.6.3.

4.2.2.8 Adiabatic and Isentropic Transformations

An adiabatic transformation is a transformation without heat exchange. The temperature can vary during such a transformation! An isentropic process is a process in which the entropy of the system remains constant.

4.2.2.9 Isothermal Transformation

An isothermal process is a change of a system, in which the temperature remains constant.

4.2.2.10 Isobaric Transformation

An isobaric process is a thermodynamic process in which the pressure stays constant.

4.2.2.11 Isochoric Transformation

An isochoric process, also called *a constant volume process*, is a thermodynamic process during which the volume of the closed system remains constant.

4.2.2.12 State Equation

In a system in equilibrium, all the variables are not independent but linked by the so-called equations of state. These equations can be of type $f(P,V,T) = 0$. We can also have other quantities than P, V, and T and not consider a gas but a solid or a liquid. Many state equations exist to describe real fluids. Among these let us quote the van der Waals equation, which has the merit of qualitatively reproducing isotherms and describing the liquid-gas transition:

$$\left(P + \frac{n^2a}{V^2}\right)(V - nb) = nRT \qquad\qquad (4.7)$$

where a and b are positive real characteristics of the substance. A physical discussion shows that the coefficients a, b describe some effects of the interaction between gas molecules. The case where $a = 0$ and $b = 0$ describes the ideal gas; this is a model where molecules do not interact with each other.

4.2.2.13   Ideal Gas

The ideal gas is a model which describes the behavior of a rarefied gas. In this model, gas molecules are sufficiently distant from each other so that we can neglect the electrostatic interactions, which depend on the nature of gas. These molecules behave like small billiard balls in a chaotic movement which bounce against each other without the dissipation of energy by friction (Fig. 4.4).

The ideal gas law is an equation of state for an ideal gas, given by:

$$Pv = rT. \qquad\qquad (4.8)$$

This equation shows that for an ideal gas, the multiple Pv is proportional to T where P is the pressure of gas in pascals; $v = V/m$, the specific volume in $\mathrm{m}^3.\mathrm{kg}^{-1}$; T the temperature in kelvins; and r the constant of the considered gas in $\mathrm{J.K}^{-1}.\mathrm{kg}^{-1}$. Equation (4.8) can also be expressed as a function of the mass:

$$PV = mrT \qquad\qquad (4.9)$$

where V is the volume occupied by gas in m^3 and m its mass in kg. Equation (4.9) shows that one of these three state variables (pressure, volume, or temperature) depends on the other two variables:

Fig. 4.4 Chaotic movement without energy dissipation of an ideal gas

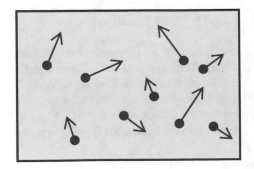

$$P = f(V, T) \quad \text{or} \quad V = g(P, T) \quad \text{or} \quad T = h(P, V). \tag{4.10}$$

It is still possible to express (4.8) according to the matter in moles:

$$PV = Nk_B T = \frac{N}{N_A} N_A k_B T = nRT \tag{4.11}$$

where T is the temperature in kelvins, V the volume in m^3, $k_B = 1.38 \times 10^{-23}$ J.K^{-1} Boltzmann's constant, N the number of molecules of the considered gas, and $n = N/N_A$, the number of moles with $N_A = 6.022 \times 10^{23}$ the number of Avogadro. In (4.11), R denotes the constant of ideal gas with $R = N_A k_B = 8.314$ J.K^{-1}.mol^{-1}. Let us recall that 1 mole of ideal gas at 1 atm and 0 °C occupies 22.4 liters. From (4.11), we also obtain:

$$PV = Nk_B T \quad \Rightarrow \quad P = \frac{N}{V} k_B T = n_V k_B T \tag{4.12}$$

where $n_V = N/V$ represents the number of molecules per unit volume. From (4.4), we deduce:

$$P = \frac{1}{3} m \, n_V \left\langle \vec{v}^2 \right\rangle = \frac{2}{3} \times \frac{1}{2} m \, n_V \left\langle \vec{v}^2 \right\rangle. \tag{4.13}$$

By equating (4.13) and (4.12), we obtain:

$$P = \frac{2}{3} \times \frac{1}{2} m \, n_V \left\langle \vec{v}^2 \right\rangle = n_V k_B T \quad \Rightarrow \quad \frac{1}{2} m \left\langle \vec{v}^2 \right\rangle = \frac{3}{2} k_B T. \tag{4.14}$$

Equation (4.14) allows interpreting the temperature as a measurement of the kinetic energy. It is related to the molecular agitation of an ideal gas. We can define the concept of temperature by using the zero principle of thermodynamics, which states that two systems in a prolonged contact converge to a state of thermal equilibrium and that two systems in thermal equilibrium with a third are themselves in thermal equilibrium. Temperature therefore appears as an intensive variable that takes the same value for all systems in thermal equilibrium.

4.2.2.14   Graphical Representation of the Evolution of a System

The variations of the state of a system, as a consequence of transformations, can be represented by various diagrams which allow to follow the evolution of the system such as the diagram of Clapeyron (P,V) represented in Fig. 4.5, the diagram of Amagat (PV,P), isentropic diagrams (T,S), the diagram (H,S), the diagram of Mollier (P,H), etc. The variables H and S will be formally defined later.

Fig. 4.5 Diagram of
Clapeyron: it is a diagram in
which we carry the volume
in abscissa and the pressure
in ordinate

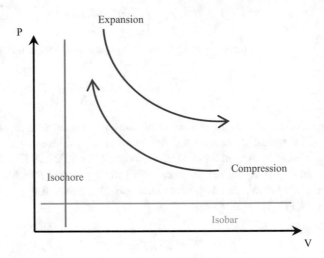

4.2.2.15 Functions of State

A system can perform a transformation that brings it from the initial equilibrium state
1 to the final equilibrium state 2 by taking different paths. In general, the variation
ΔX of a physical quantity X depends on the path followed during a transformation.
But, in thermodynamics, there are functions f related to variables of state, of which
the variations during a transformation are independent of the followed path. Such
functions are called *functions of state*. The variation Δf of these functions is an exact
differential:

$$\Delta f = f_2 - f_1. \tag{4.15}$$

The internal energy U, the enthalpy H, and the entropy S, which will be introduced later, are state functions because they do not depend on the followed path
between the initial and final states. The work W and the heat Q, also introduced later,
are not state functions because they depend on the followed path.

4.2.2.16 Partial Pressure

Let us consider a mixture of several gases, which occupy a closed chamber of the
volume V at the temperature T. The partial pressure P_i of gas i is the pressure, which
it would have if it occupied alone the volume V at the temperature T. The pressure of
the gaseous mixture is the sum of the partial pressures of its constituents:

$$P = \sum_i P_i.$$ (4.16)

Equation (4.16) defines the Dalton empirical law. The mixture, which verifies it, is called *ideal*.

4.3 First Principle of Thermodynamics

4.3.1 Introduction

A transformation causes the variation of state variables of a system as a result of an exchange of the mass and the energy with the external environment. In physics, the energy characterizes the ability to modify a state, to produce a work generating a movement, a light, or a heat. Any action or change of the state requires energy to be exchanged. This energy appears under multiple forms.

4.3.2 Internal Energy

4.3.2.1 Definition

The internal energy of a system, denoted U, is its contents in energy. Each solid, liquid, or gaseous system is formed of particles (atoms or molecules) which incessantly vibrate. To this microscopic movement is associated a kinetic energy Ec_i for each particle. Moreover, between these particles there exist interaction forces with which is associated a potential energy Ep_i for each particle. The concept of the internal energy brings together all the kinetic and potential energies of n particles that make up this system. It represents the total amount of the mechanical energy stored inside a system.

$$U = \sum_{i=1}^{n} Ec_i + \sum_{i=1}^{n} Ep_i.$$ (4.17)

4.3.2.2 Properties

At thermal equilibrium, the internal energy U of a system is a state function which depends only on the initial state and the final state; it is identified on the macroscopic scale and has a well-defined value; it is expressed in joules or in calories (1 cal = 4.1855 J). The internal energy of a system is modified as a result of energy exchanges with the outside in the form of heat and work. Let us note that the internal energy of an ideal gas depends only on the temperature. Its variation is therefore zero for any isothermal transformation. The demonstration can be found in Exercise 25.

4.3.3   Heat

Heat, denoted Q, is a form of energy that is not a state function because it depends on the followed path to go from the initial state to the final state. It is exchanged by shocks between particles in disordered and incessant movement. It is a chaotic transfer, associated with a potential: the temperature. Such spontaneous heat transfer always occurs from a region of higher temperature to another region of lower temperature. Do not confuse temperature and heat. Indeed, temperature is a state function of a system that characterizes its average atomic or molecular excitation level, whereas heat is the energy that must be brought or removed from the previous system to change its temperature. Thus, it is necessary to bring a quantity of the heat Q to a pan filled with water at the temperature T_1 to raise its temperature to a value T_2. In thermodynamics, there are two types of heat, the sensible heat and the latent heat.

4.3.3.1   Sensitive Heat

Sensitive heat is related to a temperature change ΔT of the system following a warming or a cooling without the phase transition of the system. It is proportional to the amount of matter expressed in kilograms or moles and to the difference in temperature. For an infinitesimal transformation, the quantity of heat exchanged between the system and the outside has the following expression:

$$dQ = m\,C\,dT. \tag{4.18}$$

Here m is the mass of matter that makes up the system, expressed in kg, and C, the mass heat capacity in $J.kg^{-1}.K^{-1}$. It depends on the nature of matter that forms the system and dt the infinitesimal variation of the temperature between two neighboring states, expressed in K. This quantity of heat can also be expressed as a function of the number of moles n of the system:

$$dQ = n\,C\,dT. \tag{4.19}$$

The molar heat capacity is expressed in this case in $J.mol^{-1}.K^{-1}$. For a finite transformation between the equilibrium states 1 and 2, the exchanged heat Q has for expression, by starting from (4.18) and by considering the mass heat capacity independent of the temperature:

$$Q = \int_1^2 dQ = \int_1^2 m\,C\,dT = m\,C \int_1^2 dT = m\,C\,(T_2 - T_1) = m\,C\Delta T. \quad (4.20)$$

4.3.3.2 Latent Heat

Latent heat is the exchanged heat Q with the external environment during a change of the physical state of matter that composes the system (solid, liquid, gas). It is denoted by L. When it is expressed for 1 kg of matter, we speak about *the mass latent heat*, and when it is expressed for 1 mole, it is *the latent heat molar*:

$$Q = m\,L \quad (4.21)$$

with L in J/kg and

$$Q = n\,L \quad (4.22)$$

with L in mol/kg.

For each physical state of matter, there are six types of latent heats related to six changes of the physical state: melting, solidification, vaporization, liquefaction, sublimation, and condensation (Fig. 4.6).

4.3.3.3 Calorimetry

- *Definition of calorie*:

The usual unit of the quantity of heat is the calorie (in abbreviation cal), of which the definition is as follows: calorie is the quantity of heat necessary to raise the temperature of 1 g of liquid water from 14.5 to 15.5 °C under the atmospheric pressure. Its multiple units are the kilocalorie (1 kcal $= 10^3$ cal) and the therm (1 thm $= 10^6$ cal). The calorimetric measurements allow determination of the quantities of heat, heat capacities, and latent heats. They are based on three

Fig. 4.6 Changes of the physical state of matter

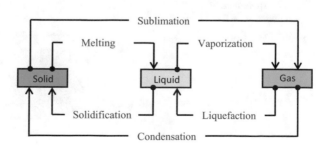

principles: the principle of equality of heat exchanges, the principle of inverse transformations, and the principle of the equilibrium temperature.

– *Principle of equality of heat exchanges*:

When system 1 exchanges only heat with system 2, the quantity of heat gained by one of the two systems is equal to the quantity of heat lost by the other system.

– *Principle of inverse transformations*:

The quantity of heat which is necessary to supply to a system to pass from state 1 to state 2 is equal to the quantity of heat that it restores to the outside environment when it comes back from state 2 to state 1.

– *Principle of the equilibrium temperature*:

When a system 1 is in contact with a system 2, of respective temperatures T_1 and T_2, heat flows from the warmer system to the colder system until these two systems reach a common equilibrium temperature T.

4.3.4 Work

The work of pressure forces, denoted W, is a transfer associating a force to a displacement. It is a form of energy expressed in joules or in calories. The work depends on the followed path; it most often results from a volume variation of a cylinder-piston set of a thermal engine. Let us consider for this purpose a cylinder containing a fluid and separated from the outside by a piston which moves along the Ox-axis under the action of an external force $F_{ext} = P_{ext}\, S$ where P_{ext} is the external pressure and S the circular surface of the piston (Fig. 4.7). The elementary work exercised by this force is:

$$\delta W = \vec{F}_{ext} \cdot \vec{dx} = F_{ext} \cdot dx = P_{ext}\, S\, dx = P_{ext}\, dV_{ext} \qquad (4.23)$$

Fig. 4.7 Movement of a piston under the action of an external force $\vec{F}_{ext}$

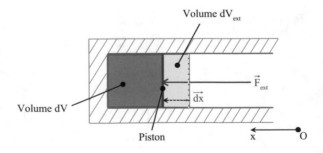

Volume dV_{ext}

$\vec{F}_{ext}$

$\vec{dx}$

Volume dV

Piston

Fig. 4.8 The work
W represents the area under
the curve (P,V)

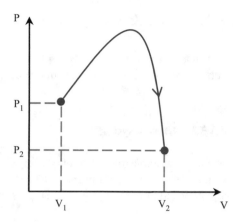

where dV_{ext} is the differential increase of the volume of the external environment and thus opposed to the lost volume dV by the cylinder, from where:

$$\delta W = P_{ext}\ dV_{ext} = -P_{ext}\ dV. \tag{4.24}$$

The elementary work δW is an algebraic quantity:

- If $\delta W > 0$: the system receives work from the outside environment by undergoing a compression $(dV < 0)$.
- If $\delta W < 0$: the system provides work to the outside environment by undergoing a relaxation $(dV > 0)$.

For a finite transformation from state 1 to state 2, that gives:

$$W = -\int_{1}^{2} P_{ext}\ dV. \tag{4.25}$$

Moreover, in the case of a quasi-static or reversible transformation, we have at each instant $P_{ext} = P$ so that $\delta W = -P\ dV \Rightarrow -W$ represents the area under the curve (P,V) in the diagram of Clapeyron (Fig. 4.8).

In physics, power is the rate of doing work, the amount of energy transferred per unit time. Having no direction, it is a scalar quantity. Its unit in the international system is the watt (W) with $1\ W = 1\ J/s$. Other units are also used as the horsepower with $1\ hp = 735.5\ W$ (metric horsepower, French system) or $1\ hp = 745.7\ W$ (imperial horsepower, Anglo-Saxon system). This unit is used in the automotive industry. The creation of this unit is due to the Scottish engineer James Watt (1736–1819).

4.3.5 *Foundations of the First Principle*

The first principle of thermodynamics, as any principle in physics, is not proved; it is verified by experiment. It states that spontaneous creation of energy does not exist.

4.3.5.1 Closed System

A closed system exchanging only energy to the external environment, the variation of its internal energy during a transformation from state 1 to state 2 is equal to the algebraic sum of the exchanged energies in the form of work W and heat Q:

$$\Delta U = U_2 - U_1 = W + Q. \tag{4.26}$$

For an infinitesimal transformation, we write:

$$dU = \delta W + \delta Q. \tag{4.27}$$

Equation (4.26) reveals that when a system evolves from a state of equilibrium 1 to a state of equilibrium 2, the algebraic sum of mechanical and calorific energies received or transferred by the system to the outside depends on the initial state and the final state and not on the followed path to go from 1 to 2. The internal energy is thus a state function.

4.3.5.2 Isolated System

An isolated system not exchanging either matter or energy with the outside environment, its internal energy is preserved:

$$\Delta U = U_2 - U_1 = W + Q = 0 \quad \Rightarrow \quad U_1 = U_2 = U = \text{constant.} \tag{4.28}$$

Equation (4.28) implicitly assumes that mechanical work and heat are counted in the same unit with 1 cal = 4.1852 J.

4.3.5.3 Open System

An open system exchanging matter and energy with the external environment, the variation of its internal energy will include a contribution related to the matter flow.

4.4 Enthalpy

Enthalpy, denoted H, comprises the internal energy of a system to which is added the work which this system must exert against the external pressure to occupy its own volume. Enthalpy is a chemical potential which predicts the evolution and the equilibrium of a thermodynamic system, and from which we can deduce all its properties. It is an extensive state function which plays a privileged role in the isobaric transformations which are very useful in chemistry. It is defined by:

$$H = U + PV. \tag{4.29}$$

Here the term PV is the expansion or the compression energy of the system. H is always greater than U. Enthalpy is expressed in joules or in calories. For an infinitesimal transformation, we have, starting from (4.24) and (4.27):

$$dU = \delta Q + \delta W = dQ - PdV. \tag{4.30}$$

By introducing the enthalpy function, we get:

$$dH = dU + d(PV) = dU + PdV + VdP = dQ - PdV + PdV + VdP. \tag{4.31}$$

In the end, we obtain:

$$dH = dQ + VdP. \tag{4.32}$$

4.5 Heat Capacities

Let us consider an infinitesimal transformation that transforms a system from a state P,V,T to a very close state $P + dP$, $V + dV$, $T + dT$ by receiving a quantity of heat δQ. The heat capacity characterizes the response in temperature of the system to an influx of heat. This is the energy that must be brought to a system to increase its temperature by 1 kelvin. It is expressed in joules by kelvins (J/K). It is an extensive quantity: the more the quantity of matter is important, the more the thermal capacity is big. The constant volume capacity characterizes a system that undergoes an isochoric transformation. It is given by:

$$C_V = \left(\frac{\partial U}{\partial T} \right)_V. \tag{4.33}$$

The constant pressure capacity characterizes a system that undergoes an isobaric transformation. It is given by:

$$C_P = \left(\frac{\partial H}{\partial T}\right)_P .$$ (4.34)

It is sometimes useful to use the mass thermal capacity, expressed in $J.kg^{-1}.K^{-1}$, or the molar thermal capacity, expressed in $J.mol^{-1}.K^{-1}$. For an ideal gas, Eqs. (4.33) and (4.34) lead to the equation of Mayer:

$$C_P - C_V = nR$$ (4.35)

where n is the number of moles and R is the constant of ideal gases in $J.K^{-1}.mol^{-1}$. The heat capacities C_P and C_V are in $J.K^{-1}$. By taking:

$$\gamma = \frac{C_P}{C_V}$$ (4.36)

we always get for an ideal gas:

$$C_V = \frac{nR}{\gamma - 1}, \quad C_P = \frac{\gamma nR}{\gamma - 1}, \quad PV^\gamma = Cte \text{ (adiabatic relaxation).}$$ (4.37)

For liquid water under a pressure of 1 atmosphere, the heat capacity at 15 °C is exactly 1 $cal.g^{-1}.K^{-1}$.

4.6   Second Principle of Thermodynamics

4.6.1   Introduction

The second principle of thermodynamics gives the direction of the evolution of transformations and introduces a distinction between the reversible and irreversible transformations.

4.6.2   Statements of the Second Principle

4.6.2.1   Clausius's Statement

The heat transfer towards a higher temperature cannot be done without the energy input. This postulate is an expression of the general principle of the irreversibility of real transformations which take place in a definite direction and are always irreversible, in particular because of the existence of friction forces.

4.6.2.2   Kelvin's Statement

It is impossible to completely transform heat into work. This postulates the impossibility of building an engine by taking a quantity of heat from a single source of heat to provide an equivalent quantity of work. Thus, so that an engine works continuously, it is necessary, in addition to a high-temperature source to capture heat, to have a low-temperature sink to reject the heat it no longer needs. To achieve this enigmatic principle, let us imagine we want to create work by taking heat from a hot system at the temperature T. To this end, we attach to this system a cylinder-piston set filled with a gas of volume V_1 and of temperature $T_1 < T$ and let this gas push the piston as it receives heat (Fig. 4.9a). Once a work has been supplied to the external environment, the fluid increases in volume and reaches a volume V_2 and a temperature T_2 between T and T_1 (Fig. 4.9b). If we wish to continue to uninterruptedly transform heat into work in this way, we have to reduce the volume of gas to its initial level V_1 by cooling it. The only way to extract some heat from this gas is to put it in contact with a colder system of temperature $T' < T_2$. This generates a non-zero compressive work in the opposite direction with the difference being that this time the heat and the work transfers are lower than on the outward journey; this process has an industrial interest, hence the notion of the yield of a thermal engine (Fig. 4.9c).

4.6.2.3   Mathematical Statement

Let us imagine a heat engine which takes a heat Q from a reservoir at low temperature T_2 and yield it to a reservoir at higher temperature T_1. This transfer being impossible without a contribution of energy, this cycle is therefore unrealistic in practice. The energy balance of such a transformation is written mathematically as follows:

$$\sum \frac{dQ}{T} = \frac{Q}{T_2} - \frac{Q}{T_1} > 0 \quad \text{because} \quad T_1 > T_2. \tag{4.38}$$

In Eq. (4.38), we have complied with the sign convention mentioned in the previous paragraphs, namely, that the received energy by the system is positively counted, while the energy yielded by it is negatively counted. Since the previous process is impossible according to the Clausius statement, we deduce that for a thermal engine to actually work, its energy balance must imperatively verify the following conditions:

$$\sum \frac{dQ}{T} \leq 0. \tag{4.39}$$

The conditions described in Eq. (4.39) with the sign ($=$) for a reversible cycle and the sign ($<$) for an irreversible cycle are the mathematical formulation of the second

a

Heat transfer

Hot source of temperature T

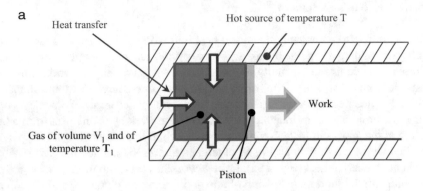

Work

Gas of volume V$_1$ and of
temperature **T**$_1$

Piston

b

Hot source of temperature T

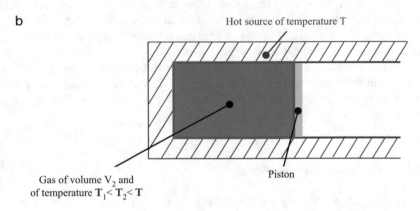

Gas of volume V$_2$ and
of temperature **T**$_1$< **T**$_2$< **T**

Piston

Cold sink of temperature T′

c   Release of heat

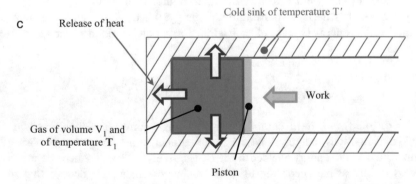

Work

Gas of volume V$_1$ and
of temperature **T**$_1$

Piston

Fig. 4.9 (**a**) To create work, we take heat from a hot system of temperature T. To this end, we attach to this system a cylinder-piston set filled with a gas of volume V_1 and of temperature $T_1 < T$, and we let this gas push the piston as it receives heat. (**b**) Once a work has been provided to the external environment, the fluid increases in volume and reaches a volume V_2 and a temperature T_2 between T and T_1. (**c**) To continue to uninterruptedly convert heat into work in this way, we must reduce the volume of gas to its initial level V_1 by cooling it. The only way to extract some heat from this gas is to put it in contact with a colder system of the temperature $T' < T_2$. This generates a non-zero compressive work in the opposite direction with the difference being that this time the heat and the work transfers are lower than on the outward journey; this process has an industrial interest, hence the notion of the yield of a thermal engine

principle of thermodynamics that summarizes the irreversibility postulates of Clausius and Kelvin.

4.6.3 Entropy

Entropy, denoted S, is the last of five physical quantities which characterizes the state of a thermodynamic system, namely, the temperature T, the pressure P, the volume V, the internal energy U, and finally the entropy S. We will use Eq. (4.39) as a starting point to introduce this new quantity.

4.6.3.1 Reversible Cycle

Let us consider a reversible thermodynamic cycle composed of two reversible transformations, the one going from A to B and the other from B to A (Fig. 4.10).

Equation (4.39) can be replaced by an integral while respecting the postulates of the irreversibility of Clausius and Kelvin. The energy balance of this process is:

$$\int_A^B \frac{dQ(1)}{T} + \int_B^A \frac{dQ(2)}{T} = \int_A^B \frac{dQ(1)}{T} - \int_A^B \frac{dQ(2)}{T} = 0 \;\Rightarrow\; \int_A^B \frac{dQ(1)}{T}$$

$$= \int_A^B \frac{dQ(2)}{T} = \int_A^B \frac{dQ_{\text{rev}}}{T}. \tag{4.40}$$

Equation (4.40) reveals that the integral for a reversible transformation does not depend on the followed path. It only depends on the final state and the initial state. It is an extensive state function. This integral stems from the variation of a quantity S called *entropy*; it is defined by:

Fig. 4.10 Reversible cycle

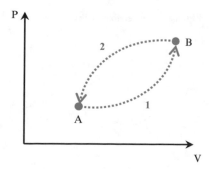

$$\Delta S = S_B - S_A = \int_A^B dS = \int_A^B \frac{dQ_{\text{rev}}}{T}. \tag{4.41}$$

4.6.3.2   Irreversible Cycle

Let us consider an irreversible cycle formed by an irreversible transformation AB and a reversible transformation BA (Fig. 4.11).

Equation (4.39) can be replaced by an integral while respecting the postulates of the irreversibility of Clausius and Kelvin. The energy balance of this process is:

$$\int_A^B \frac{dQ_{\text{irr}}}{T} + \int_B^A \frac{dQ_{\text{rev}}}{T} = \int_A^B \frac{dQ_{\text{irr}}}{T} - \int_A^B \frac{dQ_{\text{rev}}}{T} < 0 \Rightarrow \int_A^B \frac{dQ_{\text{rev}}}{T} > \int_A^B \frac{dQ_{\text{irr}}}{T}$$

$$\Rightarrow \Delta S > \int_A^B \frac{dQ_{\text{irr}}}{T}. \tag{4.42}$$

For an elementary transformation, we have:

$$dS > \frac{dQ_{\text{irr}}}{T} \quad \Rightarrow \quad dS = \frac{dQ_{\text{irr}}}{T} + \sigma \tag{4.43}$$

where σ represents a source of entropy characterizing the irreversibility of the transformation $\Rightarrow$ there is a creation of entropy. Equation (4.43) shows that the entropy of a thermodynamic system can only be positive or zero. It constitutes the most general statement of the second principle of thermodynamics with $\sigma = 0$ for a reversible transformation and $\sigma \neq 0$ for an irreversible transformation.

Fig. 4.11 Irreversible cycle

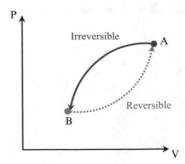

4.6.3.3 Isolated System

As an isolated system does not exchange either matter or energy with the outside environment, we have $dQ = 0 \Rightarrow dS = 0$. During the evolution of a closed system from state 1 to state 2, the entropy can only increase or remain constant.

$$\Delta S = S_2 - S_1 \geq 0. \tag{4.44}$$

4.6.3.4 Spontaneous Heat Transfer and Disorder Notion

Let us consider an isolated system, separated by an insulating partition into two compartments at different temperatures T_A and T_B such as $T_A > T_B$. Let us remove the partition and look at what direction the heat will flow. To do this, let us suppose that an infinitesimal quantity of heat dQ passes from compartment A towards compartment B and let us determine the sign of dQ (Fig. 4.12).

For each compartment taken apart, we have:

$$dS_A = \frac{dQ}{T_A} \quad \text{and} \quad dS_B = \frac{dQ}{T_B}. \tag{4.45}$$

The total variation of entropy of the enclosure after the removal of the partition is:

$$dS = dQ \left(\frac{1}{T_B} - \frac{1}{T_A} \right) > 0. \tag{4.46}$$

We have for an isolated system $dS > 0$, and as $T_A > T_B$, it follows that $dQ > 0$, that is to say that compartment B well receives heat as we supposed at the beginning. We see that the second principle well explains the privileged and the irreversible sense of the flow of heat from high temperatures to lower temperatures. This irreversible process of heat will continue until the temperatures in both compartments become equal. This example well illustrates, on the one hand, that the second principle is intimately linked to the notion of disorder and, on the other hand, the natural tendency of the thermodynamic system to increase its disorder. Indeed, in the final state, the gas molecules that were initially confined in each compartment were well mixed and distributed homogeneously throughout the enclosure. The disorder has thus reached its maximal threshold. The notion of entropy was introduced in 1865 by

Fig. 4.12 Spontaneous transfer of heat between two isolated compartments

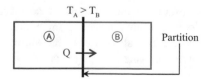

the German physicist Rudolf Clausius. This quantity is expressed in J.kg^{-1}. At the absolute zero where nothing moves, the entropy is zero.

4.6.3.5 Thermodynamic Identity

The entropy equation which we have just introduced does not allow specifying the exact form of the function S. This one is determined by very precise physical considerations. For a system in the thermodynamic equilibrium, we have:

$$\frac{1}{T} = \left(\frac{\partial S}{\partial U}\right)_{V,N} \quad \text{and} \quad \frac{P}{T} = \left(\frac{\partial S}{\partial V}\right)_{U,N} \tag{4.47}$$

where N is the number of molecules contained in the system. At N fixed, the system is isolated. Equation (4.47) is stored more easily in the following form:

$$dU = T\, dS - P\, dV. \tag{4.48}$$

This equation is called *thermodynamic identity*. Compared with the infinitesimal form of the first principle of thermodynamics:

$$dU = \delta W + \delta Q. \tag{4.49}$$

We see that in the reversible case, we have a term-by-term equality:

$$\delta W_{\text{rev}} = -P\, dV \tag{4.50}$$

and therefore:

$$\delta Q_{\text{rev}} = T\, dS. \tag{4.51}$$

This brings us back to a different path to Eq. (4.40).

4.6.4   The Gibbs Paradox

4.6.4.1 Introduction

The physical paradoxes go against the common sense. They are nevertheless indispensable to the progress of science. The Gibbs paradox is at the origin of the indiscernibility notion from which ensues all quantum mechanics. It intervenes during the calculation of the entropy of mixing of two ideal gases. It was named after the American physicist and chemist Josiah Willard Gibbs who discovered it in 1861.

4.6.4.2   Entropy of an Ideal Gas

Let us consider n moles of an ideal gas enclosed in an enclosure, of which the entropy depends on the temperature T and the volume V. The thermodynamic identity defined in Eq. (4.48) is written as:

$$dU = T\, dS - P\, dV \quad \Rightarrow \quad dS = \frac{dU}{T} + P\frac{dV}{T} \tag{4.52}$$

and by starting from (4.11) and (4.34):

$$P = \frac{nRT}{V} \quad \text{and} \quad U = n\, C_V T. \tag{4.53}$$

From (4.52) and (4.53), we deduce:

$$dS = n\left(C_V \frac{dT}{T} + R\, \frac{dV}{T} \right). \tag{4.54}$$

- *The volume V remains constant* $\Rightarrow dV = 0$:

In this case, the entropy S is a function of T only, which brings us to the derivative:

$$dS = n\, C_V \frac{dT}{T}. \tag{4.55}$$

Knowing that $\ln T$ is a primitive of $1/T$, we deduce:

$$S = n\, C_V\, ln(T) + \text{Cte.} \tag{4.56}$$

In fact, we consider as a constant all that does not vary with the studied problem. So, in this case, any function of V and n is necessarily a constant. For this reason, we write the previous solution in the form:

$$S = n\, C_V\, ln(T) + f(V, n) \tag{4.57}$$

where f is an unknown function of V and n.

- *The temperature T remains constant* $\Rightarrow dT = 0$:

In this case, the entropy S is a function of V only, of which we know the derivative:

$$\frac{dS}{dV} = \frac{nR}{V}.$$

(4.58)

We deduce:

$$\frac{dS}{dV} = \frac{d}{dV}\left[n\,C_V\,ln(T) + f(V,n),\right] = \frac{d}{dV}f(V,n) = \frac{nR}{V} \Rightarrow f(V,n)$$

$$= nR\,ln(V) + \text{Cte}'$$

(4.59)

As before, the constant Cte' may be a function of T and n because they are constants in this case. But, we know that $f(V,n)$ is not a function of T. The entropy is thus written as:

$$S = n\,C_V\,ln(T) + nR\,ln(V) + s_0(n)$$

(4.60)

where $s_0(n)$ is a function of n only. To determine its form, let us use the extensive nature of entropy. Let us consider for this purpose a new system formed of $n' = \lambda n$ moles occupying a volume $V' = \lambda V$ at the same temperature T as the previous system. Let us use the entropy expression obtained above for the new system:

$$S' = n'\,C_V\,ln(T) + n'R\,ln(V') + s_0(n')$$
$$= \lambda n\,C_V\,ln(T) + \lambda nR\,ln(\lambda V) + s_0(\lambda n).$$

(4.61)

Since entropy is an extensive quantity, its value for the new system is:

$$S' = \lambda S.$$

(4.62)

This allows us to write the expression of S' in the following form:

$$S' = \lambda S = \lambda\left[nC_V\,ln(T) + nR\,ln(V) + s_0(n)\right].$$

(4.63)

We deduce from (4.61) and (4.63):

$$S' = \lambda n\,C_V\,ln(T) + \lambda nR\,ln(\lambda V) + s_0(\lambda n)$$
$$= \lambda\left[nC_V\,ln(T) + nR\,ln(V) + s_0(n)\right].$$

(4.64)

In the end, we arrive at:

$$\lambda n\,R\,ln(\lambda) + s_0(\lambda n) = \lambda s_0(n).$$

(4.65)

Equation (4.65) is a satisfied mathematical equation whatever λ and n. Let us put:

$$\lambda = \frac{1}{n}. \tag{4.66}$$

We find:

$$-R\,ln(n) + s_0(1) = \frac{1}{n}\,s_0(n) \quad \Rightarrow \quad s_0(n) = ns^0 - nR\,ln(n) \tag{4.67}$$

where $s_0(1) = s^0$ is a constant. By putting this expression in (4.60), we obtain:

$$S = n\left[C_V\,ln(T) + R\,ln\left(\frac{V}{n}\right) + s^0\right] \tag{4.68}$$

where V/n is the molar volume, whereas s^0 is a constant which may depend on the nature of gas, but stays without big meaning because it is the entropy variations that present a practical interest in most of the cases. Equation (4.68) reveals that when the temperature T tends towards 0, entropy does not become zero. The third principle of thermodynamics is thus put into default. This means that the ideal gas is a model with a limited range of validity.

4.6.4.3   Entropy of Mixing

To study the Gibbs paradox, let us consider an isolated chamber separated into two compartments of the same volume V by a movable partition. In one compartment there are n moles of an ideal gas 1 and in the other, n moles of an ideal gas 2 at the same temperature. The law of ideal gases implies the equality of pressures in both compartments. The outer walls are adiabatic and non-deformable so that the internal energy of the system formed by all two gases remains constant. We are interested in the evolution of entropy when we remove the partition, allowing the mixing of the two gases. Depending on whether the gases are identical or not, we expect two different results.

- *Different gases*: the process is irreversible, and according to the second principle of thermodynamics, entropy must increase.

Let us consider the case where the gases in the two compartments are different. Before the removal of the separating partition, gas 1 is defined by (V,T,n) and gas 2 by (V,T,n). After the removal of the partition, gases 1 and 2 are mixed and each gas occupies all the volume that is offered. The entropy of the enclosure is the sum of entropies of the two gases. In the initial state, the entropy S_i of the enclosure is worth by starting from (4.68):

$$S_i = S_i^1(V,T,n) + S_i^2(V,T,n)$$

$$S_i = n\left[C_V \, ln(T) + R \, ln\left(\frac{V}{n}\right) + s_1^0\right] + n\left[C_V \, ln(T) + R \, ln\left(\frac{V}{n}\right) + s_2^0\right]$$
$$= 2n\left[C_V \, ln(T) + R \, ln\left(\frac{V}{n}\right)\right] + n\left(s_1^0 + s_2^0\right). \tag{4.69}$$

In the final state, each gas occupies the volume 2 V. The entropy S_f of the enclosure is:

$$S_f = n\left[C_V \, ln(T) + R \, ln\left(\frac{2V}{n}\right) + s_1^0\right] + n\left[C_V \, ln(T) + R \, ln\left(\frac{2V}{n}\right) + s_2^0\right]. \tag{4.70}$$

The entropy variation between the final state and the initial state is:

$$S_f - S_i = 2nR \, ln \, 2 > 0. \tag{4.71}$$

This is a positive variation of the entropy of the isolated system (1 + 2): there has been an increase in entropy due to the mixing of the two gases. Since the mixing process is irreversible, it is not surprising that entropy increases, according to the second principle of thermodynamics.

- *Identical gases*: the process is reversible and entropy must remain constant.

If the two gases are identical, we have $s_1^0 = s_2^0$ in (4.69) and (4.70) which does not change the result found in (4.71). We arrive at a positive variation of the entropy of the mixture, whereas we should find a variation of the entropy null because the process is, this time, reversible. Indeed, to put back the partition to separate the total volume in two compartments only restores the initial state (no change in compartments A and B). This result is therefore a paradox!

4.6.4.4 Paradox Resolution

The obtained expression for the entropy of a discernable ideal gas is not extensive:

$$S(2V, T, 2n) \neq 2 \, S(V, T, n). \tag{4.72}$$

The origin of this problem comes from the bad count of the number of microscopic states of a gas (the microscopic state of a system is specified at a given moment by the knowledge of the various microscopic variables of the system such as the position, speed, energy, etc. of each of the particles which constitute it as opposed to the macroscopic state, which is specified by the knowledge of the various macroscopic state variables of the system such as the pressure, temperature, etc.). Indeed, the indiscernibility principle of the particles of a gas in quantum mechanics states that when we permute two particles of a microscopic state, we obtain the same microscopic state because the particles are indistinguishable. Thus, when the mixture of the two identical gases is carried out, we are in the presence of $2n$ moles of a single gas which occupies the volume 2 V. In these conditions, the expression of the final entropy takes the following form:

$$S_f = 2n\left[C_V \ln(T) + R \ln\left(\frac{2V}{n}\right) + s^0\right] \tag{4.73}$$

with $s^0 = s_1^0 = s_2^0$. We verify that the entropy of the enclosure before and after the removal of the partition remains unchanged:

$$S_f - S_i = 0. \tag{4.74}$$

4.7 Heat Machines

4.7.1 Introduction

A heat engine is a machine which uses energy exchanges to produce work or consumes work to achieve energy exchanges, naturally impossible. Two reservoirs of heat to high and low temperatures are required for its operation. In what follows, we will use the following notations:

- T_H and T_L for the high and low temperatures of hot and cold reservoirs, respectively.
- Q_H is the exchanged heat between the hot source and the machine and Q_L is the exchanged heat between the machine and the cold sink. These two heats can each be of positive or negative sign.
- W is the produced or consumed work. It can be positive or negative.

Whatever the mode of operation and the yield of the engine, it cannot create or destroy any energy in accordance with the first principle of thermodynamics, and therefore, we shall always have:

$$Q_H + Q_B + W = 0. \tag{4.75}$$

4.7.2 Thermodynamic or Driving Machines

Thermodynamics or driving machines are heat machines, which produce work; it is the case of the steam engines (steam locomotives, steamboats, etc.), the gasoline or diesel combustion engines (cars, planes, etc.), and the power plants (nuclear reactor, gas or coal boiler, etc.). A thermodynamic machine takes some heat from a high-temperature source ($Q_H > 0$) and produces work ($W < 0$). As we saw in Sect. 4.6.2, if

Fig. 4.13 Energy transfers of a thermodynamic machine

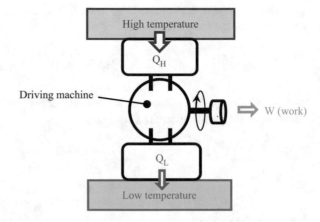

we want to perform this transformation, continuously, we have no choice but to reject some heat in a low-temperature sink ($Q_L < 0$) (Fig. 4.13). In power plants, the two temperature reservoirs are easily identifiable: water vapor draws some heat from the heart of the plant and rejects the unused heat through large cooling chimneys. In the automotive and aerospace engines, the air that they use as a working fluid is drained away because of the combustion products which prevent its reuse. The cooling is directly done in the atmosphere, outside the engine casing. The cold sink is therefore not easily identifiable.

4.7.3   Dynamo-Thermal or Receiving Machines

The dynamo-thermal or receiving machines are heat transfer machines such as the refrigerator, air conditioner, and gas liquefier. These machines have a reverse operation to that of the thermodynamic machines. Indeed, these machines extract some heat from a reservoir at low temperature ($Q_L > 0$) to reject it in a reservoir at higher temperature ($Q_H < 0$). An unavoidable consequence is that they consume work ($W > 0$) (Fig. 4.14).

4.7.4   Carnot's Cycle

(i) *Introduction*

The Carnot cycle, named after the French physicist and engineer Nicolas Léonard Sadi Carnot (1796–1832) who discovered it, is a theoretical thermodynamic cycle of a heat engine, consisting of four reversible processes: an isothermal relaxation, an adiabatic relaxation, an isothermal compression, and an adiabatic compression. When it comes to a thermodynamic machine, this cycle is the most efficient of all

Fig. 4.14 Energy transfers of a dynamo-thermal machine

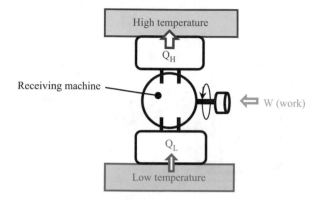

existing cycles to get a work from two constant temperature heat sources. It is described in a clockwise direction in the diagram of Clapeyron. The reversibility of Carnot's cycle allows the inversion of the direction of transformations and leads to a reversal of all the signs of exchanged energies. This is the case of a dynamo-thermal machine. The efficiency of other cycles and actual machines is compared to that of Carnot's cycle via the yield, a dimensionless number between 0 (zero efficiency) and 1 (perfect efficiency).

(ii) *Cycle description for a thermodynamic machine*

The cycle is composed of four successive transformations which occur schematically in a cylinder-piston system (Fig. 4.15):

- *1 → 2: Reversible isothermal relaxation at T_1 with the supplied Q_1:*

The gas is at the temperature T_1; it captures some heat Q_1 at the high-temperature source. The pressure of gas decreases and its volume increases. The piston performs the useful positive work.

- *2 → 3: Reversible adiabatic relaxation:*

The high-temperature source is suppressed, and the pressure of gas continues to decrease and its volume to increase. At point 3, the temperature is equal to T_2.

- *3 → 4: Isothermal compression at T_2 with heat release Q_2:*

The gas is at the temperature $T_2 < T_1$, and it gives up some heat to the low-temperature source. The pressure of gas increases and its volume decreases. This time, the piston is doing negative work.

- *4 → 1: Adiabatic compression:*

The low-temperature source is removed. The pressure of gas increases and its volume decreases. The temperature increases until it reaches T_1. Work, denoted W, is equal to the surface between the points 1234. It represents the total work done by the piston. It is equal to the quantity of heat involved during this cycle.

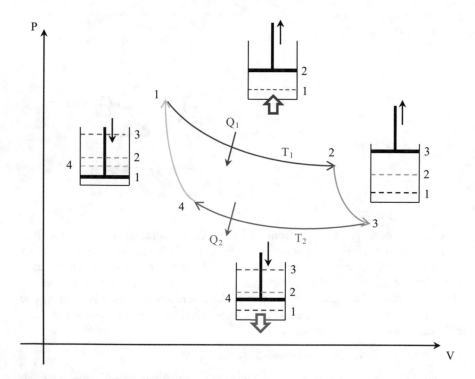

Fig. 4.15 Carnot's cycle in the Clapeyron diagram

4.7.5   Yield of a Thermal Machine

The yield of a thermal machine characterizes the efficiency with which the energy conversion is performed. It is defined by the ratio between the useful energy and therefore which interests us and the energy to be converted and thus to be paid.

4.7.5.1   Thermodynamic or Driving Machines

The yield η of a thermodynamic machine, also called *efficiency*, is always lower than a limit value which depends on the temperatures of sources at high and low temperatures:

$$\eta \le 1 - \frac{T_\mathrm{L}}{T_\mathrm{H}}. \tag{4.76}$$

The maximal yield is obtained for the limiting case of a reversible thermodynamic machine:

$$\eta_{\text{rev}} = 1 - \frac{T_{\text{L}}}{T_{\text{H}}}. \tag{4.77}$$

This is the case of the Carnot machine. Equation (4.77) shows that even for a reversible cycle, an efficiency equal to 1 requires either a low-temperature source at $T_{\text{L}} = 0$ K or a high-temperature source at $T_{\text{H}} \to \infty$ or both, which is practically impossible.

4.7.5.2 Dynamo-Thermal or Receiving Machines

In the receiving machines, we distinguish two types of machines according to the field of use: the low-temperature reservoir is the system to be cooled (refrigerator, air conditioner, cold room, etc.) or it is the high-temperature reservoir which is the system to be heated (swimming pool, house, etc.).

- *The low-temperature reservoir is the useful source*:

The yield of such a machine is lower than a limit value which depends on the temperatures of the reservoirs at low and high temperatures:

$$\eta \leq \frac{T_{\text{L}}}{T_{\text{H}} - T_{\text{L}}}. \tag{4.78}$$

The maximal yield is obtained for the limit case of a reversible machine. Equation (4.78) shows that efficiency is not necessarily less than 1. It is all the greater and tends towards infinity as the temperatures of the reservoirs are close to each other.

- *The high-temperature reservoir is the useful source*:

The yield of such a machine is lower than a limit value which depends on the temperatures of the reservoirs at low and high temperatures:

$$\eta \leq \frac{T_{\text{H}}}{T_{\text{H}} - T_{\text{L}}}. \tag{4.79}$$

The maximal yield is obtained for the limit case of a reversible machine. Equation (4.79) shows that efficiency is not necessarily less than 1. It is all the greater and tends to infinity as the temperatures of the reservoirs are close to each other.

4.7.6 Cycle of Beau de Rochas or Cycle of Otto

The Beau de Rochas cycle, named after the French engineer Alphonse Eugène Beau (1815–1893), is a four-stroke cycle. It is a theoretical thermodynamic cycle of the spark ignition internal combustion engines such as the gasoline engines of

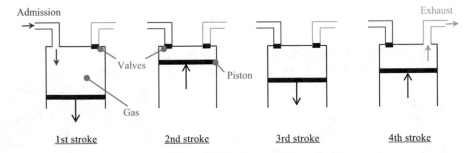

Fig. 4.16 The four-stroke cycle of Beau de Rochas

automobiles. A stroke refers to the full travel of the piston along the cylinder, in either direction. The cycle is thus characterized by four linear movements of the piston (Fig. 4.16):

1. *Admission or first stroke*: The cycle starts at the top dead center. The exhaust valve is closed, the intake valve opens, the piston descends, and a mixture of air and vaporized gasoline coming from the carburetor or the injection is inhaled into the cylinder. At the end of this phase, the intake valve closes, and we work with the locked air at T_1.
2. *Compression or second stroke*: The exhaust valve remains closed and the intake valve is also closed. The piston goes back up, compressing the admitted mixture. This phase is fast and since the heat exchanges are slow, this transformation is supposed adiabatic, quasi-static, and isentropic.
3. *Combustion-relaxation or third stroke*: The two valves remain closed. Around the second top dead center, the air-fuel mixture is ignited, usually by a spark plug. The combustion of the air-fuel mixture causes a strong increase of the pressure in the cylinder. The gas expansion forces the piston to descend, providing to the outside environment a useful work $\delta W = -P\, dV < 0$. This transformation is assumed isentropic.
4. *Exhaust or fourth stroke*: The exhaust valve opens and the pressure drops instantly to the atmospheric pressure. The flue gas is evacuated by the rise of the piston.

The agitation of gas molecules is several hundred meters per second, whereas the piston has a slow speed of a few meters per second. This is the reason why the cycle is considered quasi-static. This cycle is modeled by the following transformations (Fig. 4.17):

1. The admission is modeled by the isobar $0 \rightarrow 1$.
2. The compression is modeled by the adiabatic transformation $1 \rightarrow 2$.
3. The explosion takes place at constant volume according to the $2 \rightarrow 3$ transformation, followed by the adiabatic relaxation in $3 \rightarrow 4$.
4. The opening of the valve is modeled by the isochore $4 \rightarrow 5$, and the gas exhausts by the isobar $5 \rightarrow 0$.

Fig. 4.17 Theoretical cycle of Beau de Rochas

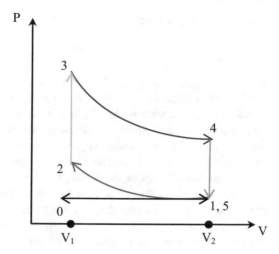

The theoretical yield calculated for an ideal cycle is maximal. It is the ratio between the provided work and the heat transfer. It is given by:

$$\eta = 1 - \frac{1}{\tau^{\gamma-1}}. \tag{4.80}$$

In this expression, the gas is supposed ideal. The parameter τ represents the volumetric compression ratio. It is a key parameter of engines. It is given by:

$$\tau = \frac{V_2}{V_1} \tag{4.81}$$

where V_1 and V_2 represent the volumes of the isochoric transformations $2 \rightarrow 3$ and $4 \rightarrow 5$, respectively, and γ is the ratio of heat capacities at constant pressure and volume:

$$\gamma = \frac{C_P}{C_V} \approx 1.4. \tag{4.82}$$

If the parameter τ increases, the yield η also increases, but a too strong compression favors explosions, and to remedy it, it is necessary to add an anti-knock. The area of the cycle corresponds to the work actually provided by the engine. The actual cycle departs significantly from the ideal cycle.

4.7.7 Diesel's Cycle

(i) *Introduction*

Initially planned for pulverized coal, the diesel engine quickly used the fuels such as the fuel oil, diesel, or, more rarely, vegetable oils. Unlike a gasoline engine, the diesel engine does not have spark plugs, since the ignition is spontaneous due to the phenomenon of auto-ignition. The basic principle is the presence of the pistons which slide inside the cylinders connected to the intake and exhaust manifolds. The functioning of the diesel engine consists in five strokes: the admission of the air allowed by the descent of the piston, the compression of the air produced by the inverse movement, the injection, the combustion and the relaxation, and the exhaust of the burned gas. This operating cycle is the most used for motor vehicles equipped with this type of engine. There are, however, two-stroke diesel engine cycles used by the ships. The diesel engine knew a significant number of improvements such as the overeating by the turbo-compressor which allowed increasing its performances after the 1980s. By way of this innovation, it possesses a better yield than its competitor, the gasoline engine. Diesel or gasoline, both engines emit pollutants such as carbon dioxide or nitrogen monoxides; except that in theory, the gasoline cars produce much more CO_2 than the diesel engines with a difference of 25%, even with a catalytic converter. In addition, the diesel engines have the particularity of reducing the diesel consumption, so, one point for the diesel engine. However, CO_2 is not the only cause of air pollution. Nitrogen oxides such as NO_2 and other greenhouse gases are also guilty of this growing pollution. And this is here that the diesel engine can keep a low profile because it rejects much more than the gasoline engine because the gasoline engine benefits from the contribution of the catalyst that allows it to significantly reduce the pollutants. In addition, the diesel engine, through its blackish smokes, emits a lot of nitrogen oxides but also benzo(a)pyrenes. These are fine particles, known to be carcinogenic and suspected to be responsible for the allergies and cardiopulmonary diseases.

(ii) *Diesel's Cycle*

The Diesel cycle, named after the German engineer Rudolf Christian Karl Diesel (1858–1913), is a four- or two-stroke cycle. It is a theoretical thermodynamic cycle of spontaneous ignition engines. This cycle is characterized by five linear movements of the piston (Fig. 4.18):

1. The air intake is modeled by the isobar $0 \rightarrow 1$.
2. The compression of the air is modeled by the isentropic transformation $1 \rightarrow 2$. This is the phase of the rising piston.
3. The fuel injection and the air-fuel mixture combustion are modeled by the isobar $2 \rightarrow 3$. The piston begins its descent with a progressive release of fuel.
4. The isentropic relaxation is modeled by the transformation $3 \rightarrow 4$. The piston finishes its descent after the release of all the fuel.

Fig. 4.18 Theoretical cycle
of Diesel

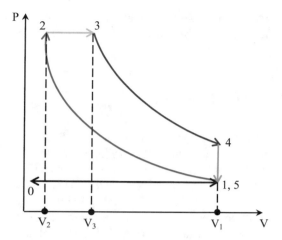

5. The isochoric cooling of the cylinder-piston system is modeled by the transformation $4 \rightarrow 5$ under the action of the atmosphere and the exhaust of the burned gas by the isobar $5 \rightarrow 0$.

The theoretical yield, calculated for an ideal cycle, is maximal. It is the ratio between the provided work and the heat transfer. It is given by:

$$\eta = 1 - \frac{\beta^{-\gamma} - \alpha^{-\gamma}}{\gamma(\beta^{-\gamma} - \alpha^{-\gamma})}. \tag{4.83}$$

In this expression, gas is supposed ideal. The parameters α and β are given by:

$$\alpha = \frac{V_1}{V_2} \quad \text{and} \quad \beta = \frac{V_1}{V_3}. \tag{4.84}$$

The parameter α is the volumetric ratio of the engine with $\alpha > \beta$, and γ the ratio of the heat capacities at constant pressure and volume:

$$\gamma = \frac{C_P}{C_V}. \tag{4.85}$$

The compression ratios of gasoline engines gradually increased, reaching 9 or 10 today. The thermodynamic efficiency of these engines is of the order of 60%. From the beginning, diesel engines had compression ratios higher than 10. They can reach 20 today with an average between 14 and 16. β is worth about 3/4 of the parameter α. The greater the β, the higher the yield of the engine. With a compression ratio of 16, the efficiency of the diesel engine reaches 65%. The theoretical efficiency of the diesel engines is therefore slightly higher than that of gasoline engines. But, from a technical point of view, the diesel engines are more suitable for the supercharging, which increases their actual efficiency, while the gasoline engines have an effective efficiency that rarely exceeds 30%.

4.7.8 Rankine's Cycle

The Rankine cycle of the name of the Scottish engineer and physicist William John Macquorn Rankine (1820–1872) is a cycle which takes place in wet vapor (term to be defined later). It comprises two isobars with an isothermal change of state and two adiabatic. This is nothing more than a Carnot's cycle applied to condensable vapors.

In practice, this cycle is not possible because it is difficult to compress in the isentropic way a two-phase mixture $(1 \rightarrow 2_{liq})$ as it is difficult to control the condensation $(3 \rightarrow 1)$ (Fig. 4.19). The actual cycle is therefore different from the theoretical cycle in that it must verify the following two properties:

- The surface of the cycle in the diagram (T,S) must be maximal. This surface represents the works of relaxation and compression.
- The compression work must be minimal.

The constituent elements of a steam engine are the boiler, turbine, condenser, and circulation pump (Fig. 4.20). The functioning of such a machine can be schematically described as follows:

1. Vaporization at constant pressure P_1 of the water in the boiler.
2. Isentropic relaxation of the just saturation vapor in the turbine (expression to be defined later). During this step, some energy is provided in the form of work to the outside of the steam engine until a pressure $P_2 < P_1$.
3. Condensation of water vapor at a constant pressure in the condenser.

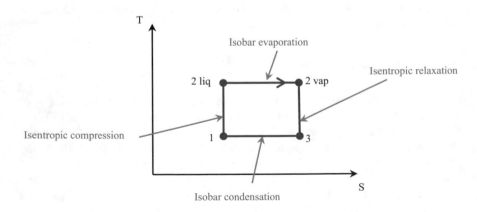

Fig. 4.19 The Rankine cycle in wet vapor. This is a Carnot cycle applied to condensable vapors

Fig. 4.20 Diagram of
operation of a steam engine

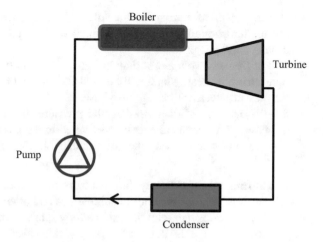

Fig. 4.21 Stirling cycle

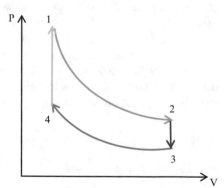

4. Isentropic compression of the just saturation liquid in the pump (expression to be
   defined later).
5. Heating of water at a constant pressure P_1 in the boiler.

4.7.9   Stirling's Cycle

The cycle of Stirling of the name of Pastor Christian Robert Stirling (1790–1868)
who invented it has the characteristic of being reversible, that is to say if a mechanical work is provided, it can function as a heat pump and provide heat or cold. It also
has the advantage to be closed because the used fluid always remains in the heat
engine. As most of the theoretical thermal cycles, it comprises four stages
(Fig. 4.21):

- 1 → 2: *Isothermal relaxation*: the volume increases while the pressure decreases, but the temperature remains constant. It is during this transformation that the driving energy is produced.
- 2 → 3: *Constant volume cooling*: the gas passes into the regenerator and cools by transferring its heat, which will be used for the next cycle. The temperature and the pressure decrease during this phase.
- 3 → 4: *Isothermal compression*: the pressure of gas increases as its volume decreases. A mechanical energy is supplied to the gas during this stage.
- 4 → 1: *Isochoric heating*: the gas circulates in the regenerator and takes some heat.

The Stirling cycle is slightly different from the Carnot cycle. As it produces few vibrations, it is ideal for the nuclear submarines and other military applications. The absence of gas exchange with the external environment makes it particularly useful in polluted environments. It presents a good yield, which can reach 40%, whereas the efficiency of a combustion engine for the automotive use rarely reaches 35% for the gasoline and 42% for the diesel. Among the disadvantages, the sealing of pistons must be greater than in an internal combustion engine because of very strong temperature variations and the need to use a gas the least viscous possible to minimize the friction losses. It is of delicate conception and of a high price because it has only a few applications in large series. This situation could favorably evolve with the development of research on renewable energy production.

4.7.10   Brayton's Cycle

The Brayton cycle is the basic thermodynamic cycle of the gas turbine functioning which is primarily used for electricity production in gas-producing countries such as Algeria. It takes its name from the American engineer George Brayton (1830–1892) who developed it, although its invention was attributed to Barber in 1791. It is also known as the Joule cycle. Another similar cycle, the Ericsson cycle, uses an external heat source and a regenerator.

The theoretical cycle comprises two adiabatic processes and two isobaric processes (Fig. 4.22):

- 4 → 1: Adiabatic compression in the turbo-compressor
- 1 → 2: Isobaric heating in the combustion chamber
- 2 → 3: Adiabatic relaxation in the turbine
- 3 → 4: Isobaric cooling in the atmosphere

This cycle integrates the irreversibility of compression and relaxation (the difference between the reversible point 1s and the real point 1 at the output of the compressor and the difference between the point 3s and the point 3 at the output of the turbine). There are two types of the Brayton cycle, depending on whether it is

Fig. 4.22 Brayton cycle including irreversibility in the diagram (T,S)

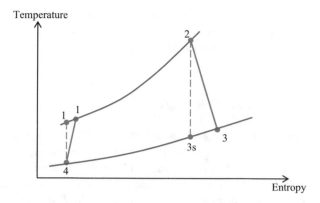

open or closed to the atmosphere using internal or closed combustion with a heat exchanger. This is the first variant which is used in the gas turbine power plants.

4.8   Properties of Pure Bodies

4.8.1   Introduction

To understand the functioning of heat machines, it is fundamental to have a minimum of knowledge on the properties of the used fluids.

4.8.2   Notion of Phase

4.8.2.1   Usual Definitions

A pure body is a body composed of a single chemical species. Water is a pure body, while air is not. A pure body can be in the solid, liquid, and gaseous (vapor) phases. Both solid and liquid phases are called *condensed phases*. At the microscopic level, the solid and liquid phases are differentiated by the fact that the molecules can move relative to each other in the liquids, while they are rigidly bound in the solids. At the macroscopic level, the solids are the bodies which have a volume and a defined form, little compressible for the usual values of the pressure. The liquids are also slightly compressible fluids that have a well-defined volume, but take the form of the bottom of volumes they occupy. The volume of a liquid is practically constant at a constant temperature, therefore independent of the pressure. Both gaseous and liquid phases are called *fluid phases*. The difference between these two phases at the microscopic level lies in the average distance between a molecule and its closest neighbors. This distance is small in a liquid and large in a gas. At the macroscopic level, the gases are fluids without shape or volume. They occupy all the space available to them. Let us note to close this paragraph that the different phases of a pure body do not mix. In other words, when two phases of the same pure body are present at the same time, they are distinguishable from each other.

4.8.2.2   Clear Definitions

The notion of phase is in fact more extensive than the simple distinction between solid, liquid, and gas. Indeed, in any phase, all the intensive variables must vary continuously. In other words, to distinguish two different phases, at least one of intensive variables must undergo a discontinuity. In a solid, for example, there are as many different phases as possible crystallizations because each type of molecular arrangement corresponds to a different density. The different ways of crystallization of a body are called *allotropic varieties*. Irons α and β on the one hand and carbon, graphite, and diamond on the other hand are examples of allotropic varieties.

4.8.3   Diagram (P,T)

The isobaric cooling curves of the pure body are characterized by the presence of plateaus where the temperature remains constant as long as two phases coexist. The temperature of these plateaus depends on the pressure, and at a fixed pressure, the coexistence of two phases of a pure body is made at a temperature thermodynamically imposed (Fig. 4.23).

The diagram (P,T) of a pure body represents the most stable phase of a pure body as a function of pressure and temperature. Figure 4.24 gives the representation of a usual pure body, of which the characteristics are:

- The melting, vaporization, and sublimation curves are of positive slope.
- The curve separating the S and L domains is quasi-vertical.
- The domains S, L, and V have a common point called *triple point III*.

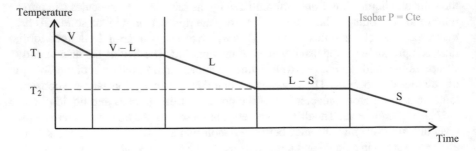

Fig. 4.23 Typical isobaric cooling curve of a pure body

Fig. 4.24 Diagram (P,T) of
a pure body

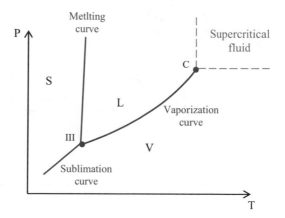

- Only the curve, which separates L and V, stops at a point C called *critical point*. The other curves which delimit two domains only stop for reasons of experimental feasibility.

The triple point III of a pure body is the unique pressure and the unique temperature for which the pure body can be found in the three phases simultaneously. This point is experimentally very important for the adjusting of thermometers because it is a universal benchmark of the temperature. For water, the triple point is defined by a temperature $T = 273.16$ K and a pressure $P = 611$ Pa. The critical point C of a pure body is the unique pressure and the unique temperature beyond which there is no possible distinction between the liquid and the solid. The body is said to be in *supercritical fluid phase*. For water, the critical point C is defined by a temperature $T = 647$ K and a pressure $P = 221 \times 10^5$ Pa. The diagram (P,T) of water has a notable difference with respect to the other usual bodies. Indeed, the melting curve that separates the S and V domains has a negative slope, signifying that the water can pass from the solid phase to the liquid phase by increasing the pressure at a constant temperature. This characteristic is linked to the fact that ice is less dense than liquid water, a rare property of pure bodies. It helps explain why ice cubes float on water.

4.8.4   Saturation Vapor Pressure

At a given temperature T, the equilibrium of a liquid with its vapor is only possible under a pressure P called *saturation vapor pressure* (Fig. 4.25). The curve which represents the saturation vapor pressure as a function of temperature distinguishes the following cases:

- If $P < P_{sat}$: the fluid is in the gaseous state called *dry vapor*.
- If $P > P_{sat}$: the fluid is in the liquid state.

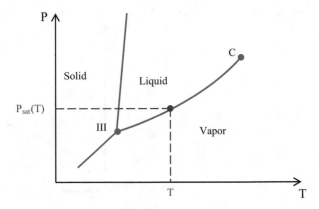

Fig. 4.25 The saturation vapor pressure P_{sat} depends on the temperature T

- If $P = P_{sat}$: the fluid is in the liquid-vapor equilibrium. In this case, the vapor is called *just saturation vapor* or *dry saturation vapor*. The set V–L is called *saturation vapor* or *wet vapor*. For a liquid, we are talking about *just saturation liquid*.

Numerous formulas have been proposed for representing algebraically the curve of the saturation vapor pressure. The Duperray formula gives a good approximation for water between 100 and 200 °C:

$$P = \left(\frac{t}{100}\right)^4 \tag{4.86}$$

where P is expressed in atmospheres and t in degrees Celsius. The value of the saturation pressure of water at 100 °C is 1 atm.

4.8.5 Clapeyron's Diagram

Clapeyron's diagram represents the state in which a body exists as a function of the pressure and the mass volume rather than the volume because the mass volume is an intensive quantity. It is composed of two curves: the boiling curve, which separates the zone where the pure body is liquid of the zone where there is a vapor-liquid coexistence, and the curve of dew which separates the zone where the pure body is a vapor of the zone where there is a liquid-vapor coexistence. These curves meet at critical point C (Fig. 4.26).

Fig. 4.26 Diagram of
Clapeyron

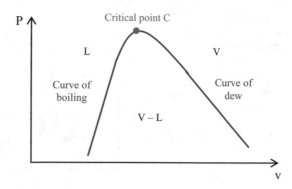

4.8.6   Theorem of Moments

(i) *Vapor Mass Fraction*

The vapor mass fraction of a pure body represents the proportion of the vapor mass
of this body:

$$x_V = \frac{\text{vapor mass}}{\text{total mass}} = \frac{m_V}{m}. \tag{4.87}$$

(ii) *Liquid Mass Fraction*

The liquid mass fraction of a pure body represents the proportion of the liquid
mass of this body:

$$x_L = \frac{\text{liquid mass}}{\text{total mass}} = \frac{m_L}{m}. \tag{4.88}$$

(iii) *Vapor Molar Fraction*

The vapor molar fraction of a pure body represents the proportion of the vapor
quantity of this body:

$$x_V = \frac{\text{vapor quantity}}{\text{total quantity}} = \frac{n_V}{n}. \tag{4.89}$$

(iv) *Liquid Molar Fraction*

The liquid molar fraction of a pure body represents the proportion of the liquid
quantity of this body:

Fig. 4.27 Schematic representation of the theorem of moments

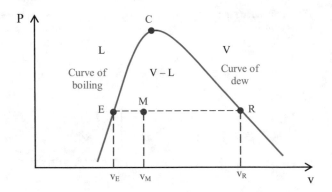

$$x_L = \frac{\text{liquid quantity}}{\text{total quantity}} = \frac{n_L}{n}. \qquad (4.90)$$

For a pure body, the mass fraction and the molar fraction have the same value. The sum of the vapor and the liquid fractions is worth 1:

$$x_V + x_L = 1. \qquad (4.91)$$

(v) *Theorem of Moments*

By noting M the representative point of a pure body and E and R the points of the boiling and the dew curves at the same pressure, we have (Fig. 4.27):

$$x_V = \frac{EM}{ER} \quad \text{and} \quad x_L = \frac{MR}{ER}. \qquad (4.92)$$

The advantage of representing the liquid-vapor transition in Clapeyron coordinates is that the reading of a point M on the change of the state plateau allows determination the composition of the pure two-phase body and its liquid and vapor mass fraction.

4.8.7   Variations of State Functions During a Transition

4.8.7.1   Enthalpy of Change of State or Latent Heat

A change of state is accompanied by a variation of the internal energy U and the enthalpy H of a pure body, even if the pressure and the temperature remain constant. This is a novelty compared to single-phase bodies for which the internal energy U and the enthalpy H do not vary if the pressure and temperature do not vary. This result is general; it is valid for any change of state and for any pure body. This variation of the energy stored during a change of state is interpreted at the microscopic scale by the reorganization of the structure of the pure body: the interactions between the molecules or the atoms of the body change in nature. To determine the enthalpy change of the pure body when it undergoes a change of state, we must refer to data tables. As enthalpy is an extensive quantity, these tables give the variation of enthalpy per unit of mass so that they can be exploitable whatever the mass of the studied pure body. The mass change state enthalpy $\Delta h_{1\to2}(T)$ or the latent heat of change of state $L_{1\to2}(T)$ is the mass enthalpy change of the pure body during the transition of phase $1 \to 2$. This quantity is tabulated according to temperature because it depends only on it. The variation of enthalpy $\Delta H_{1\to2}(T)$ due to the change of state is therefore:

$$\Delta H_{1\to2} = m\, L_{1\to2}(T). \tag{4.93}$$

The latent heat of change of state $L_{1\to2}(T)$ is expressed in J/kg, where m is the mass of the pure body. The latent term refers to the fact that energy is well contained somewhere and ready to serve. Equation (4.93) leads us to make the following comments:

- The considered pressure here is the plateau pressure at temperature T.
- $L_{1\to2}(T)$ calculates the enthalpy change of the pure body between the initial state and the final state. In the case of solidification, for example, the initial state is the liquid at the temperature T at the limit of solidification, that is to say, at the appearance of the first crystal of the solid. The final state is the solid at the temperature T at the limit of liquefaction, that is to say, at the disappearance of the last drop of the liquid. It is therefore a state function.
- It is clear that $L_{1\to2}(T) = -L_{2\to1}(T)$. The data tables only contain the positive latent heat values. The latent heat of melting, vaporization, and sublimation is positive. The inverse transformations have therefore the negative latent heat:

$$L_{S\to L} > 0, \quad L_{L\to V} > 0, \quad L_{S\to V} > 0. \tag{4.94}$$

The latent heat of vaporization of water at $T = 100\,°C$ is $L_{vap} = 2.25 \times 10^3$ kJ/kg.
- If the transformation is isobaric, thus isothermal because a change of state at a constant pressure implies that the temperature is thermodynamically imposed, the variation of enthalpy during the change of state is equal to the heat received by the body:

$$\Delta H_{1\to2} = Q. \tag{4.95}$$

This equality is not verified for any transformation.

4.8.7.2 Internal State Change Energy

To describe the evolution of a pure body during a change of state, we choose enthalpy rather than the internal energy because the changes of state currently realized are isobars. But, we can also define a mass internal energy of change of state. It is a function of the temperature only. So, by definition of enthalpy, we can write:

$$U = H - PV \quad \Rightarrow \quad \Delta U = \Delta H - \Delta(PV). \tag{4.96}$$

According to the first principle of thermodynamics, the PV product is negligible for the solid and liquid phases and not negligible for the vapor phase, so we can write:

$$\Delta U_{\text{melting}} = \Delta H_{\text{melting}} \quad \text{and} \quad \Delta U_{\text{solidification}} = \Delta H_{\text{solidifiaction}}. \tag{4.97}$$

As for the vapor phase, it becomes:

$$\Delta U_{\text{vaporization}} = \Delta H_{\text{vaporization}} - \left[(PV)_{\text{vapor}} - (PV)_{\text{liquid}} \right]$$
$$\approx m\, L_{\text{vaporization}} - (PV)_{\text{vapor}}. \tag{4.98}$$

4.8.7.3 Entropy of Change of State

For an isothermal and an isobaric transformation, we can determine the relationship between the entropy variation and the enthalpy variation, starting from the thermodynamic identity (see Eq. 4.48):

$$dH = T\, dS + V dP = T\, dS \quad \Rightarrow \quad dS = \frac{dH}{T}. \tag{4.99}$$

This is made possible because the transformation is by hypothesis isobaric ($P = \text{Cte} \Rightarrow dP = 0$). Taking into account that the transformation is, by hypothesis, isothermal and by integration of (4.99), it becomes:

$$\Delta S = \int \frac{\Delta H}{T} = \frac{1}{T} \int dH = \frac{\Delta H}{T}. \tag{4.100}$$

Equation (4.100) shows that a change of state also results in a variation of the entropy of the pure body. The mass entropy variation is thus related to the latent heat and to the considered temperature by:

$$\Delta s_{1\to2}(T) = \frac{L_{1\to2}(T)}{T}.$$ (4.101)

This variation is called *entropy of change of state*. It depends only on temperature. The entropy variation of a pure body from state 1 to state 2 is given by:

$$\Delta S_{1\to2}(T) = m\,\Delta s_{1\to2}(T).$$ (4.102)

Equation (4.102) leads us to make the following comments:

- The entropy of the pure body being a function of state, this variation of the mass entropy does not depend on the transformation undergone to pass from the initial state to the final state.
- Signs of entropy of the state change and the latent heat are identical. During a transition from a more ordered phase (solid) to a less ordered phase (liquid), the entropy of the pure body increases and the latent heat is positive. This is the theoretical argument that explains the sign of the latent heat. So:

  - For a melting: $\Delta s_{S\to L} > 0$
  - For a solidification: $\Delta s_{L\to S} < 0$
  - For a vaporization: $\Delta s_{L\to V} > 0$
  - For a liquefaction: $\Delta s_{V\to L} < 0$

4.8.8   Supercooling and Overheating

When we very slowly cool an extremely pure body in the liquid phase, there is a field in which the liquid should not exist because it is at a temperature where only the solid should be present. The liquid is in a metastable phase. It is said to be *supercooled* because it is very energetically expensive to show a solid-liquid interface rather than to grow an already existing solid. Similarly, by an isobaric heating of an extremely pure body in the liquid phase, we reach a point on the vaporization curve and normally at this point the beginning of the vaporization of the liquid should appear. However, no vaporization takes place; the body remains in the liquid state beyond this point. The body is in a metastable state. It is said to be *overheated* because the formation of the first bubble is energetically disadvantaged. In these two cases, the infinitesimal modification of pressure or temperature by the intrusion of an external element destroys this equilibrium, and the transition starts.

4.9 Solved Exercises

4.9.1 Fundamental Notions

Exercise 1
For each of the systems defined below, specify whether it is an isolated, closed, or open system.
1. *A burning candle*
2. *A car in functioning*
3. *A glass of boiling tea*
4. *A calorimeter*
5. *A light bulb lit*

Solution
An isolated system does not exchange either matter or energy with the outside environment. An open system exchanges matter and energy with the outside. A closed system exchanges only the energy with the outside.

1. A burning candle is an open system because the wax that composes it disappears in the form of the hot blackish smoke. So, there is an exchange of matter and energy with the outside.
2. A running car is an open system because the outside air and fuel enter the engine and the burned gas comes out. The engine is cooled by giving up heat to the ambient air. There is therefore the exchange of matter (air sucked by the engine + released burned gases) and energy (cooling of the engine in the open air).
3. A glass of boiling tea is an open system because the water it contains goes into the vapor state and leaves the system. The tea cools by contact with the cold ambient air. There is thus exchange of matter (water vapor leaving the glass) and energy (cooling of the tea in the open air).
4. A calorimeter is an isolated system because there is neither exchange of matter nor exchange of energy with the external environment.
5. A lit bulb is a closed system because there is the exchange of energy in the form of light and no exchange of matter.

Exercise 2
For each of the systems defined below, specify whether it is a homogeneous or a heterogeneous system
1. *A glass of water with ice cubes*
2. *A cup of coffee*
3. *Vapor which escapes from a pan of heated milk*

Solution
1. Water with ice cubes in the glass is in two states, liquid and solid at the same temperature and pressure. However, the density is not the same for the liquid and

the solid. We are in the presence of two phases, and this system is therefore heterogeneous.

2. The coffee in the cup is a homogeneous two-component system (ground coffee and water).
3. The vapor that escapes from a pan of heated milk is some water in gaseous form. It is a homogeneous system with a single constituent.

Exercise 3

What is a system in thermodynamic equilibrium? Define the stable, unstable, and metastable equilibrium.

Solution

- A system is in thermodynamic equilibrium if the variables which define it are constant. Among these variables, let us quote the temperature, the pressure, its volume, and its structure, that is to say, the arrangement of the atoms within the molecule and the molecules between them.
- A system is in stable equilibrium if, moved away substantially from this equilibrium, it returns to it as soon as it is left to itself.
- A system is in unstable equilibrium if an infinitesimal perturbation is sufficient to shift it to a more stable equilibrium state.
- Thermodynamics tells us about the evolution of a system from an initial state to a more stable final state. The thermodynamic stability is characterized by the invariance of the parameters which define this system. The kinetics informs us about the time taken by this system to reach this final state. The kinetic stability is reached if this time is very long. For example, carbon in the diamond state is thermodynamically unstable under the usual temperature and pressure conditions. On the other hand, the kinetics of transformation into graphite is infinitely slow. This makes it a kinetically stable system. The metastability is the property for a state to be kinetically stable but not thermodynamically. The speed of the transformation leading to the stable state is very low, even almost zero. If we consider a physicochemical system represented by its potential energy, a metastable state corresponds to a local minimum of energy. In order for the system to reach the state of the absolute minimum of the energy corresponding to the state of the stable equilibrium, it must be supplied with a quantity of energy called *the activation energy*.

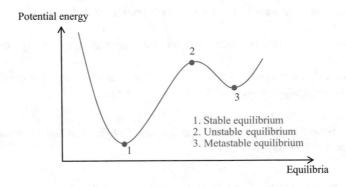

Exercise 4

Indicate on each of the following cases whether the system state is in a state of equilibrium.

1. *A material point in a low vertical position.*

2. *A material point in a high vertical position.*

3. *An extremely pure liquid phase water cooled very slowly below 0 °C.*
4. *An extremely pure liquid phase water heated very slowly above 100 °C.*
5. *A mixture of dihydrogen and dioxygen.*

Solution

1. The material point in the low vertical position is a position of equilibrium because as soon as it is removed, it returns to this position which represents the absolute minimum of the potential energy.
2. The material point in the high vertical position is not a position of equilibrium because as soon as it is removed from it, it moves further away. This position represents the local maximum of the potential energy.
3. When we very slowly cool extremely pure water in the liquid phase below 0 °C, the liquid should not exist because it is at a temperature where only the solid should be present. The liquid is in a metastable phase. It is said to be *supercooled* because it is very energetically expensive to show up a solid-liquid interface than to grow an already existing solid.
4. In the same way, by an isobaric heating of an extremely pure water in the liquid phase, we reach the point 100 °C on the curve of vaporization, and normally at this point, the beginning of vaporization of the liquid must appear. However, no vaporization takes place; the body remains in the liquid state beyond this point. The body is in a metastable state. It is said to be *overheated* because the formation of the first bubble is energetically disadvantageous. In the case of supercooling and superheating, the infinitesimal modification of pressure or temperature by the intrusion of an external element destroys this equilibrium, and the transition starts.
5. A mixture of dihydrogen and dioxygen is not a state of equilibrium because it is the high activation energy which is at the origin of the non-evolution of the system. It is enough to approach an inflamed match to the mixture so that the reaction of the formation of the water occurs.

Exercise 5

What is a thermodynamic transformation? Define open and closed transformations, and also the thermodynamic cycle.

Solution

- A thermodynamic transformation corresponds to a variation of the variables which define a thermodynamic state.
- An open transformation is a transformation where the final state is different from the initial state.
- A closed transformation is a transformation where the final state is the same as the initial state.
- A thermodynamic cycle is a series of open transformations.

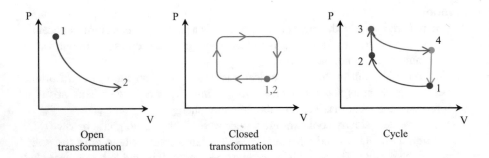

Open
transformation

Closed
transformation

Cycle

Exercise 6

Illustrate with a simple example the notions of reversible and irreversible transformations.

Solution

Let us consider the example of a thread, at the end of which we add each time a small mass. At any time, we can remove a mass to return to the previous state. In this case, the transformation is reversible. But, when we arrive by the addition of successive masses to the breaking point of the thread, it breaks, and even if we remove the last small mass that caused this break, the thread will not be reconstituted. The transformation is irreversible.

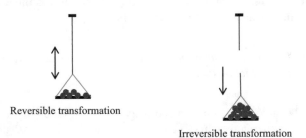

Reversible transformation

Irreversible transformation

Exercise 7

What is an isentropic process, and what are the causes of the entropy variation of a system?

Solution

In thermodynamics, an isentropic process is a process in which the system entropy remains constant. The variation of entropy of a system during a transformation has two causes:

- The creation of entropy due to the transformation irreversibility
- The exchange of entropy between the system and the external environment that surrounds it, through a heat transfer

We can therefore consider two types of isentropic transformations:

- An irreversible transformation, of which the creation of entropy is compensated by the entropy ceded by the system to the external environment, due to a heat transfer
- A reversible transformation (no creation of entropy) and an adiabatic transformation (no heat exchange)

Exercise 8
Liquid nitrogen is dinitrogen gas cooled below its boiling point at $-195.79\,°C$. It has the appearance of a limpid liquid from where white vapors escape. What is its boiling point in kelvins, and what purposes can it serve?

Solution
The Kelvin scale is deduced from the Celsius scale by the following affine function:

$$T(K) = T(°C) + 273.15, \tag{1}$$

from which we deduce the boiling point of nitrogen:

$$T(\text{Nitrogen}) = -195.79\,°C + 273.15 = 77.36\ K. \tag{2}$$

It is a cryogenic liquid commonly used both in the field of the scientific research and in the industry because of its low cost.

Exercise 9
Helium 4 can be liquefied at ambient pressure and at a temperature of $-269\,°C$, i.e., 4.13 K or $-452.2\,°F$. What is its boiling point in K and $°F$, and for what is it used?

Solution
The Kelvin scale is deduced from the Celsius scale by the following affine function:

$$T(K) = T(°C) + 273.15, \tag{1}$$

from which we deduce the boiling point of helium:

$$T(\text{Helium}\,4) = -269\,°C + 273.15 = 4.15\ K. \tag{2}$$

The Fahrenheit temperature scale is deduced from the Celsius scale by the following affine function:

$$T(°F) = T(°C) \times 1.8 + 32, \tag{3}$$

from which we deduce the boiling point of helium:

$$T(\text{Helium4}) = -269\,°C \times 1.8 + 32 = -452.2\,°F. \tag{4}$$

Liquid helium is used in the cooling of superconducting coils which create very high magnetic fields. These superconducting coils are used in medical context, such as magnetic resonance imaging (MRI), nuclear magnetic resonance (NMR), and magneto-encephalography (MEG). Liquid helium is also essential for the vast majority of experiments in low-temperature physics.

Exercise 10
A woman weighing 65 kg stands on the heel of her right shoe of circular shape and the radius r = 1 cm.
1. *What pressure does she exert on the ground?*
2. *Compare this pressure to the atmospheric pressure. We give the acceleration due to gravity g = 10 m/s², and the atmospheric pressure P_0 = 1.013 × 10⁵ Pa.*

Solution
1. The pressure exerted by the heel on the ground is:

$$P = \frac{mg}{\pi r^2} = \frac{(65\,\text{kg})\,(10\ \text{m.s}^{-2})}{(3.14)\,(10^{-2}\,\text{m})^2} = 207 \times 10^4\,\text{Pa}. \tag{1}$$

2. With an atmospheric pressure P_0 = 1.03 × 10⁵ Pa, the comparison between the two pressures gives:

$$\frac{P}{P_0} = \frac{207 \times 10^4\,\text{Pa}}{1.013 \times 10^5\,\text{Pa}} = 20.4. \tag{2}$$

The pressure exerted by the woman is about 20 times greater than the atmospheric pressure.

Exercise 11
What is the pressure exerted on the bottom of a container of height h = 76 cm when it is filled with (1) water? (2) Mercury? We give:
- *Acceleration due to gravity g = 9.8 m/s²*
- *Density of water ρ(H₂O) = 1000 kg/m³*
- *Density of Mercury ρ(Hg) = 13.6 × 10³ kg/m³*

Solution

1. *Water pressure*:

$$P(H_2O) = \rho_{H_2O}gh = \left(1000 \text{ kg.m}^{-3}\right)\left(9.8 \text{ m.s}^{-2}\right)(0.76 \text{ m}) = 7.45 \times 10^3 \text{ Pa}. \quad (1)$$

2. *Mercury pressure*:

$$P\,(Hg) = \rho_{Hg}gh = \left(13.6 \times 10^3 \text{ kg.m}^{-3}\right)\left(9.8 \text{ m.s}^{-2}\right)(0.76 \text{ m})$$
$$= 1.013 \times 10^5 \text{ Pa}. \quad (2)$$

Exercise 12

The surface S_1 of the large piston of a hydraulic press is 1300 cm², while the surface S_2 of the small piston has a value of 26 cm². Find the force F_1 supported by the large piston when a force F_2 of 100 N is applied on the small piston.

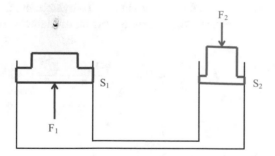

Solution

As the water contained in the press is incompressible, the pressure exerted on the liquid by the large piston is equal to the pressure exerted on the liquid by the small piston:

$$P = \frac{F_1}{S_1} = \frac{F_2}{S_2}. \quad (1)$$

The force F_1 is therefore:

$$F_1 = \frac{S_1}{S_2} \times F_2 = \frac{1300 \text{ cm}^2}{26 \text{ cm}^2} \times 100 \text{ N} = 5 \text{ kN}. \quad (2)$$

Exercise 13

A syringe of internal diameter $D = 1$ cm is filled with a liquid and in which can slide without friction a piston of the same diameter.

1. *What minimal force F must we exert on the piston to transmit the liquid through a needle where there is a relative pressure $P = 2 \times 10^5$ Pa with respect to the atmospheric pressure?*
2. *What work do we need to do to transmit 10 cm^3 of this liquid?*

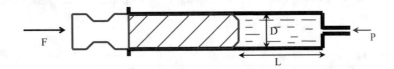

Solution

1. The pressures are integrally transmitted because the liquid is incompressible; it is therefore exerted on the piston of a pressure P. To overcome this pressure P, it is necessary to exert a minimal force F on the piston of diameter D such that:

$$\frac{F}{S} = \frac{F}{\frac{\pi D^2}{4}} = P \quad \Rightarrow \quad F = \frac{\pi D^2}{4} \times P = \frac{(3.14)\,(0.01\text{ m})^2}{4} \times (2 \times 10^5 \text{ Pa})$$

$$= 15.4 \text{ N}. \tag{1}$$

2. To transmit a volume $V = 10$ cm^3 of the liquid, the piston moves by a length L:

$$L = \frac{V}{\frac{\pi D^2}{4}}. \tag{2}$$

   The exerted force is constant and the frictions are negligible. It is therefore necessary to provide a work $W = F \times L$, because the force F and the displacement L have the same direction:

$$W = F \times L = \left(\frac{\pi D^2}{4} \times P\right)\left(\frac{V}{\frac{\pi D^2}{4}}\right) = PV = (2 \times 10^5 \text{ Pa})\,(10^{-5}\text{ m}^3) = 2\,\text{J}. \tag{3}$$

Exercise 14

Let us consider 1 mole of a gas occupying a volume V_m under a pressure P and at a temperature T. We suppose that these quantities are bound by the equation:

$$\left(P + \frac{a}{V_m^2}\right)(V_m - b) = RT.$$

1. *Establish the corresponding equation for n moles.*
2. *By putting $A = n^2 a$ and $B = nb$, which of the two quantities is additive?*

Solution

1. *Equation for n moles*:

   V_m being the molar volume, we can write:

$$V_m = \frac{V}{n}.$$ (1)

By injecting (1) into the expression given in the statement, we find:

$$\left[P + \frac{a}{\left(\frac{V}{n}\right)^2}\right]\left(\frac{V}{n} - b\right) = RT \quad \Rightarrow \quad \left(P + \frac{n^2 a}{V^2}\right)(V - nb) = nRT.$$ (2)

2. *Additive quantity*:

   Let us put $A = n^2 a$ and $B = nb$, and (2) takes the form:

$$\left(P + \frac{A}{V^2}\right)(V - B) = nRT$$ (3)

where B is an additive extensive quantity. Indeed, if $n = n_1 + n_2$, then:

$$B = nb = (n_1 + n_2)b = n_1 b + n_2 b = B_1 + B_2.$$ (4)

Here A is also an extensive quantity, but it is not an additive quantity because:

$$A = n^2 a \neq n_1^2 a + n_2^2 a.$$ (5)

Exercise 15

1. *Calculate the number of molecules per cm^3 in an ideal gas at $27\,°C$ under a pressure of 10^{-6} atmosphere.*
2. *Calculate the occupied volume per mole of an ideal gas at a temperature of $0\,°C$ under the normal atmospheric pressure.*

Solution

1. *Number of molecules per cm^3 in an ideal gas*:

   According to Eq. (4.12) of ideal gases given in the course, we have:

$$PV = Nk_BT \quad \Rightarrow \quad P = \frac{N}{V}k_BT = n_Vk_BT. \tag{1}$$

Here N is the number of molecules of the considered gas, $n_V = N/V$ represents the number of molecules per unit volume, T the temperature in kelvins, V the volume of the gas in m^3, and $k_B = 1.38 \times 10^{-23}$ J.K^{-1}, Boltzmann's constant. Knowing from Eq. (4.4) of the course that 1 atm = 101 325 Pa, we deduce the number of molecules per cubic meter:

$$n_V = \frac{P}{k_BT} = \frac{(101\ 325\ \text{Pa})\ (10^{-6}\ \text{atm})}{(1.38 \times 10^{-23}\ \text{J.K}^{-1})\ (27\ ^\circ\text{C} + 273.15)\text{K}}$$

$$= 2.45 \times 10^{19}\ \text{molecules.m}^{-3}. \tag{2}$$

The number of molecules per cubic centimeter is:

$$n_V = \frac{2.45 \times 10^{19}\ \text{molecules.m}^{-3}}{10^6} = 2.45 \times 10^{13}\ \text{molecules.cm}^{-3}. \tag{3}$$

2. *Volume occupied per mole of an ideal gas*:

It is still possible to express Eq. (1) according to the matter in moles:

$$PV = Nk_BT = \frac{N}{N_A}N_Ak_BT = nRT. \tag{4}$$

Here T is the temperature in K; V the volume in m^3; $k_B = 1.38 \times 10^{-23}$ J.K^{-1}, Boltzmann's constant; N the number of molecules of the considered gas; $n = N/N_A$, the number of moles with $N_A = 6.022 \times 10^{23}$ Avogadro's number; and $R = N_Ak_B = 8.314$ J.K^{-1}.mol^{-1}, the ideal gas constant. For 1 mole ($n = 1$ mol), Eq. (4) is transformed into:

$$PV_m = nRT. \tag{5}$$

In (5), V_m is the molar volume, that is to say the volume occupied by 1 mole of an ideal gas, and n in this case is worth 1; from which we deduce the molar volume at the temperature of 0 °C under the normal atmospheric pressure:

$$V_m = \frac{nRT}{P} = \frac{(1\ \text{mol})\ (8.314\ \text{J.K}^{-1}.\text{mol}^{-1})\ (273.15\ \text{K})}{101\ 325\ \text{Pa}} = 22.4 \times 10^{-3}\ \text{m}^3$$

$$= 22.4\ \text{liters}. \tag{6}$$

Exercise 16
What is the value of the ideal gas constant when it is expressed in:

1. $L.atm.mol^{-1}.K^{-1}$
2. $J.mol^{-1}.K^{-1}$
3. $cal.mol^{-1}.K^{-1}$

Solution

One mole of an ideal gas occupies a volume of 22.4 liters under normal conditions of pressure ($P = 1$ atm $= 1.013 \times 10^5$ Pa) and temperature ($T = 273$ K).

1. *Value of the ideal gas constant R, expressed in* $L.atm.mol^{-1}.K^{-1}$:

   The ideal gas equation is given by:

   $$PV = nRT \tag{1}$$

   where T is the temperature, V the volume, n the number of moles, and R the constant of ideal gases, from which we deduce:

   $$R = \frac{PV}{nT} = \frac{(1 \text{ atm}) (22.4 \text{ L})}{(1 \text{ mol}) (273 \text{ K})} = 0.0821 \, L.atm.mol^{-1}.K^{-1}. \tag{2}$$

2. *Value of the ideal gas constant R, expressed in* $J.mol^{-1}.K^{-1}$:

   Knowing that 1 joule amounts to 1 pascal cubic meter ($1 \text{ J} = 1$ Pa.m^3), we find:

   $$R = \frac{PV}{nT} = \frac{\left(1.013 \times 10^5 \text{ Pa}\right) \left(22.4 \times 10^{-3} \text{ m}^3\right)}{(1 \text{ mol}) (273 \text{ K})} = 8.31 \, J.mol^{-1}.K^{-1}. \tag{3}$$

3. *Value of the ideal gas constant R, expressed in* $cal.mol^{-1}.K^{-1}$:

   Knowing that 1 cal is equivalent to 4.18 J and knowing the value of R in $J.mol^{-1}.K^{-1}$, a rule of three is sufficient to determine the constant of ideal gases in $cal.mol^{-1}.K^{-1}$:

   $$R = \frac{\left(8.31 \, J.mol^{-1}.K^{-1}\right) (1 \text{ cal})}{4.18 \, J} = 1.99 \, cal.mol^{-1}.K^{-1}. \tag{4}$$

Exercise 17

A container contains an ideal gas, of which the pressure is $P_1 = 1.1 \times 10^5$ Pa and the temperature $T_1 = 50 \,°C$. The gas is cooled to constant volume until the temperature $T_2 = 10 \,°C$.

1. *What is the pressure P_2 of the gas?*
2. *What is the quantity of matter of the gas if its volume is 1 liter? We give the ideal gas constant $R = 8314 \, J.mol^{-1}.K^{-1}$.*

Solution

1. *Gas pressure*:

It is an isochoric transformation from state (P_1,V_1,T_1,n_1) to state (P_2,V_1,T_2,n_2). The ideal gas equation is written for state 1:

$$P_1V_1 = n_1RT_1 \tag{1}$$

and for state 2:

$$P_2V_1 = n_1RT_2. \tag{2}$$

By making the ratio (1)/(2), we obtain:

$$\frac{P_1}{P_2} = \frac{T_1}{T_2} \quad \Rightarrow \quad P_2 = \frac{T_2}{T_1} \times P_1 = \frac{(10\,^\circ C + 273.15)\,K}{(50\,^\circ C + 273.15)\,K}\left(1.1 \times 10^5\,Pa\right)$$
$$= 9.64 \times 10^4\,Pa. \tag{3}$$

2. *Quantity of matter of gas*:

Since we have the pressure and the temperature values which define states 1 and 2 and since the transformation is isochoric, the quantity of matter can be determined from Eq. (1) or (2). From (1), it becomes:

$$P_1V_1 = n_1RT_1 \quad \Rightarrow \quad n_1 = \frac{P_1V_1}{R\,T_1}. \tag{4}$$

For $V_1 = 1$ liter $= 10^{-3}\,m^3$, we obtain:

$$n_1 = \frac{\left(1.1 \times 10^5\,Pa\right)\left(10^{-3}\,m^3\right)}{\left(8.314\,J.K^{-1}.mol^{-1}\right)(50\,^\circ C + 273.15)\,K} = 0.041\,mol. \tag{5}$$

Exercise 18

A car tire is inflated with air at a temperature of 20 °C under a pressure of 2.1 bars. Its internal volume, supposed constant, is 30 liters. The air is supposed an ideal gas.
1. *How much air does it contain?*
2. *After driving for a while, a pressure check is carried out leading to a pressure value of 2.3 bars. By assuming that the volume of the tire has not changed during the journey, what is the temperature of the air enclosed in the tire? Express the result in the Celsius scale.*
3. *Are the pressure values recommended by the manufacturers for inflating with air different for inflating with nitrogen? We give the ideal gas constant $R = 8.314$ J. $mol^{-1}\ K^{-1}$, the air molar mass $M = 29$ g/mol, the molar mass of nitrogen $M' = 28$ g/mol, and 1 bar $= 10^5$ Pa.*

Solution

1. *Air quantity in the tire*:

The ideal gas equation is given as a function of the matter in moles by:

$$PV = nRT \tag{1}$$

where T is the temperature in K, V the volume in m^3, n the number of moles, and R the constant of ideal gases in mol^{-1}.K^{-1}. From (4), we deduce the quantity of air in the tire, in moles:

$$n = \frac{PV}{RT} = \frac{(2.1 \text{ bars}) \left(10^5 \text{ Pa}\right) \left(30 \times 10^{-3}\text{m}^3\right)}{\left(8.314 \text{ J.K}^{-1}.\text{mol}^{-1}\right) (20 \,^\circ\text{C} + 273.15)\text{K}} = 2.59 \text{ moles}. \tag{2}$$

Assuming that 1 mole weighs 29 g, we end up with the quantity of air m in the tire, in grams:

$$m = nM = (2.59 \text{ mol}) \left(29 \text{ g.mol}^{-1}\right) = 75.1 \text{ g}. \tag{3}$$

2. *Air temperature enclosed in the tire*:

As the volume of the tire did not vary, the number of moles of air contained in the tire did not also vary (isochoric transformation between two thermodynamic states). Starting from (1), it becomes:

$$T' = \frac{P'V}{n^*R} = \frac{(2.3 \text{ bars}) \left(10^5 \text{ Pa}\right) \left(30 \times 10^{-3}\text{m}^3\right)}{(2.59 \text{ mol}) \left(8.314 \text{ J.K}^{-1}.\text{mol}^{-1}\right)} = 320.6 \text{ K}. \tag{4}$$

The Celsius scale is deduced from the Kelvin scale by the following affine function:

$$T(\,^\circ\text{C}) = T(\text{K}) - 273.15 \text{ K} = 320.6 \text{ K} - 273.15 \text{ K} = 47.9 \,^\circ\text{C}. \tag{5}$$

3. *Difference between inflating of the tires with air and nitrogen*:

As the molar masses of air (29 g/mol) and nitrogen (28 g/mol) are very close to each other, the pressure values recommended by the car manufacturers are almost the same for both gases.

Exercise 19

An ideal gas is considered in the three following successive states:
- *State 1 characterized by P_1, V_1, T_1*
- *State 2 characterized by P_2, V_2, T_2*
- *State 3 characterized by P_3, V_3, T_3*

We give $P_1 = 1.0 \times 10^5$ Pa, $V_1 = 2$ liters, and $T_1 = 300$ K.

1. *The transition from state 1 to state 2 is carried out at a constant pressure by a temperature rise of 20 K. Determine P_2, V_2, and T_2.*
2. *The transition from state 2 to state 3 takes place at a constant temperature by a pressure increase of 1×10^4 Pa. Determine P_3, V_3, and T_3.*

Solution

1. *Transformation from state 1 to state 2*:

• *Determination of the pressure P_2*:

The transition from state 1 to state 2 takes place at a constant pressure (isobaric transformation):

$$P_2 = P_1 = 1.0 \times 10^5 \, \text{Pa}. \tag{1}$$

• *Determination of the temperature T_2*:

The temperature increases by 20 K:

$$T_2 = T_1 + 20 \, \text{K} = 300 \, \text{K} + 20 \, \text{K} = 320 \, \text{K}. \tag{2}$$

• *Determination of the volume V_2*:

The ideal gas equation applied to state 1 gives:

$$P_1 V_1 = nRT_1 \tag{3}$$

where T_1 is the temperature in K, P_1 the pressure in Pa, V_1 the volume in m^3, n the number of moles, and R the constant of ideal gases in mol^{-1} K^{-1}. The ideal gas equation applied to state 2 gives:

$$P_1 V_2 = nRT_2 \tag{4}$$

where T_2 is the temperature in K, $P_2 = P_1$ the pressure in Pa, V_2 the searched volume in m^3, n the number of moles, and R the constant of ideal gases in mol^{-1} K^{-1}. By making the ratio (2)/(1), we obtain:

$$\frac{V_2}{V_1} = \frac{T_2}{T_1} \quad \Rightarrow \quad V_2 = \frac{T_2}{T_1} \times V_1 = \frac{320 \, \text{K}}{300 \, \text{K}} \left(2 \times 10^{-3} \, \text{m}^3 \right) = 2.1 \times 10^{-3} \, \text{m}^3$$

$$= 2.1 \text{ liters.} \tag{5}$$

2. *Transformation from state 2 to state 3*:

- *Determination of the temperature T_3*:

The transition from state 2 to state 3 is done at a constant temperature (isothermal transformation):

$$T_3 = T_2 = 320 \text{ K.} \tag{6}$$

- *Determination of the pressure P_3*:

During the transition from state 2 to state 3, the pressure increases by 10^4 Pa:

$$P_3 = P_2 + 10^4 \text{ Pa} = 1.0 \times 10^5 \text{ Pa} + 10^4 \text{ Pa} = 1.1 \times 10^5 \text{ Pa.} \tag{7}$$

- *Determination of the volume V_3*:

The ideal gas equation applied to state 2 gives:

$$P_2 V_2 = nRT_2 \tag{8}$$

where T_2 is the temperature in K, P_2 the pressure in Pa, V_1 the volume in m^3, n the number of moles, and R the constant of ideal gases in mol^{-1}.K^{-1}. The ideal gas equation applied to state 3 gives:

$$P_3 V_3 = nRT_2 \tag{9}$$

where $T_3 = T_2$ is the temperature in K, P_3 the pressure in Pa, V_3 the searched volume in m^3, n the number of moles, and R the constant of ideal gases in mol^{-1}.K^{-1}. By making the ratio (3)/(2), we obtain:

$$\frac{P_3 V_3}{P_2 V_2} = 1 \quad \Rightarrow \quad V_3 = \frac{P_2}{P_3} \times V_2 = \frac{1.0 \times 10^5 \text{ Pa}}{1.1 \times 10^5 \text{ Pa}} \text{ (2.1 liters)}$$

$$= 1.91 \text{ liters.} \tag{10}$$

Exercise 20

The coefficients a and b of the van der Waals equation of state of carbon dioxide have the respective values 0.366 kg.m^5.s^{-2}.mol^{-2} and 4.29 m^3.mol^{-1}. We place 2 moles of this gas in an enclosure of volume V = 1 liter at a temperature T = 300 K. Compare the pressures given by the equations of state of the ideal gas (GP) and the Van der Walls gas (VdW) knowing that the exact value of the pressure is P = 38.5 bars. Conclude.

Solution

The carbon dioxide pressure is given within the framework of the ideal gas model by:

$$P_{GP}V = nRT \tag{1}$$

where T is the temperature in kelvins, V the volume in m^3, and $R = 8.314$ J.K^{-1}. mol^{-1}, the constant of ideal gases, from which we deduce the value of the pressure:

$$P_{GP} = \frac{nRT}{V} = \frac{(2 \text{ moles}) \left(8.314 \text{ J.K}^{-1}.\text{mol}^{-1}\right) (300 \text{ K})}{10^{-3} \text{ m}^3} = 4.99 \times 10^6 \text{ Pa.} \tag{2}$$

In relation to the value of the pressure assumed to be exact by hypothesis, the relative uncertainty made in the calculation of pressure within the framework of the ideal gas model is:

$$\left| \frac{P - P_{GP}}{P} \right| = \left| \frac{38.5 \times 10^5 \text{ Pa} - 4.99 \times 10^6 \text{ Pa}}{38.5 \times 10^5 \text{ Pa}} \right| = 29.6\%. \tag{3}$$

The carbon dioxide pressure is given within the framework of the Van der Waals model by:

$$\left(P_{VdW} + \frac{n^2 a}{V^2} \right) (V - nb) = nRT \quad \Rightarrow \quad P_{VdW} = \frac{nRT}{V - nb} - \frac{n^2 a}{V^2}$$

$$= \frac{(2 \text{ mol}) \left(8.314 \text{ J.K}^{-1}.\text{mol}^{-1}\right) (300 \text{ K})}{10^{-3} \text{ m}^3 - (2 \text{ mol}) \left(4.29 \times 10^{-5} \text{ m}^3.\text{mol}^{-1}\right)} \tag{4}$$

$$- \frac{(2 \text{ mol})^2 \left(0.366 \times 10^{-5} \text{ kg.m}^5.\text{s}^{-2}.\text{mol}^{-2}\right)}{\left(10^{-3} \text{ m}^3\right)^2} = 3.99 \times 10^6 \text{ Pa}$$

where T is the temperature in kelvins, V the volume in m^3, and $R = 8.314$ J.K^{-1}. mol^{-1} the constant of ideal gases and a and b two constants which depend on the nature of the considered gas. Compared to the value of the pressure assumed to be exact by hypothesis, the relative uncertainty made in the calculation of pressure within the framework of the Van der Waals model is:

$$\left| \frac{P - P_{VdW}}{P} \right| = \left| \frac{38.5 \times 10^5 \text{ Pa} - 3.99 \times 10^6 \text{ Pa}}{38.5 \times 10^5 \text{ Pa}} \right| = 3.6\%. \tag{5}$$

On the basis of the results mentioned in (3) and (5), we can conclude that the ideal gas model is unacceptable, whereas the Van der Waals gas model gives a better precision.

Exercise 21
A supposed ideal gas mixture of 0.2 g of H_2, 0.21 g of N_2, and 0.51 g of NH_3 is subjected to a pressure of 1 atmosphere and a temperature of 27 °C. Calculate:
1. *The molar fractions*
2. *The partial pressure of each gas*
3. *The total volume*

We give gas the ideal constant $R = 8.314 \ J.mol^{-1}.K^{-1}$, and the molar mass of hydrogen $M(H) = 1 \ g.mol^{-1}$ and that of nitrogen $M(N) = 14 \ g.mol^{-1}$.

Solution

1. *Molar fractions*:

Let m_i be the mass of gas i, M_i its molar mass, and n_i the number of moles it contains with:

$$n_i = \frac{m_i}{M_i} \tag{1}$$

from which we deduce the number of moles of:

- *Hydrogen H_2*:

$$n_{H_2} = \frac{m_{H_2}}{M_{H_2}} = \frac{0.2 \ g}{(1+1) \ g.mol^{-1}} = 0.1 \ mol. \tag{2}$$

- *Nitrogen N_2*:

$$n_{N_2} = \frac{m_{N_2}}{M_{N_2}} = \frac{0.21 \ g}{(14+14) \ mol.g^{-1}} = 0.0075 \ mol. \tag{3}$$

- *Ammonia (hydrogen nitride) NH_3*:

$$n_{NH_3} = \frac{m_{NH_3}}{M_{NH_3}} = \frac{0.51 \ g}{(14+1+1+1) \ g.mol^{-1}} = 0.03 \ mol. \tag{4}$$

The molar fraction of the gas i is, by definition:

$$x_i = \frac{n_i}{\sum_i n_i} = \frac{n_i}{n_{H_2} + n_{N_2} + n_{NH_3}} = \frac{n_i}{0.1 \ mol + 0.0075 \ mol + 0.03 \ mol}$$
$$= \frac{n_i}{0.1375 \ mol}. \tag{5}$$

with $\sum n_i$ the total number of moles, and $\sum x_i = 1$. The molar fractions of the three gases are therefore:

- *Hydrogen H_2*:

$$x_{H_2} = \frac{n_{H_2}}{0.1375 \ mol} = \frac{0.1 \ mol}{0.1375 \ mol} = 0.727. \tag{6}$$

- *Nitrogen N_2:*

$$x_{N_2} = \frac{n_{N_2}}{0.1375 \text{ mol}} = \frac{0.0075 \text{ mol}}{0.1375 \text{ mol}} = 0.055. \tag{7}$$

- *Ammonia (hydrogen nitride) NH_3:*

$$x_{NH_3} = \frac{n_{NH_3}}{0.1375 \text{ mol}} = \frac{0.03 \text{ mol}}{0.1375 \text{ mol}} = 0.218. \tag{8}$$

We verify well that the sum of molar fractions is equal to unity:

$$\sum_i x_i = 0.727 + 0.055 + 0.218 = 1. \tag{9}$$

2. *Partial pressure of each gas*:

The partial pressure P_i of each gas i is given by:

$$P_i = x_i P \tag{10}$$

with:

$$P = \sum_i P_i = 1 \text{ atm.} \tag{11}$$

- *Hydrogen H_2:*

$$P_{H_2} = x_{H_2} P = (0.727)(1 \text{ atm}) = 0.727 \text{ atm.} \tag{12}$$

- *Nitrogen N_2:*

$$P_{N_2} = x_{N_2} P = (0.055)(1 \text{ atm}) = 0.055 \text{ atm.} \tag{13}$$

- *Ammonia (hydrogen nitride) NH_3:*

$$P_{NH_3} = x_{NH_3} P = (0.218)(1 \text{ atm}) = 0.055 \text{ atm.} \tag{14}$$

3. *Total volume*:

The gas mixture being supposed ideal, its equation of state is written as:

$$PV = \sum_i n_i RT = nRT \tag{15}$$

where T is the temperature, V the volume of the mixture, P the total pressure, n the total number of moles, and R the constant of ideal gases, from which we deduce:

$$V = \frac{nRT}{P} = \frac{(0.1375\,\text{mol})\,(8.314\,\text{J.K}^{-1}.\text{mol}^{-1})\,(27\,^\circ\text{C}+273.15)\text{K}}{10^5\,\text{Pa}} = 3.43 \times 10^{-3}\,\text{m}^3$$
$$= 3.43\,\text{liters}.$$

$$\tag{16}$$

4.9.2   First Principle of Thermodynamics

Exercise 22
1. *Determine the work put at stake by 2 liters of an ideal gas maintained at 25 °C under the pressure of 5 atmospheres (state 1) which undergoes an isothermal relaxation to occupy a volume of 10 liters (state 2):*
   (a) *In a reversible way*
   (b) *In an irreversible way*
2. *At the same temperature, the gas is brought back from state 2 to state 1. Determine the work of compression in the two following cases:*
   (a) *Reversible*
   (b) *Irreversible*

Solution
The supposed ideal gas starts from state 1 characterized by ($P_1 = 5$ atm, $V_1 = 2$ liters, $T_1 = 298$ K) to state 2 defined by (P_2, $V_2 = 10$ liters, $T_2 = T_1 = 298$ K). Both temperatures T_1 and T_2 are equal because the transformation is by, hypothesis, isothermal. Let us start with the value of the pressure P_2 by first applying the ideal gas equation to state 1:

$$P_1 V_1 = n_1 R T_1 \tag{1}$$

and, a second time, to state 2:

$$P_2 V_2 = n_1 R T_1. \tag{2}$$

By making the ratio (2)/(1), we obtain:

$$\frac{P_2}{P_1} \times \frac{V_2}{V_1} = 1 \quad \Rightarrow \quad P_2 = \frac{V_1}{V_2} \times P_1 = \frac{2 \text{ liters}}{10 \text{ liters}} \times 5 \text{ atm} = 1 \text{ atm}$$

$$= 101\,325 \text{ Pa.} \tag{3}$$

1. (a) *The work involved in the isothermal reversible relaxation*:

The relaxation being reversible, the transformation is very slow so that at each moment we have $P_{\text{ext}} = P_{\text{gaz}}$. We find:

$$W_{\text{rev}}(1 \rightarrow 2) = -\int_1^2 P_{\text{ext}} \, dV = -\int_1^2 P_{\text{gaz}} \, dV = -\int_1^2 \frac{nRT}{V} \, dV = -nRT \, ln\left(\frac{V_2}{V_1}\right)$$

$$= -P_1 V_1 \, ln\left(\frac{V_2}{V_1}\right)$$

$$= -\left(5 \times 101\,325 \text{ Pa}\right)\left(2 \times 10^{-3} \text{ m}^3\right) \, ln\left(\frac{10 \times 10^{-3} \text{ m}^3}{2 \times 10^{-3} \text{ m}^3}\right)$$

$$= -1630.8 \,\text{J.}$$

$$\tag{4}$$

This work is negative because it is provided to the outside environment.

(b) *The work involved in the isothermal irreversible relaxation*:

An irreversible isothermal relaxation is a rapid transformation from state 1 to state 2. This transformation can be treated as an isochoric transformation followed by an isobaric transformation where the external pressure is equal to the pressure of state 2 (final state). Under these conditions, the work done by this relaxation is that of the isobaric transformation:

$$W_{\text{irr}}(1 \rightarrow 2) = -\int_1^2 P_{\text{ext}} \, dV = -\int_1^2 P_{\text{gaz}} \, dV = -\int_1^2 P_2 \, dV = -P_2 \int_1^2 dV \tag{5}$$

$$= -P_2\left(V_2 - V_1\right) = -(101\,325 \text{ Pa})\,(10 - 2)\,10^{-3} \text{ m}^3$$

$$= -810.6 \text{ J.}$$

This result shows that we get back less useful work when the gas relaxes in an irreversible way compared to the reversible case treated in (1a).

2. (a) *The work involved in the isothermal reversible compression*:

The compression being reversible, the transformation is very slow so that at each moment we have $P_{\text{ext}} = P_{\text{gaz}}$. We obtain:

$$W_{rev}(2 \rightarrow 1) = -\int_2^1 P_{ext}\, dV = -\int_2^1 P_{gaz}\, dV = -\int_2^1 \frac{nRT}{V}\, dV = -nRT\, ln\left(\frac{V_1}{V_2}\right)$$

$$= P_1 V_1\, ln\left(\frac{V_2}{V_1}\right) = (5 \times 101\,325\,\text{Pa})\,(2 \times 10^{-3}\,\text{m}^3)\, ln\left(\frac{10 \times 10^{-3}\,\text{m}^3}{2 \times 10^{-3}\,\text{m}^3}\right)$$

$$= +1630.8\,\text{J}.$$

$$(6)$$

This work is positive because it is provided to the considered system by the external environment. It is equal in absolute value to the work done by the relaxation because the transformation is reversible.

(b) *The work involved in the isothermal irreversible compression*:

The irreversible isothermal compression is a rapid transformation from state 2 to state 1. This transformation can be treated as an isochoric transformation followed by an isobaric transformation where the external pressure is equal to the pressure of state 1 (final state). Under these conditions, the work done by this compression is that of the isobaric transformation:

$$W_{irr}(2 \rightarrow 1) = -\int_2^1 P_{ext}\, dV = -\int_2^1 P_{gaz}\, dV = -\int_2^1 P_1\, dV = -P_1 \int_2^1 dV$$

$$= -P_1\,(V_1 - V_2) = -(5 \times 101\,325\,\text{Pa})\,(2 - 10\,)\,10^{-3}\,\text{m}^3$$

$$= 4053\,\text{J}.$$

$$(7)$$

This result shows that the irreversible compression requires much more work from the outside environment than the reversible compression.

Exercise 23

An ideal gas undergoes a transformation from state 1 to state 2 according to three different paths a, b, and c such that:
- *Path a: an isochore followed by an isobar*
- *Path b: an isobar followed by an isochore*
- *Path c: an isotherm defined by PV = Cte*

States 1 and 2 are respectively defined by ($P_1 = 1$ bar, $V_1 = 3$ liters, T_1) and ($P_2 = 3$ bars, $V_2 = 1$ liter, T_1).

1. *Represent the three transformations in the coordinates of Clapeyron.*
2. *Calculate the variation of the internal energy between states 1 and 2.*
3. *Calculate the work done in the three aforementioned cases.*
4. *Deduct the exchanged heat. Are these heats received or transferred by the studied system?*

Solution

1. *Representation of the three transformations in the coordinates of Clapeyron*:

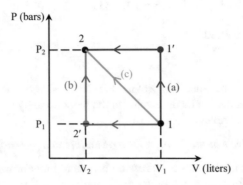

2. *Calculation of the variation of the internal energy ΔU between states 1 and 2*:

Since the internal energy is a state function, its variation does not depend on the followed path to go from state 1 to state 2. To find its value, let us choose the isothermal path (c). For an ideal gas, of which the internal energy U depends only on the temperature (see the course), we have:

$$\Delta U = 0 \text{ J}. \tag{1}$$

The variation of the internal energy is zero whatever the chosen path to go from 1 to 2.

3. *Calculation of the work done*:

• *Path (a)*:

$$W_a = W_{11'} + W_{1'2} = 0 - P_2(V_2 - V_{1'}) = -P_2(V_2 - V_1) = P_2(V_1 - V_2)$$
$$= \left(3 \times 10^5 \text{ Pa}\right) (3 - 1) \, 10^{-3} \text{ m}^3 = 600 \text{ J}. \tag{2}$$

In this equation $W_{11'} = 0$ because the transformation is done at constant volume ($\Delta V = 0$). With regard to $W_{1'2}$, the transformation is carried out at constant pressure P_2.

• *Path (b)*:

$$W_b = W_{12'} + W_{2'2} = -P_1(V_{2'} - V_1) + 0 = -P_1(V_2 - V_1) = P_1(V_1 - V_2)$$
$$= \left(10^5 \text{ Pa}\right) (3 - 1) 10^{-3} \text{ m}^3 = 200 \text{ J}. \tag{3}$$

In this equation $W_{2'2} = 0$ because the transformation is carried out at constant volume ($\Delta V = 0$). As for $W_{12'}$, the transformation is carried out at constant pressure P_1.

- *Path (c)*:

$$W_c = -\int_1^2 P\, dV = -\int_1^2 \frac{nRT}{V}\, dV = -nRT\, \ln\left(\frac{V_2}{V_1}\right) = 330\,\text{J}. \qquad (4)$$

4. Exchanged heats:

Since $\Delta U = W + Q = 0 \Rightarrow W = -Q$, we deduce for the three paths:

- *Path (a)*:

$$Q_a = -600\ \text{J}. \qquad (5)$$

- *Path (b)*:

$$Q_b = -200\ \text{J}. \qquad (6)$$

- *Path (c)*:

$$Q_c = -330\ \text{J}. \qquad (7)$$

These heats are negative; they are therefore transferred by the system to the outside environment.

Exercise 24

One mole of an ideal gas at an initial temperature $T_1 = 298$ K expands from a pressure $P_1 = 5$ atm to a pressure $P_2 = 1$ atm. In each of the following cases:
  I. *Isothermal and reversible relaxation*
 II. *Isothermal and irreversible relaxation*
III. *Adiabatic and reversible relaxation*
IV. *Adiabatic and irreversible relaxation*
   Calculate:
1. *The final temperature of gas*
2. *The variation of the internal energy of gas*
3. *The work done by gas*
4. *The quantity of heat put at stake*
5. *The enthalpy variation of gas*

We give for this gas $\Delta U = C_V \Delta T$, the constant volume heat capacity $C_V = 3R/2$, and the constant pressure capacity $C_P = 5R/2$ with $R = 8.314$ J.mol^{-1}.K^{-1} the constant of ideal gases.

Solution

I. *Isothermal and Reversible Relaxation*

1. *Final gas temperature*:

Since the transformation from state 1 (P_1, T_1, V_1) to state 2 (P_2, T_2, V_2) is isothermal, the initial and final temperatures are equal:

$$T_2 = T_1 = 298 \text{ K.} \tag{1}$$

2. *Variation of the internal energy of gas*:

The internal energy is a state function; its variation does not depend on the followed path to go from state 1 to state 2. For an isothermal transformation from state 1 to state 2, the internal energy U of ideal gas depends only on temperature (see the course). Therefore:

$$\Delta U = 0 \text{ J.} \tag{2}$$

3. *Work done by gas*:

The relaxation being reversible, the transformation is very slow so that at each moment we have $P_{ext} = P_{gaz}$. It becomes:

$$W_{rev}(1 \rightarrow 2) = -\int_1^2 P_{ext} \, dV = -\int_1^2 P_{gaz} \, dV = -\int_1^2 \frac{nRT}{V} \, dV$$

$$= -nRT \, ln\left(\frac{V_2}{V_1}\right). \tag{3}$$

Since the ratio V_2/V_1 is unknown, let us express it according to the known variables P_1 and P_2 by first applying the ideal gas equation time to state 1:

$$P_1 V_1 = n_1 R T_1 \tag{4}$$

and a second time to state 2:

$$P_2 V_2 = n_1 R T_1. \tag{5}$$

By making the ratio (2)/(1), we obtain:

$$\frac{P_2}{P_1} \times \frac{V_2}{V_1} = 1 \quad \Rightarrow \quad \frac{V_2}{V_1} = \frac{P_1}{P_2} \tag{6}$$

By injecting (6) into (3), we get:

$$W_{rev}(1 \rightarrow 2) = -nRT \, ln\left(\frac{P_1}{P_2}\right) = -(1 \, mol) \, (8.314 \, J.K^{-1}.mol^{-1}) \, (298 \, K) \, ln\left(\frac{5 \, atm}{1 \, atm}\right)$$
$$= -3987.5 \, J. \tag{7}$$

This work is negative because it is provided to the outside environment.

4. *Quantity of heat put at stake*:

A closed system exchanging only the energy to the external environment, the variation of its internal energy during a transformation from state 1 to state 2 is equal to the algebraic sum of the exchanged energies in the form of mechanical work W and heat Q:

$$\Delta U = U_2 - U_1 = W + Q. \tag{8}$$

According to (2), we have $\Delta U = 0$. The heat put at stake during the isothermal and reversible relaxation $1 \rightarrow 2$ is:

$$U = W + Q = 0 \quad \Rightarrow \quad Q = -W = 3987.5 \, J. \tag{9}$$

This energy is positive. It is therefore received by the system.

5. *Gas enthalpy variation*:

Enthalpy, denoted H, comprises the internal energy of a system to which is added the work which this system must exert against the external pressure to occupy its own volume. Enthalpy is a chemical potential which predicts the evolution and the equilibrium of a thermodynamic system, and from which we can deduce all its properties. It is an extensive state function which plays a privileged role in the isobaric transformations which are very useful in chemistry. It is defined by:

$$H = U + PV. \tag{10}$$

Here the term PV corresponds to the energy of expansion or compression of the system. H is always greater than U. Enthalpy is in joules or in calories. For an infinitesimal transformation, we have, by starting from (4.24) and (4.27) of the course:

$$dU = \delta Q + \delta W = dQ - P dV. \tag{11}$$

By introducing the enthalpy function, it becomes:

$$dH = dU + d(PV). \tag{12}$$

The relaxation being isothermal and by starting from (6), we deduce that $d(PV) = 0$. The variation of enthalpy is thus:

$$dH = dU \quad \Rightarrow \quad \Delta H = \Delta U = 0. \tag{13}$$

II. *Isothermal and Irreversible Relaxation*

1. *Final gas temperature*:

Since the transformation from state 1 (P_1, T_1, V_1) to state 2 (P_2, T_2, V_2) is isothermal, the initial and final temperatures are equal:

$$T_2 = T_1 = 298 \text{ K}. \tag{14}$$

2. *Variation of the internal energy of gas*:

The internal energy is a state function; its variation does not depend on the followed path to go from state 1 to state 2. For an isothermal transformation from state 1 to state 2, the internal energy U of an ideal gas depends only on temperature (see the course):

$$\Delta U = 0 \text{ J} \tag{15}$$

3. *Work done by gas*:

An irreversible isothermal relaxation is a rapid transformation from state 1 to state 2. This transformation can be treated as an isochoric transformation followed by an isobaric transformation where the external pressure is equal to the pressure of state 2 (final state). Under these conditions, the work done by this relaxation is that of the isobaric transformation:

$$W_{\text{irr}}(1 \rightarrow 2) = -\int_1^2 P_{\text{ext}}\, dV = -\int_1^2 P_{\text{gaz}}\, dV = -\int_1^2 P_2\, dV = -P_2 \int_1^2 dV \tag{16}$$
$$= -P_2\, (V_2 - V_1).$$

Since the difference $V_2 - V_1$ is unknown; let us express it according to the known variables P_1 and P_2 by applying the ideal gas equation to 1 mole of gas a first time to state 1:

$$P_1 V_1 = nRT_1 \quad \Rightarrow \quad V_1 = \frac{nRT_1}{P_1} \tag{17}$$

and a second time to state 2:

$$P_2 V_2 = nRT_1 \quad \Rightarrow \quad V_2 = \frac{nRT_1}{P_2}. \tag{18}$$

By injecting (17) and (18) into (16), we obtain:

$$W_{irr}(1 \to 2) = -P_2 (V_2 - V_1) = -P_2 \left(\frac{nRT_1}{P_2} - \frac{nRT_1}{P_1} \right) = -nRT_1 \left(1 - \frac{P_2}{P_1} \right)$$

$$= -(1 \text{ mol}) \left(8.314 \text{ J.K}^{-1}.\text{mol}^{-1} \right) (298 \text{ K}) \left(1 - \frac{1 \text{ atm}}{5 \text{ atm}} \right)$$

$$= -1982.1 \text{ J}.$$

$$\tag{19}$$

This work is negative because it is yielded by the system during this relaxation. It is an absolute value lower than the work yielded by the system during a reversible relaxation.

4. *Quantity of heat put at stake*:

A closed system exchanging only the energy to the external environment, the variation of its internal energy during a transformation from state 1 to state 2 is equal to the algebraic sum of the exchanged energies in the form of mechanical work W and heat Q:

$$\Delta U = U_2 - U_1 = W + Q. \tag{20}$$

According to (15), we have $\Delta U = 0$. The heat put at stake during the isothermal and irreversible expansion $1 \to 2$ is:

$$\Delta U = W + Q = 0 \quad \Rightarrow \quad Q = -W = 1982.1 \text{ J}. \tag{21}$$

The energy is positive. It is therefore received by the system.

5. *Gas enthalpy variation*:

Enthalpy, denoted H, comprises the internal energy of a system to which is added the work which this system must exert against the external pressure to occupy its own volume. Enthalpy is a chemical potential which predicts the evolution and the equilibrium of a thermodynamic system, and from which we can deduce all its properties. It is an extensive state function which plays a privileged role in the isobaric transformations which are very useful in chemistry. It is defined by:

$$H = U + PV. \tag{22}$$

Here the term PV corresponds to the energy of expansion or compression of the system. H is always greater than U. Enthalpy is in joules or in calories. For an infinitesimal transformation, we have, by starting from (4.24) and (4.27) of the course:

$$dU = \delta Q + \delta W = dQ - PdV. \tag{23}$$

By introducing the enthalpy function, we arrive at:

$$dH = dU + d(PV). \tag{24}$$

The relaxation being isothermal and by starting from (15), we deduce that $d(PV) = R \, d(T) = 0$. The variation of enthalpy is thus:

$$dH = dU \quad \Rightarrow \quad \Delta H = \Delta U = 0. \tag{25}$$

III. *Adiabatic and Reversible Relaxation*

1. *Final gas temperature*:

Since the transformation from state 1 (P_1, T_1, V_1) to state 2 (P_2, T_2, V_2) is adiabatic, there is no heat exchange between the system and the external medium $\Rightarrow \delta Q = 0$. By starting from Eq. (4.33) of the course, we obtain:

$$dU = C_V \, dT = \delta W + \delta Q = \delta W + 0 = \delta W. \tag{26}$$

By applying the ideal gas equation to our one-mole system, we deduce from (26):

$$C_V \, dT = \delta W = -P \, dV = -\frac{RT}{V} \, dV. \tag{27}$$

Moreover, Eqs. (4.35) and (4.36) of the course give us:

$$C_P - C_V = R \quad \text{and} \quad \gamma = \frac{C_P}{C_V}. \tag{28}$$

By injecting (28) into (27), we obtain:

$$C_V \, dT = -\frac{RT}{V} \, dV = -\frac{(C_P - C_V) \, T}{V} \, dV \quad \Rightarrow \quad \frac{dT}{T} = -\frac{C_P - C_V}{C_V} \times \frac{dV}{V}$$

$$= -\left(\frac{C_P}{C_V} - 1\right) \times \frac{dV}{V} = -(\gamma - 1) \times \frac{dV}{V} = (1 - \gamma) \, \frac{dV}{V}. \tag{29}$$

By integrating (29), we find:

$$\int \frac{dT}{T} = (1 - \gamma) \int \frac{dV}{V}. \tag{30}$$

We finally get:

$$lnT = (1 - \gamma) \ lnV \quad \Rightarrow \quad T = \text{Cte } V^{1-\gamma} \quad \Rightarrow \quad TV^{\gamma-1} = \text{Cte.} \tag{31}$$

As $PV = RT$, we deduce from (31):

$$\left(\frac{PV}{R}\right) V^{\gamma-1} = \text{Cte} \quad \Rightarrow \quad PV^{\gamma-1+1} = R \times \text{Cte} = \text{Cte} \quad \Rightarrow \quad PV^{\gamma} = \text{Cte.} \tag{32}$$

By replacing V with RT/P, deduced from the ideal gas equation, we arrive at an equation which links this time the temperature and the pressure:

$$PV^{\gamma} = P\left(\frac{RT}{P}\right)^{\gamma} = P^{1-\gamma} \, T^{\gamma} = \frac{\text{Cte}}{R} = \text{Cte.} \tag{33}$$

By applying (33) to states 1 and 2, we deduce:

$$P_1^{1-\gamma} T_1^{\gamma} = P_2^{1-\gamma} T_2^{\gamma} \quad \Rightarrow \quad T_2 = T_1 \left(\frac{P_1}{P_2}\right)^{\frac{1-\gamma}{\gamma}}. \tag{34}$$

On the basis of the hypotheses of the exercise, we can write:

$$\begin{aligned} C_V &= \frac{3R}{2} \\ C_P &= \frac{5R}{2} \end{aligned} \quad \Rightarrow \quad \gamma = \frac{\frac{5R}{2}}{\frac{3R}{2}} = \frac{5}{3}. \tag{35}$$

By injecting (35) into (34), we determine the value of the temperature T_2 of state 2:

$$T_2 = T_1 \left(\frac{P_1}{P_2}\right)^{\frac{1-\gamma}{\gamma}} = (298 \text{ K}) \left(\frac{5 \text{ atm}}{1 \text{ atm}}\right)^{\frac{1-\frac{5}{3}}{\frac{5}{3}}} = 156.5 \text{ K.} \tag{36}$$

2. *Variation of the internal energy of gas*:

The internal energy is a state function; its variation does not depend on the followed path to go from state 1 to state 2. For an adiabatic transformation from state 1 to state 2, we have by starting from Eq. (4.33) of the course:

$$\Delta U = C_V\,\Delta T = \frac{3R}{2}\,(T_2 - T_1) = \frac{3\left(8.314\ \mathrm{J.K^{-1}.mol^{-1}}\right)}{2}\ (156.5\ \mathrm{K} - 298\ \mathrm{K})$$
$$= -1761.7\ \mathrm{J.mol^{-1}}.$$

$$(37)$$

3. *Work done by gas*:

Since the closed system only exchanges energy to the external environment, the variation of its internal energy during a transformation from state 1 to state 2 is equal to the algebraic sum of the exchanged energies in the form of mechanical work W and heat Q. Since the transformation is adiabatic, there is no heat exchange between the system and the external environment $\Rightarrow Q = 0$:

$$\Delta U = U_2 - U_1 = W + Q = W = -1761.7\ \mathrm{J.mol^{-1}}. \qquad (38)$$

4. *Quantity of heat put at stake*:

For an adiabatic relaxation, there is no heat exchange $\Rightarrow Q = 0$ J.

5. *Gas enthalpy variation*:

By starting from Eq. (4.34) of the course, we obtain:

$$\Delta H = C_P\,\Delta T = \frac{5R}{2}\,(T_2 - T_1) = \frac{5\left(8.314\ \mathrm{J.K^{-1}.mol^{-1}}\right)}{2}\ (156.5\ \mathrm{K} - 298\ \mathrm{K})$$
$$= -2941.1\ \mathrm{J.mol^{-1}}.$$

$$(39)$$

IV. *Adiabatic and Irreversible Relaxation*

1. *Final gas temperature*:

Since the transformation from state 1 (P_1,T_1,V_1) to state 2 (P_2,T_2,V_2) is adiabatic, there is no heat exchange between the system and the external medium $\Rightarrow \delta Q = 0$. By starting from Eq. (4.33) of the course, it becomes:

$$dU = C_V\,dT = \delta W + \delta Q = \delta W + 0 = \delta W. \qquad (40)$$

We deduce:

$$\Delta U = C_V \, \Delta T = W \quad \Rightarrow \quad C_V \, (T_2 - T_1) = -P_2 \, (V_2 - V_1). \tag{41}$$

By applying the ideal gas equation $PV = RT$ to the system formed of 1 mole a first time in state 1 and a second time in state 2, we replace the volumes V_1 and V_2, which are unknown, by the pressures P_1 and P_2, of which the values are given in the statement of this exercise. That is:

$$C_V \, (T_2 - T_1) = -P_2 \left(\frac{RT_2}{P_2} - \frac{RT_1}{P_1} \right) \quad \Rightarrow \quad T_2 = T_1 \times \frac{C_V + \dfrac{RP_2}{P_1}}{(C_V + R)} = T_1 \times \frac{\dfrac{3R}{2} + \dfrac{RP_2}{P_1}}{\left(\dfrac{3R}{2} + R \right)}$$

$$= T_1 \times \frac{3P_1 + P_2}{5P_1} = (298 \text{ K}) \frac{3 \, (5 \, \text{atm}) + 1 \, \text{atm}}{5 \, (5 \, \text{atm})} = 190.7 \text{ K}.$$

$$\tag{42}$$

2. *Variation of the internal energy of gas*:

The internal energy is a state function; its variation does not depend on the followed path to go from state 1 to state 2. For an adiabatic transformation from state 1 to state 2, we have by starting from Eq. (4.33) of the course, we obtain:

$$\Delta U = C_V \, \Delta T = \frac{3R}{2} \, (T_2 - T_1) = \frac{3 \, (8.314 \, \text{J.K}^{-1}.\text{mol}^{-1})}{2} \, (190.7 \text{ K} - 298 \text{ K})$$

$$= -1338.1 \text{ J.mol}^{-1}.$$

$$\tag{43}$$

3. *Work done by gas*:

Since the closed system only exchanges energy to the external environment, the variation of its internal energy during a transformation from state 1 to state 2 is equal to the algebraic sum of the exchanged energies in the form of mechanical work W and heat Q. Since the transformation is adiabatic, there is no heat exchange between the system and the external environment $\Rightarrow Q = 0$:

$$\Delta U = U_2 - U_1 = W + Q = W = -1338.1 \text{ J.mol}^{-1}. \tag{44}$$

4. *Quantity of heat put at stake*:

For an adiabatic relaxation, there is no heat exchange $\Rightarrow Q = 0$ J.

5. *Gas enthalpy variation*:

By starting from Eq. (4.34) of the course, we obtain:

$$\Delta H = C_{\mathrm{P}}\,\Delta T = \frac{5R}{2}\,(T_2 - T_1) = \frac{5\left(8.314\ \mathrm{J.K^{-1}.mol^{-1}}\right)}{2}\,(190.7 - 298\ \mathrm{K}) \quad (45)$$
$$= -2230.2\ \mathrm{J.mol^{-1}}.$$

Exercise 25

I. *Let us consider an isothermal compression of 1 mole of an ideal gas contained in a vertical cylinder of the section S. We suppose that the weight of the piston is negligible compared to the other forces involved in the experiment. The temperature T is kept constant during this experiment by means of a thermostat. P_1 and P_2 are the initial and final pressures with P_1 equal to the atmospheric pressure.*

    1. *What is an isothermal compression and how can we experimentally do it? If gas is ideal, on which parameter its internal energy depends?*

    2. *Represent this transformation in the Clapeyron coordinates.*

    3. *Calculate the work W_1 supplied by the external environment to 1 mole of an ideal gas.*

II. *We now realize a sudden compression by putting on the piston S a sandbag of mass M so that the final equilibrium pressure is P_2 and the temperature of the cylinder keeps its value T.*

    1. *What happens thermodynamically under these conditions?*

    2. *Calculate the work W_2 supplied by the external environment to 1 mole of an ideal gas.*

    3. *Represent graphically the work provided in the reversible and the irreversible cases by plotting $y = W_1/(P_1 V_1)$ as a function of $x = P_2/P_1$. We will verify that the work supplied to the gas in the irreversible transformation is always superior to the work of the reversible compression.*

III. *We make this time a sudden compression from P_1 to $2P_1$ then from $2P_1$ to P_2 with $P_1 < 2P_1 < P_2$. Compare the results of this experiment with the results found in I and II.*

Solution
Part I

1. • *Experimental realization of an isothermal compression*:

During an isothermal process, the temperature of the system is well defined, which implies that the transformation is quasi-static and reversible. These two conditions are experimentally realized if the transition from the initial state 1 to the final state 2 takes place very slowly. To do this, we can deposit on the piston some sand in very small quantities so that the external pressure, and therefore that of the system, passes almost continuously from the value P_1 to the value P_2. At any moment, it is possible to reverse the direction of this evolution and the system will exactly return by the same thermodynamic states. In addition, the heated gas by the compression is warmer than the environment, and to maintain isothermy, the heat must leave the system. The additional work observed for the isothermal compression corresponds to the heat energy lost by the system.

• Internal energy of ideal gases:

The internal energy of a system, denoted U, is its contents in energy. Each solid, liquid, or gaseous system is formed of the particles (atoms or molecules) which incessantly vibrate. To this microscopic movement is associated a kinetic energy Ec_i for each particle. Moreover, between these particles there exist the interaction forces with which is associated a potential energy Ep_i for each particle. The concept of internal energy brings together all the kinetics and the potential energies of the n particles which make up this system. It represents the total amount of the mechanical energy stored inside a system.

$$U = \sum_{i=1}^{n} Ec_i + \sum_{i=1}^{n} Ep_i. \tag{1}$$

As there is no interaction between the particles of the ideal gas, the second term is therefore zero. We distinguish two cases:

– *Monatomic ideal gas (MIG)*: the molecules are assimilated to their mass center and only the translational kinetic energies are to be taken into account:

$$U_{\text{MIG}} = \sum_{i=1}^{n} Ec_i = \sum_{i=1}^{n} \frac{1}{2} m_i v_i^2 = N \langle E_C \rangle = \frac{3}{2} Nk_B T \tag{2}$$

where $\langle E_C \rangle$ is the average kinetic energy of the N particles which make up the system, k_B the Boltzmann constant, and T the temperature.

– *Ideal polyatomic gas (IPG)*: it is necessary not only to take into account the kinetic energy of translation of the mass center of each molecule of gas, but also the barycentric kinetic energy, denoted E_V, relative to the disordered movement of atoms in the barycentric frame of the molecule. In the case of an ideal diatomic gas, the internal energy of this system is expressed as:

$$U_{\text{IPG}} = \frac{5}{2} Nk_B T. \tag{3}$$

In any case, we retain that the internal energy of an ideal gas (IG) depends only on temperature:

$$U_{\text{IG}} = f(T). \tag{4}$$

In particular, for any isothermal transformation, we have:

$$\Delta U_{\text{IG}} = 0. \tag{5}$$

2. *Representation of an isothermal compression in the Clapeyron coordinates*:

During an isothermal transformation of 1 mole of an ideal gas, the equation of state can be written in the following form:

$$PV = RT = \text{Cte} \quad \Rightarrow \quad P = \frac{\text{Cte}}{V}. \tag{6}$$

Equation (6) shows that the pressure varies inversely with volume. Its representation $P = f(V)$, given in the Clapeyron diagram, shows a decrease in the volume of gas from V_1 to V_2 following an increase of pressure from P_1 to P_2. This transformation is carried out at a constant temperature.

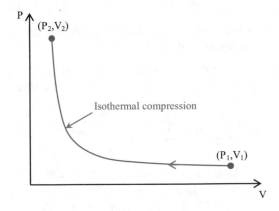

3. *Work W_1 supplied by the external environment to 1 mole of an ideal gas*:

As the compression is reversible, the transformation is very slow so that at each moment we have $P_{ext} = P_{gaz}$. The elementary work δW_1 performed during this isothermal process is therefore:

$$\delta W_1 = -P_{ext}\, dV = -P_{gas}\, dV = -RT\, \frac{dV}{V}. \tag{7}$$

As the process is isothermal, we deduce:

$$PV = RT = \text{Cte} \quad \Rightarrow \quad d(PV) = 0 \quad \Rightarrow \quad PdV + VdP = 0 \quad \Rightarrow \quad \frac{dV}{V}$$

$$= -\frac{dP}{P}. \tag{8}$$

By injecting (8) into (7), we find:

$$\delta W_1 = -RT \, \frac{dV}{V} = RT \, \frac{dP}{P}. \tag{9}$$

By integrating (9), we obtain the work W_1 supplied by the external environment to one ideal mole of gas:

$$W_1 = \int_1^2 RT \, \frac{dP}{P} = RT \int_1^2 \frac{dP}{P} = RT \, \ln\left(\frac{P_2}{P_1}\right). \tag{10}$$

This work is positive because it is provided to the considered system by the external environment.

Part II

1. *Description of the process*:

By putting a bag of sand on the piston, it abruptly descends. This process is no longer reversible because the gas pressure is not defined during the transition from state 1 to state 2. This compression cannot be represented by a curve in the Clapeyron diagram. Only the points corresponding to the initial state and the final state can be included. Gas undergoes a monobaric evolution at pressure $P_2 = P_1 + Mg/S$ and a monothermal evolution at the temperature T.

2. *Calculation of the work W_2 supplied by the external environment to 1 mole of an ideal gas*:

The work W_2 provided is that of the monobaric process. For an infinitesimal evolution, we obtain:

$$\delta W_2 = -P_{ext} \, dV = -P_2 \, dV. \tag{11}$$

By integrating between states 1 and 2, we get:

$$W_2 = - \int_1^2 P_{ext} \, dV = -P_2 \int_1^2 dV = -P_2 \, (V_2 - V_1) = P_2 V_2 \left(\frac{V_1}{V_2} - 1\right). \tag{12}$$

By starting from (8), we arrive at the provided work W_2 according to the exercise data:

$$W_2 = RT \, \left(\frac{P_2}{P_1} - 1\right). \tag{13}$$

This work is positive because it is provided to the considered system by the external environment.

3. *Graphic representation of the provided work in the reversible and irreversible*
   cases:

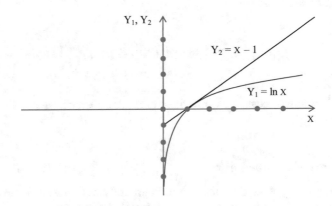

By starting from (10), we obtain:

$$W_1 = RT \ ln\left(\frac{P_2}{P_1}\right) \quad \Rightarrow \quad \frac{W_1}{RT} = ln\left(\frac{P_2}{P_1}\right). \tag{14}$$

Let us put:

$$Y_1 = \frac{W_1}{RT} = ln\left(\frac{P_2}{P_1}\right) = lnX. \tag{15}$$

By starting of (13), we find:

$$W_2 = RT \ \left(\frac{P_2}{P_1} - 1\right) \quad \Rightarrow \quad \frac{W_2}{RT} = \frac{P_2}{P_1} - 1. \tag{16}$$

Let us put:

$$Y_2 = \frac{W_2}{RT} = \frac{P_2}{P_1} - 1 = X - 1 \tag{17}$$

The plot of $Y_1 = f(X)$ and $Y_2 = h(X)$ reveals that the provided work by an irreversible
isothermal compression is always greater than the provided work by a reversible
isothermal compression.

Part III

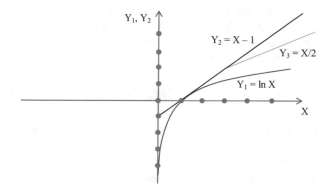

In this third and last experiment, the compression of gas is always performed in an isothermal and irreversible way, but in two stages, a first time from P_1 to $2P_1$ and a second time from $2P_1$ to P_2 with $P_1 < 2P_1 < P_2$. The given work to gas will be the sum of the works done during these two stages:

– *The work W performed during the first stage from P_1 to $2P_1$:*

By starting from (13), we immediately deduct:

$$W = RT \left(\frac{2P_1}{P_1} - 1 \right). \tag{18}$$

– *The work W' carried out during the second stage from $2P_1$ to P_2:*

$$W' = RT \left(\frac{P_2}{2P_1} - 1 \right). \tag{19}$$

– *The provided total work W_3:*

$$W_3 = W + W' = RT + RT \left(\frac{P_2}{2P_1} - 1 \right) = RT \frac{P_2}{2P_1}. \tag{20}$$

Let us put:

$$Y_3 = \frac{W_3}{RT} = \frac{P_2}{2P_1} = \frac{X}{2}. \tag{21}$$

Knowing by hypotheses that:

$$P_2 > 2P_1 \quad \Rightarrow \quad \frac{P_2}{2P_1} > 1 \quad \Rightarrow \quad \frac{X}{2} > 1 \quad \Rightarrow \quad X > 2 \tag{22}$$

inequality (22) forces us to draw the part $X > 2$ of the line $Y_3 = X/2$. The plot of this curve at the same time as the curves of the experiments described in parts I and II reveals us that a process in stages, one monobaric and the other monothermal, allows getting closer to the reversible evolution.

Exercise 26

A vertical cylindrical tube, transparent and heat-insulated with a diameter $D = 3$ cm and a height $H = 1.1$ m, contains a mass $M = 1$ kg of mercury at the initial temperature T_1. This tube is returned 50 times leading to a rise of its temperature of ΔT.

1. *Calculate the work performed by liquid mercury.*
2. *Calculate the variation of the internal energy of mercury.*
3. *Calculate the temperature variation ΔT knowing that all the work was used to heat liquid mercury. We give the density of mercury $\rho = 13\,600$ kg.m^{-3}, its specific heat $C = 138$ J.kg^{-1}.K^{-1}, and the acceleration due to gravity $g = 9.81$ m.s^{-2}.*

Solution

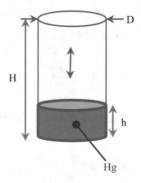

1. *Work performed by liquid mercury*:

The work performed by liquid mercury is equal to the work of its weight when it moves of a distance $H - h$ multiplied by the number of the reversals of the tube:

$$W = Mg\,(H - h) \times 50. \tag{1}$$

In Eq. (1), h is unknown. Let us determine its expression based on the exercise data. The basic surface of the circular tube has the following expression:

$$S = \pi \left(\frac{D}{2}\right)^2. \tag{2}$$

The density of mercury ρ is equal to:

$$\rho = \frac{M}{V} = \frac{M}{h \times S} \quad \Rightarrow \quad h = \frac{M}{\rho S}. \tag{3}$$

By injecting (2) into (3), we obtain h according to the exercise data:

$$h = \frac{M}{\rho S} = \frac{M}{\rho \pi \left(\frac{D}{2}\right)^2}. \tag{4}$$

By starting from (1), the work performed by liquid mercury is:

$$W = Mg \left[H - \frac{M}{\rho \pi \left(\frac{D}{2}\right)^2} \right] \times 50 = (1 \text{ kg}) \left(9.81 \text{ m.s}^{-2} \right)$$

$$\times \left[(1.1 \text{ m}) - \frac{(1 \text{ kg})}{(13\,600 \text{ kg.m}^{-3})\,(3.14)\left(\frac{0.03 \text{ m}}{2}\right)^2} \right] \times 50 = 488.5 \text{ J}. \tag{5}$$

2. *Variation of the mercury internal energy*:

The variation of the mercury internal energy is given by the first principle of thermodynamics:

$$\Delta U = W + Q = W = 488.5 \text{ J}. \tag{6}$$

The tube being insulated, there is no heat exchange to the external environment $Q = 0$.

3. *Temperature variation ΔT*:

The rise of mercury temperature is due to the viscous forces of mercury as a result of the 50 tube reversals and not to any heat exchange to the external environment because the tube is insulated. It is therefore the work of these friction forces that has turned into heat. This heat is proportional to the quantity of matter expressed in kilograms and to the temperature difference in kelvins:

$$Q = M\,C\,\Delta T = W \quad \Rightarrow \quad \Delta T = \frac{W}{MC} = \frac{488.5\ \text{J}}{(1\ \text{kg})\,(138\ \text{J.kg}^{-1}.\text{K}^{-1})} = 3.54\ \text{K}. \quad (7)$$

Exercise 27

1. Calculate in joules the quantity of heat needed to raise the temperature of 300 liters of water from 20 to 100 °C.
2. Express the result in kilojoules, kilocalories, and kilowatt-hours. We give:
   - Mass thermal capacity of water $C = 4.18\ kJ.kg^{-1}.°C^{-1}$
   - Density of water: $\rho = 1000\ kg/m^3$
   - 1 cal = 4.18 J
   - $1\ kJ = 10^3\ J$
   - 1 Wh = 3600 J

Solution

1. *Heat quantity in joules*:

The quantity of heat required to carry 300 liters of water from 20 to 100 °C is given by:

$$Q = m\,C\,\Delta T \qquad\qquad (1)$$

where Q is expressed in the international system in J, m in kg, C in J.kg^{-1}.K^{-1}, and T in K. In this equation, ΔT is a temperature difference, so it can express itself indifferently in °C or in K. The other quantities can be expressed in other units, generally imposed by the unit of the thermal capacity. From (1), we obtain:

$$Q = \left[\frac{(300 \times 10^{-3}\ \text{m}^3)\,(1000\ \text{kg})}{1\ \text{m}^3}\right] (4.18 \times 10^3\ \text{J.kg}^{-1}.°\text{C}^{-1})\,(100 - 20\,°\text{C})$$
$$= 100\ 320\ 000\ \text{J}.$$

$$(2)$$

2. *Quantity of heat in kilojoules, calories, and kilowatt-hours*:

- *Quantity of heat in kilojoules*:

Knowing by hypotheses that $1\ \text{kJ} = 10^3\ \text{J}$, it becomes:

$$Q = \frac{(100\ 320\ 000\ \text{J})\,(1\ \text{kJ})}{10^3\ \text{J}} = 100\ 320\ \text{kJ}. \qquad (3)$$

- *Quantity of heat in kilocalories*:

Knowing by hypotheses that 1 cal = 4.18 J, we obtain:

$$Q = \frac{(100\ 320\ 000\ \text{J})\ (1\ \text{cal})}{4.1855\ \text{J}} = 2.4 \times 10^7\ \text{cal} = 2.4 \times 10^4\ \text{kcal}. \qquad (4)$$

- *Quantity of heat in kilowatt-hours*:

The kilowatt-hour is a unit of energy. Knowing by hypotheses that $1\ \text{Wh} = 3600\ \text{J}$, we find:

$$Q = \frac{(100\ 320\ 000\ \text{J})\ (1\ \text{Wh})}{3600\ \text{J}} = 2.8 \times 10^4\ \text{Wh} = 28\ \text{kWh}. \qquad (5)$$

Exercise 28

Calculate the power in kilowatts to carry 500 liters of water from 10 to 60 °C in 6 hours. We give the mass thermal capacity of water $C = 4.18\ \text{kJ.kg}^{-1}.°C^{-1}$ and the density of water $\rho = 1000\ \text{kg/m}^3$.

Solution

The required quantity of heat to carry 500 liters of water from 10 to 60 °C is given by:

$$Q = m\ C\ \Delta T \qquad (1)$$

where Q is expressed in the international system in J, m in kg, C in $\text{J.kg}^{-1}.\text{K}^{-1}$, and T in K. In this equation, ΔT is a temperature difference, so it can express itself indifferently in °C or in K. The other quantities can be expressed in other units, generally imposed by the unit of the thermal capacity. From (1), we obtain:

$$Q = \left[\frac{\left(500 \times 10^{-3}\ \text{m}^3\right)\ (1000\ \text{kg})}{1\ \text{m}^3} \right]\ \left(4.18 \times 10^3\ \text{J.kg}^{-1}.°C^{-1}\right)\ (60 - 10\ °C)$$
$$= 104\ 500\ 000\ \text{J}. \qquad (2)$$

The needed power to produce 104 500 000 J in 6 hours is:

$$p = \frac{Q}{\text{time}} = \frac{104\ 500\ 000\ \text{J}}{(6\ \text{hours})(3600\ \text{secondes})} = 4837.9\ \text{W} = 4.8\ \text{kW}. \qquad (3)$$

Exercise 29

Hundred tons of scraps are heated in an electric oven to obtain some liquid iron at 1535 °C. The initial temperature is 20 °C. This operation lasts 5 hours and the efficiency of the oven is 70%.
1. *What is the required energy for this operation?*
2. *Deduce the power of the oven. We give:*
    - *The mass thermal capacity of iron $C_{iron} = 450\ \text{J.kg}^{-1}.K^{-1}$*
    - *The latent heat of the phase change of iron $L_{melting} = 270\ \text{kJ.kg}^{-1}$*

Solution

1. *Energy required for the transformation scrap → liquid iron*:

- *Heating of iron, solid → solid*:

The required energy to carry the solid iron from 20 to 1535 °C without phase transition is:

$$Q_1 = m \, C_{\text{iron}} \, \Delta T \qquad (1)$$

where Q_1 is expressed in the international system in J, m in kg, C in $\text{J.kg}^{-1}.\text{K}^{-1}$, and T in K. In this equation, ΔT is a temperature difference, so it can express itself indifferently in °C or in K. The other quantities can be expressed in other units, generally imposed by the unit of the thermal capacity. From (1), we obtain:

$$Q_1 = \left(\frac{100 \text{ tons} \times 1000 \text{ kg}}{1 \text{ ton}} \right) \left(450 \times 10^3 \text{ J.kg}^{-1}.\text{K}^{-1} \right) (1535 - 20 \,^{\circ}\text{C})$$
$$= 6.8 \times 10^{10} \text{ J.} \qquad (2)$$

- *Phase transition of iron, solid → liquid*:

The heat which accompanies the iron transition from the solid state to the liquid state without a change of the temperature (1535 °C) is the latent heat. It is the exchanged heat with the external environment during a change of the physical state of the matter that composes the system. It is noted L. When it is expressed for 1 kg of matter, we speak about the mass latent heat:

$$Q_2 = mL_{\text{melting}} = \left(\frac{100 \text{ tons} \times 1000 \text{ kg}}{1 \text{ ton}} \right) \left(2.7 \times 10^5 \text{ J.kg}^{-1} \right) = 2.7 \times 10^{10} \text{ J.} \qquad (3)$$

- The energy Q required for the transformation, scrap → liquid iron, is the sum of energies of the two previous steps:

$$Q = Q_1 + Q_2 = 6.8 \times 10^{10} \text{ J} + 2.7 \times 10^{10} \text{ J} = 9.5 \times 10^{10} \text{ J.} \qquad (4)$$

- This energy has been calculated assuming that the furnace yield is 100%. Taking into account the efficiency of the furnace, we finally end up in the sought energy:

$$Q' = \frac{Q \times 70}{100} = \frac{9.5 \times 10^{10} \text{ J} \times 70}{100} = 6.67 \times 10^{10} \text{ J.} \qquad (5)$$

2. *Oven power*:

The oven power is given by the quantity of energy actually produced by the oven on the duration of this operation:

$$p = \frac{Q'}{time} = \frac{6.67 \times 10^{10} \text{ J}}{(5 \text{ hours}) (3600 \text{ secondes})} = 3.5 \times 10^6 \text{ W} = 3.5 \times 10^3 \text{ kW}. \quad (6)$$

Exercise 30
What is the latent heat of melting of lead in J/g knowing that the impact of a standard 9 mm parabellum bullet made of this matter launched at a speed of 350 m/s partially melts it? We give the melting temperature of lead $T_{melting} = 327.46\,°C$, its thermal capacity $C = 0.129$ J/g/K, and the ambient temperature $25\,°C$.

Solution
During the impact of the bullet, its kinetic energy is transformed into heat, a part of which partially melts the lead:

$$\frac{1}{2} mv^2 = m\,C\,\Delta T + m\,L \quad \Rightarrow \quad L = \frac{1}{2}v^2 - C\,\Delta T$$
$$= \frac{1}{2}(350 \text{ m.s}^{-1})^2 - (0.129 \text{ J.g}^{-1}.\text{K}^{-1})(327.46 - 25\,°C) = 136 \text{ J.g}^{-1}. \quad (1)$$

Exercise 31
We immerse a piece of metal of mass $M = 300$ g, of thermal capacity $C = 0.06$ cal/°C/g, and of temperature $T_1 = 400\,°C$ in a negligible mass calorimeter containing 600 g of water, a third of which is frozen. How much ice remains after reaching the thermal equilibrium? We give the latent heat of melting of water $L_{melting} = 80$ cal/g.

Solution
The metal piece which cools from $T_1 = 400$ °C to the temperature of the liquid-ice coexistence $T_0 = 0$ °C provides to the calorimeter a quantity of heat Q:

$$Q = MC\,(T_0 - T_1) = (300 \text{ g})\,(0.06 \text{ cal.}°\text{C}^{-1}.\text{g}^{-1})\,|0 - 400\,°\text{C}| = 7200 \text{ cal}. \quad (1)$$

This quantity of heat is absorbed by the melting of m grams of ice:

$$m = \frac{Q}{L_{melting}} = \frac{7200 \text{ cal}}{80 \text{ cal.g}^{-1}} = 90 \text{ g}. \quad (2)$$

Initially, there was:

$$\frac{600 \text{ g}}{3} = 200 \text{ g of ice}. \quad (3)$$

90 g of the ice melted, so it remains:

$$200 - 90 \text{ g} = 110 \text{ g of ice.} \tag{4}$$

Exercise 32
Let us consider a negligible mass calorimeter. We put there a quantity of water of mass M_1, of mass heat capacity C_1, and of temperature T_1. We add there a quantity of alcohol M_2 at the temperature T_2 such that $T_2 < T_1$. After a while, the water-alcohol mixture is isothermal at an intermediate temperature T_3. What is the mass heat capacity C_2 of alcohol as a function of data? In this exercise, the masses are in kg, the mass thermal capacities in J/kg/K, and the temperatures in K.

Solution
The temperature of water has dropped from T_1 to T_3. The water has therefore provided a quantity of heat Q_1 such as:

$$Q_1 = M_1 \, C_1 \, |T_3 - T_1|. \tag{1}$$

The temperature of alcohol rose from T_2 to T_3. The alcohol has therefore received a quantity of heat Q_2 such as:

$$Q_2 = M_2 \, C_2 \, (T_3 - T_2). \tag{2}$$

At thermal equilibrium, we have $Q_1 = Q_2$:

$$M_1 \, C_1 \, |T_3 - T_1| = M_2 \, C_2 \, (T_3 - T_2) \quad \Rightarrow \quad C_2 = \frac{M_1}{M_2} \times \frac{|T_3 - T_1|}{(T_3 - T_2)}. \tag{3}$$

Exercise 33
1. *A calorimeter contains a mass $m_1 = 500$ g of water at a temperature $T_1 = 19\,°C$. We introduce a mass $m_2 = 150$ g of water at a temperature $T_2 = 25.7\,°C$. The final temperature is $T_F = 20.5\,°C$. Calculate the thermal capacity of the calorimeter knowing that the mass heat capacity of water is $C_{water} = 4180$ J.K^{-1}. kg^{-1}.*
2. *In the same calorimeter containing now $m_3 = 750$ g of water at $T_3 = 19\,°C$, a copper block of $m_4 = 550$ g and $T_4 = 92\,°C$ is immersed. The final temperature is $T_f = 23.5\,°C$. What is the mass heat capacity of copper?*

Solution
1. *Heat capacity of the calorimeter:*

The hot water transfers to the calorimeter and to the cold water which it contains a quantity of heat Q_2:

$$Q_2 = m_2 \, C_{\text{water}} \, (T_F - T_2) = (0.15 \text{ kg}) \, \left(4180 \text{ J.kg}^{-1}.\text{K}^{-1}\right) (25.7 - 20.5\,^{\circ}\text{C})$$
$$= 3260 \text{ J}.$$

$$(1)$$

- *Thermal capacity of the cold water:*

Since the thermal capacity of the cold water is an extensive quantity and expresses itself in J/K, it is worth:

$$(0.5 \text{ kg}) \left(4180 \text{ J.kg}^{-1}.\text{K}^{-1}\right) = 2090 \text{ J.K}^{-1}. \tag{2}$$

- *Heat capacity μ of the calorimeter:*

The thermal capacity in J/kg of the calorimeter-cold water set is $\mu + 2090$. The quantity of the gained heat by the cold bodies, calorimeter + cold water, is:

$$Q_1 = (\mu + 2090) \text{ J.K}^{-1} \, (T_F - T_1) = (\mu + 2090) \text{ J.K}^{-1} \, (20.5 - 19\,^{\circ}\text{C})$$
$$= (1.5\,\mu + 3135) \text{ J}. \tag{3}$$

This quantity of the gained heat by the calorimeter-cold water set is equal to the quantity of the yielded heat by the hot water:

$$Q_1 = Q_2 \quad \Rightarrow \quad (1.5\,\mu + 3135) \text{ J} = 3260 \text{ J} \quad \Rightarrow \quad \mu = 83.6 \text{ J.K}^{-1}. \tag{4}$$

The determination of the heat capacity of the calorimeter in J/K and not of its mass heat capacity in J/kg/K is due to the fact that the mass of the calorimeter is unknown.

2. *Mass thermal capacity of copper:*

The quantity of heat Q_1' received by the cold bodies formed of the calorimeter and the water which it contains is:

$$Q_1' = \left[\mu + (0.75 \text{ kg}) \left(4180 \text{ J.kg}^{-1}.\text{K}^{-1}\right)\right] \text{ J.K}^{-1} \, (T_f - T_3)$$
$$= \left[83.6 \text{ J.K}^{-1} + (0.75 \text{ kg}) \left(4180 \text{ J.kg}^{-1}.\text{K}^{-1}\right)\right] (23.5 - 19\,^{\circ}\text{C}) \tag{5}$$
$$= 14\,482 \text{ J}.$$

The quantity of heat Q_2' yielded by the hot body, formed of the immersed copper block, is:

$$Q_2' = m_4 \, C_{\text{water}} \, (T_f - T_4) = (0.55 \text{ kg}) \, C_{\text{Cu}} \, |23.5 - 92\,^{\circ}\text{C}| = (37.67 \, C_{\text{Cu}}) \text{ J}. \tag{6}$$

At thermal equilibrium, the heat acquired by the calorimeter-cold water set is equal to the heat yielded by the hot copper:

$$Q_1' = Q_2' \quad \Rightarrow \quad (37.67 \, C_{Cu}) \, \text{J} = 14\,482 \, \text{J} \quad \Rightarrow \quad C_{Cu} = \frac{14\,482 \, \text{J}}{37.67 \, \text{kg.K}} \quad (7)$$

$$= 384.4 \, \text{J.kg}^{-1}.\text{K}^{-1}.$$

Exercise 34

The water of a river at a temperature $T_1 = 16\,°C$ is used for the liquefaction of the vapor at the exit of a nuclear power plant of temperature $T_V = 130\,°C$. The temperature of the water leaving the liquefaction process is $T_2 = 60\,°C$. How much river water does it take to liquefy 1 kg of vapor? We give the latent heat of the water vaporization $L_V = 2260$ kJ/kg, the mass heat capacity of the water vapor $C_V = 2090$ J/kg/K, and the mass heat capacity of liquid water $C_V = 2090$ J/kg/K.

Solution

• *Quantity of heat Q ceded by 1 kg of vapor (hot body):*

The water vapor at the exit of the nuclear power plant being hotter than the water of the adjacent river gives up its heat. To do this, this vapor goes in three stages. At first, the water vapor cools from $T_V = 130\,°C$ to $T° = 100\,°C$ without change of state:

$$Q_1 = m_V \, C_V \, (T_V - T°) = (1 \, \text{kg}) \, (2090 \, \text{J.kg}^{-1}.\text{K}^{-1}) \, |100 - 130\,°C|$$
$$= 62\,700 \, \text{J}. \quad (1)$$

In a second step, the vapor undergoes a change of state vapor-liquid at $T° = 100\,°C$:

$$Q_2 = m_V \, L_V = (1 \, \text{kg}) \, (2260 \times 10^3 \, \text{J.kg}^{-1}) = 2260 \, 10^3 \, \text{J}. \quad (2)$$

In a third and last step, the water vapor became liquid and cools from $T° = 100\,°C$ to $T_2 = 60\,°C$ with $m_V = m_L = 1$ kg:

$$Q_3 = m_L \, C_L \, |T_2 - T°| = (1 \, \text{kg}) \, (4180 \, \text{J.kg}^{-1}.\text{K}^{-1}) \, |60 - 100\,°C|$$
$$= 167\,200 \, \text{J}. \quad (3)$$

In total, the vapor, which is the hot body, gives in to the river which represents the cold body, a quantity of heat:

$$Q = Q_1 + Q_2 + Q_3 = 62.7 \times 10^3 \, \text{J} + 2260 \times 10^3 \, \text{J} + 167.2 \times 10^3 \, \text{J}$$
$$= 2489.9 \, 10^3 \, \text{J}. \quad (4)$$

• *Quantity of heat Q' received by M kg of liquid water (cold body):*

The quantity of heat Q' received by M kg of river water, which passes from $T_1 = 16\,°C$ to $T_2 = 60\,°C$, is:

$$Q' = M \, C_{\mathrm{L}} \, |T_1 - T_2| = (M \text{ kg}) \left(4180 \text{ J.kg}^{-1}.\text{K}^{-1}\right) |16 - 60 \text{ °C}|$$
$$= \left(183.92 \times 10^3 \, M\right) \text{ J}. \tag{5}$$

- *Quantity of river water M needed to liquefy 1 kg of vapor:*

  At thermal equilibrium, we have:

$$Q = Q' \;\Rightarrow\; \left(183.92 \times 10^3 \, M\right) \text{J} = 2489.9 \times 10^3 \text{ J} \;\Rightarrow\; M = \frac{2489.9 \times 10^3 \text{ J}}{183.92 \times 10^3 \text{ J.kg}^{-1}}$$

$$= 13.54 \text{ kg}.$$
$$\tag{6}$$

It takes 13.54 kg of river water to liquefy 1 kg of water vapor coming out of a nuclear power plant. The resulting warm water causes the flora and fauna destruction.

4.9.3   Second Principle of Thermodynamics and Heat Machines

Exercise 35
Calculate the entropy variation ΔS of a system consisting of n moles of an ideal gas which undergoes an isothermal reversible transformation from state 1 (V_1,P_1,T) to state 2 (V_2,P_2,T).

Solution
As the studied gas is ideal, its internal energy U depends only on temperature and since the process which passes this gas from state 1 to state 2 is isothermal, we have:

$$\Delta U = Q + W = 0 \quad \Rightarrow \quad Q = -W. \tag{1}$$

The process being reversible, the transformation $1 \to 2$ is very slow so that at each moment we have $P_{\mathrm{ext}} = P_{\mathrm{gaz}}$. The elementary work δW performed during the isotherm is thus:

$$\delta W_{\mathrm{rev}} = -P_{\mathrm{ext}} \, dV = -P_{\mathrm{gaz}} \, dV = -RT \, \frac{dV}{V}. \tag{2}$$

By integrating (2), we obtain the work W supplied by the external environment to n moles of an ideal gas:

$$W_{\text{rev}} = \int_1^2 -nRT \frac{dV}{V} = -nRT \int_1^2 \frac{dV}{V} = nRT \, ln\left(\frac{V_1}{V_2}\right). \qquad (3)$$

The process is isothermal; we deduce:

$$PV = nRT = \text{Cte} \quad \Rightarrow \quad d(PV) = 0 \quad \Rightarrow \quad PdV + VdP = 0 \quad \Rightarrow \quad \frac{dV}{V}$$

$$= -\frac{dP}{P}. \qquad (4)$$

By injecting (4) into (3), we obtain:

$$W_{\text{rev}} = \int_1^2 nRT \frac{dP}{P} = nRT \int_1^2 \frac{dP}{P} = nRT \, ln\left(\frac{P_2}{P_1}\right). \qquad (5)$$

From (1), (3) and (5), we deduce:

$$Q_{\text{rev}} = -W_{\text{rev}} = nRT \, ln\left(\frac{V_2}{V_1}\right) = nRT \, ln\left(\frac{P_1}{P_2}\right). \qquad (6)$$

Since entropy is a state function, its variation depends only on the initial state 1 and the final state 2:

$$\Delta S = S_2 - S_1 = \int_1^2 dS = \int_1^2 \frac{dQ_{\text{rev}}}{T} = \frac{Q_{\text{rev}}}{T} = nR \, ln\left(\frac{V_2}{V_1}\right) = nR \, ln\left(\frac{P_1}{P_2}\right). \qquad (7)$$

Exercise 36
Calculate the entropy variation ΔS of a system consisting of n moles of an ideal gas which undergoes an isobaric reversible transformation from state 1 to state 2. The thermal capacity is assumed to be constant.

Solution
Since the transformation is reversible isobaric, we have:

$$dQ_{\text{rev}} = dQ_P = nC_P \, dT. \qquad (1)$$

As entropy is a state function, its variation depends only on the initial state 1 and the final state 2:

$$\Delta S = \int\limits_{1}^{2} dS = \int\limits_{1}^{2} \frac{dQ_P}{T} = \int\limits_{1}^{2} nC_P \frac{dT}{T} = nC_P \ln\left(\frac{T_2}{T_1}\right) \tag{2}$$

where C_P is the heat capacity at constant pressure.

Exercise 37

Calculate the entropy variation ΔS of a system consisting of n moles of an ideal gas which undergoes a reversible isochoric transformation from state 1 to state 2. The heat capacity is assumed to be constant.

Solution

Since the transformation is reversible isochoric, we have:

$$dQ_{rev} = dQ_V = nC_V dT. \tag{1}$$

Entropy is a state function; its variation depends only on the initial state 1 and the final state 2:

$$\Delta S = \int\limits_{1}^{2} dS = \int\limits_{1}^{2} \frac{dQ_V}{T} = \int\limits_{1}^{2} nC_V \frac{dT}{T} = nC_V \ln\left(\frac{T_2}{T_1}\right) \tag{2}$$

where C_V is the thermal capacity at constant volume.

Exercise 38

Calculate the entropy S of a system formed of 1 mole of an ideal gas according to:
1. *The temperature T and the volume V*
2. *The temperature T and the pressure P*
3. *The volume V and the pressure P*
   We give $dU = C_V dT$, $dH = C_P dT$, and $C_P - C_V = R$.

Solution

1. *Entropy S according to T and V:*

According to the first principle of thermodynamics, the infinitesimal variation of the internal energy of the system is:

$$dU = dQ_{rev} + dW_{rev}. \tag{1}$$

As the process is reversible, the passage between two neighboring states is therefore very slow so that at each moment we have $P_{ext} = P_{gaz} = P$. The elementary work δW carried out during the isotherm is thus:

$$dW_{rev} = -P_{ext}\, dV = -PdV. \tag{2}$$

According to the second principle of thermodynamics:

$$dS = \frac{dQ_{rev}}{T}. \tag{3}$$

By injecting (3) and (2) into (1) and taking into account the hypotheses, we obtain:

$$dU = C_V dT = TdS - PdV. \tag{4}$$

The equation of ideal gases gives for 1 mole of this gas:

$$PV = RT \quad \Rightarrow \quad P = \frac{RT}{V}. \tag{5}$$

By injecting (5) into (4), we obtain:

$$C_V dT = TdS - \frac{RT}{V}\, dV \quad \Rightarrow \quad dS = C_V \frac{dT}{T} + R \frac{dV}{V}. \tag{6}$$

2. *Entropy S as a function of T and P*:

Enthalpy, denoted H, comprises the internal energy of a system to which is added the work which this system must exert against the external pressure to occupy its own volume. Enthalpy is a chemical potential which predicts the evolution and the equilibrium of a thermodynamic system, and from which we can deduce all its properties. It is an extensive state function which plays a privileged role in the isobaric transformations which are very useful in chemistry. It is defined by:

$$H = U + PV. \tag{7}$$

Here the term PV corresponds to the energy of expansion or compression of the system. H is always greater than U. Enthalpy is in joules or in calories. For an infinitesimal transformation, we have, by starting from (4.24) and (4.27) of the course:

$$dU = \delta Q + \delta W = dQ - PdV. \tag{8}$$

By introducing the enthalpy function, we obtain:

$$dH = dU + d(PV) = dU + PdV + VdP. \tag{9}$$

We have shown in (4) that:

$$dU = TdS - PdV. \tag{10}$$

By injecting (10) into (9), we obtain:

$$dH = dU + d(PV) = TdS - P\,dV + PdV + VdP = TdS + VdP. \tag{11}$$

The equation of ideal gases gives for 1 mole of this gas:

$$PV = RT \quad \Rightarrow \quad V = \frac{RT}{P}. \tag{12}$$

By injecting (12) into (11) and taking into account the hypotheses, we finally obtain:

$$C_P dT = TdS + RT\frac{dP}{P} \quad \Rightarrow \quad dS = C_P\frac{dT}{T} - R\frac{dP}{P}. \tag{13}$$

3. *Entropy S according to V and P*:

By starting from (6) and (13), we obtain:

$$dS = C_V\frac{dT}{T} + R\frac{dV}{V} = C_P\frac{dT}{T} - R\frac{dP}{P} \Rightarrow R\frac{dV}{V} = (C_P - C_V)\frac{dT}{T} - R\frac{dP}{P}. \tag{14}$$

By taking into account the hypotheses ($C_P - C_V = R$), (14) is simplified:

$$\frac{dT}{T} = \frac{dV}{V} + \frac{dP}{P}. \tag{15}$$

By injecting (15) into (6), we finally deduce:

$$\begin{aligned} dS &= C_V\frac{dT}{T} + R\frac{dV}{V} = C_V\left(\frac{dV}{V} + \frac{dP}{P}\right) + (C_P - C_V)\frac{dV}{V} \\ &= C_V\frac{dV}{V} + C_V\frac{dP}{P} + C_P\frac{dV}{V} - C_V\frac{dV}{V} = C_P\frac{dV}{V} + C_V\frac{dP}{P}. \end{aligned} \tag{16}$$

We would have obtained the same result by injecting (15) into (13).

Exercise 39
1. *Calculate the entropy variation of 2 moles of an ideal gas, which isothermally relaxes from 30 to 50 liters in the following cases: (a) reversible and (b) irreversible.*
2. *Deduce the created entropy. We give $R = 8.314$ J/K/mol, the constant of ideal gases.*

Solution

1. *Entropy variation of two moles of an ideal gas*:

(a) *Variation of the reversible entropy*:

As the studied system is an ideal gas, its internal energy U depends only on temperature and since the process which passes this gas from state 1 to state 2 is isothermal, we have:

$$\Delta U = Q + W = 0 \quad \Rightarrow \quad Q = -W. \tag{1}$$

This transformation being reversible, the transformation $1 \rightarrow 2$ is very slow so that at each moment we have $P_{ext} = P_{gaz}$. The elementary work δW performed during the isotherm is thus:

$$\delta W_{rev} = -P_{ext}\, dV = -P_{gaz}\, dV = -RT\, \frac{dV}{V}. \tag{2}$$

By integration of (2), we obtain the work W_{rev} supplied by the external environment to n moles of the ideal gas:

$$W_{rev} = \int_1^2 -nRT\, \frac{dV}{V} = -nRT \int_1^2 \frac{dV}{V} = -nRT\, \ln\!\left(\frac{V_2}{V_1}\right) = -Q_{rev}. \tag{3}$$

Since entropy is a state function, its variation does not depend on the initial state 1 and the final state 2:

$$\Delta S_{rev} = S_2 - S_1 = \int_1^2 dS = \int_1^2 \frac{dQ_{rev}}{T} = \frac{Q_{rev}}{T} = nR\, \ln\!\left(\frac{V_2}{V_1}\right)$$

$$= (2\ \text{mol})\ \left(8.314\ \text{J.K}^{-1}.\text{mol}^{-1}\right)\ \ln\!\left(\frac{50\ \text{liters}}{30\ \text{liters}}\right) = 8.49\ \text{J.K}^{-1}. \tag{4}$$

(b) *Variation of the irreversible entropy*:

An irreversible isothermal relaxation is a rapid transformation from state 1 to state 2. This transformation can be treated as an isochoric transformation followed by an isobaric transformation where the external pressure is equal to the pressure of state 2 (final state). Under these conditions, the work done by this relaxation is that of the isobaric transformation:

$$W_{irr}(1 \to 2) = -\int_1^2 P_{ext}\, dV = -\int_1^2 P_2\, dV = -P_2 \int_1^2 dV = -P_2\,(V_2 - V_1). \quad (5)$$

By applying the ideal gas equation to n moles of gas at state 2:

$$P_2 V_2 = nRT \quad \Rightarrow \quad P_2 = \frac{nRT}{V_2}. \quad (6)$$

By injection of (6) into (5), we obtain:

$$W_{irr}(1 \to 2) = -\frac{nRT}{V_2}\,(V_2 - V_1). \quad (7)$$

From (1), we deduce the heat Q_{irr} exchanged with the external environment:

$$Q_{irr} = -W_{irr} = \frac{nRT}{V_2}\,(V_2 - V_1). \quad (8)$$

The variation of the irreversible entropy is therefore:

$$\Delta S_{irr} = \frac{Q_{irr}}{T} = \frac{nR}{V_2}\,(V_2 - V_1) = \frac{(2\ \text{mol})\,\left(8.314\ \text{J.K}^{-1}.\text{mol}^{-1}\right)}{50\ \text{liters}}\,(50 - 30\ \text{liters})$$
$$= 6.65\ \text{J.K}^{-1}.$$
$$(9)$$

2. *Created entropy*:

By comparing the results found in (4) and (9), we realize that the reversible entropy is superior to the irreversible entropy:

$$\Delta S_{rev} > \Delta S_{irr} \quad \Rightarrow \quad \Delta S_{rev} = \Delta S_{irr} + \sigma \quad (10)$$

where σ represents a source of entropy characterizing the irreversibility of the transformation $\Rightarrow$ there is creation of entropy. Equation (10) shows that the entropy of a thermodynamic system can only be positive or zero. It constitutes the most general statement of the second principle of thermodynamics with $\sigma = 0$ for a reversible transformation and $\sigma \neq 0$ for an irreversible transformation. The created entropy σ cannot be calculated directly; it is deduced from the reversible and irreversible entropies.

$$\sigma = \Delta S_{rev} - \Delta S_{irr} = 8.49\ \text{J.K}^{-1} - 6.65\ \text{J.K}^{-1} = 1.84\ \text{J.K}^{-1}. \quad (11)$$

Exercise 40

A mass of water m_1 at a temperature T_1 is mixed with a mass of water m_2 at a temperature T_2 in an adiabatic container. Calculate:
1. *The final temperature of the mixture*
2. *The entropy variation of the mixture during an isobar*
3. *What condition has to fill this variation so that the transformation is irreversible?*

Solution
1. *Final temperature of the mixture*:
    The final temperature of the mixture is:

$$T_F = \frac{m_1 T_1 + m_2 T_2}{m_1 + m_2}. \tag{1}$$

2. *Variation of entropy of the mixture*:

- *Entropy variation of the mass of water m_1*:

    The mass of water m_1 goes from state 1 where $T = T_1$ to state 2 where $T = T_F$ by a reversible isobaric transformation. This mass exchanges a quantity of heat:

$$\delta Q_1 = m_1 C_P dT. \tag{2}$$

    The variation of entropy is therefore:

$$\Delta S_1 = \int_{T_1}^{T_F} \frac{\delta Q_1}{T} = \int_{T_1}^{T_F} \frac{m_1 C_P dT}{T} = m_1 C_P \ ln\left(\frac{T_F}{T_1}\right). \tag{3}$$

- *Entropy variation of the mass of water m_2*:

    The mass of water m_2 passes from state 1 where $T = T_2$ to state 2 where $T = T_F$ by a reversible isobaric transformation. This mass exchanges a quantity of heat:

$$\delta Q_2 = m_2 C_P dT. \tag{4}$$

    The variation of entropy is therefore:

$$\Delta S_2 = \int_{T_2}^{T_F} \frac{\delta Q_2}{T} = \int_{T_2}^{T_F} \frac{m_2 C_P dT}{T} = m_2 C_P \ ln\left(\frac{T_F}{T_2}\right). \tag{5}$$

- *Entropy variation of the $m_1 \pm m_2$ system*:

    The variation of the mixture $m_1 + m_2$ is:

$$\Delta S = \Delta S_1 + \Delta S_2 = m_1 C_P \; ln\left(\frac{T_F}{T_1}\right) + m_2 C_P \; ln\left(\frac{T_F}{T_2}\right). \tag{6}$$

3. *Condition to be fulfilled for an irreversible transformation*:

For an irreversible transformation, the entropy variation between the final state and the initial state must be positive:

$$\Delta S > 0. \tag{7}$$

Equation (6) can be written in the form:

$$
\begin{aligned}
\Delta S &= m_1 C_P \; lnT_F - m_1 C_P \; lnT_1 + m_2 C_P \; lnT_F - m_2 C_P \; lnT_2 \\
&= (m_1 + m_2) \; C_P \; lnT_F - (m_1 C_P \; lnT_1 + m_2 C_P \; lnT_2) \\
&= (m_1 + m_2) \; C_P \; lnT_F - \left[\frac{(m_1 + m_2) \; m_1 C_P \; lnT_1}{m_1 + m_2} + \frac{(m_1 + m_2) \; m_2 C_P \; lnT_2}{m_1 + m_2} \right] \\
&= (m_1 + m_2) \; C_P \left(lnT_F - \frac{m_1 \; lnT_1 + m_2 \; lnT_2}{m_1 + m_2} \right) > 0 \\
\Rightarrow \; & lnT_F > \frac{m_1 \; lnT_1 + m_2 \; lnT_2}{m_1 + m_2}.
\end{aligned}
\tag{8}
$$

The temperature T_F is an intermediate temperature between the temperatures T_1 and T_2. The first member of inequality (8) represents the logarithm of the barycenter T_F and the second member, the barycenter of logarithms of the pairs (T_1, m_1) and (T_2, m_2) which represent the weighted temperatures T_1 and T_2 of the masses m_1 and m_2, respectively. The logarithm of the barycenter T_F is always above the barycenter of logarithms of the pairs (T_1, m_1) and (T_2, m_2). This transformation is therefore always irreversible.

Exercise 41

1. *One mole of helium atoms is enclosed in a cylinder, of which the walls are permeable to heat. This cylinder is immersed in a thermostat at $T_2 = 273$ K. Initially, the gas was at $T_1 = 300$ K. This gas, supposed ideal, is let cool at constant volume.*
   (a) *Indicate without making calculations the temperature of equilibrium. What can you say about the evolution of the gas pressure?*
   (b) *What is the value of the coefficient $\gamma = C_P/C_V$ for a monoatomic ideal gas knowing that the average value of the kinetic energy associated with the thermal agitation by degree of freedom is worth $nRT/2$?*
   (c) *Calculate the entropy variations of gas, of the thermostat, and of the set, supposed isolated.*
2. *Starting from the previous equilibrium, we halve the gas volume in an isothermal and a reversible way. Calculate the entropy variations of gas, the thermostat, and the entire experimental device, supposed isolated. We give the relation of Mayer $C_P - C_V = nR$ and $R = 8.314$ J/K/mol, the constant of ideal gases.*

Solution

1. (a) *Equilibrium temperature*:

The equilibrium state is reached when the temperature of helium in the cylinder and the temperature of the thermostat become equal. However, a thermostat is a device of which the temperature does not vary. The heat flows naturally from the hot body (gas in the cylinder at $T_1 = 300$ K) to the cold body (thermostat at $T_2 = 273$ K) until the temperatures of the two bodies are equal to the value of the temperature of the thermostat, i.e., $T_2 = 273$ K. Helium being by hypothesis an ideal gas, it verifies the equation of ideal gases:

$$PV = nRT. \tag{1}$$

The temperature of this gas has decreased at constant volume. Under (1), the pressure must also decrease in the same proportions.

(b) *Value of the coefficient $\gamma = C_P/C_V$ for a monoatomic ideal gas*:

The monatomic molecules are considered quasi-points. They have three quadratic degrees of freedom associated with the three dimensions of the translational motion. The average of the kinetic energy associated with thermal agitation is nothing other than the internal energy U of the system because for an ideal gas, the potential energy due to the interactions between the molecules is negligible:

$$U = 3 \times \frac{1}{2}\,nRT = \frac{3}{2}\,nRT. \tag{2}$$

The constant volume thermal capacity is deduced from:

$$C_V = \frac{\partial U}{\partial T} = \frac{3}{2}nR. \tag{3}$$

The enthalpy of the gas is:

$$H = U + PV = \frac{3}{2}\,nRT + nRT = \frac{5}{2}\,nRT. \tag{4}$$

The constant pressure thermal capacity is deduced from:

$$C_P = \frac{\partial H}{\partial T} = \frac{5}{2}\,nR. \tag{5}$$

From (3) and (5), we arrive at the value of the coefficient γ for a monoatomic ideal gas:

$$\gamma = \frac{C_P}{C_V} = \frac{\frac{5}{2}\,nR}{\frac{3}{2}\,nR} = \frac{5}{3}. \tag{6}$$

(c) *Entropy variations*:

To calculate the entropy variation, we can use a priori the following equation:

$$dS = \frac{dQ}{T}. \tag{7}$$

As we mentioned in the course, Eq. (7) is only valid for a reversible transformation. This is not the case here because the difference of the temperature which exists between the gas and the thermostat indicates that both systems are not in equilibrium. So, this equation cannot be used to calculate the entropy variation. Knowing that the entropy is a state function, its variation does not depend on the followed path and therefore on how to calculate it: it depends only on the initial state and the final state. From there, the requested entropy variations can be determined as follows: during the constant volume transformation, we have $\delta W = 0$, namely:

$$dU = \delta Q + \delta W = \delta Q = C_V\,dT \tag{8}$$

and consequently:

$$\Delta S_{\text{gas}} = \int \frac{\delta Q}{T} = \int_{1}^{2} \frac{C_V \, dT}{T} = C_V \, ln\left(\frac{T_2}{T_1}\right). \tag{9}$$

Knowing that:

$$
\begin{aligned}
C_P - C_V &= nR \\
\gamma &= \frac{C_P}{C_V}
\end{aligned}
\quad \Rightarrow \quad C_V = \frac{nR}{\gamma - 1}. \tag{10}
$$

By injecting (10) into (9), we obtain:

$$\Delta S_{\text{gas}} = \frac{nR}{\gamma - 1} \, ln\left(\frac{T_2}{T_1}\right) = \frac{(1 \text{ mol}) \, (8.314 \text{ J.K}^{-1}.\text{mol}^{-1})}{\frac{5}{3} - 1} \, ln\left(\frac{273 \text{ K}}{300 \text{ K}}\right)$$

$$= -1.18 \text{ J.K}^{-1}. \tag{11}$$

* *Variation of entropy of the thermostat*:

  The received heat by the thermostat is:

$$\delta Q' = -\delta Q = -C_V \, dT. \tag{12}$$

This heat is received at the constant temperature $T_2 = 273$ K:

$$\Delta S_{\text{th}} = \int \frac{\delta Q'}{T_2} = -\frac{C_V}{T_2} \int_{1}^{2} dT = \frac{nR}{\gamma - 1} \times \frac{T_1 - T_2}{T_2}$$

$$= \frac{(1 \text{ mol}) \, (8.314 \text{ J.K}^{-1}.\text{mol}^{-1})}{\frac{5}{3} - 1} \times \frac{300 - 273 \text{ K}}{273 \text{ K}} = 1.23 \text{ J.K}^{-1}. \tag{13}$$

* *Entropy variation of the thermostat+gas set*:

  The variation of the thermostat+gas set is the algebraic sum of the entropy variations of each constituent of this set:

$$\Delta S = \Delta S_{\text{gas}} + \Delta S_{\text{th}} = -1.18 \text{ J.K}^{-1} + 1.23 \text{ J.K}^{-1} = 0.05 \text{ J.K}^{-1} > 0. \tag{14}$$

The result of (14) shows that the variation of the thermostat+gas set is positive $\Rightarrow$ the transformation is undoubtedly irreversible. This calculation also reveals that:

$$\left|\Delta S_{\text{gaz}}\right| \approx \left|\Delta S_{\text{th}}\right| \gg \left|\Delta S\right|. \tag{15}$$

Equation (14) shows that the algebraic sum of the entropy variations of the gas contained in the thermostat and the thermostat itself is very small (= 0.05 J/K) compared to the value of each term of this sum as shown in equation (15). Although this transformation is irreversible, we can consider as a first approximation that the assumptions (use of $dS = dQ/T$, see resolution of question c of this exercise) we made to make our calculations of the two entropy variations are acceptable. This exercise deals in fact with the case of almost reversible or quasi-reversible transformations.

2. *Entropy variations*:

- *Variation of entropy of gas*:

Since gas is ideal, its internal energy U depends only on temperature, and since the process which passes this gas from state 2 to state 3 is isothermal, we have:

$$dU = \delta Q + \delta W = 0 \quad \Rightarrow \quad \delta Q = -\delta W. \tag{16}$$

This transformation being reversible, the passage $2 \rightarrow 3$ is very slow so that at each moment we have $P_{ext} = P_{gas}$. The elementary work δW performed during the isotherm is thus:

$$\delta W_{rev} = -P_{ext} \, dV = -P_{gas} \, dV = -RT \, \frac{dV}{V}. \tag{17}$$

By integrating (17), we obtain the work W_{rev} supplied by the external environment to n moles of the ideal gas:

$$W_{rev} = \int_{2}^{3} -nRT \, \frac{dV}{V} = -nRT \int_{2}^{3} \frac{dV}{V} = -nRT \, ln\left(\frac{V_3}{V_2}\right) = -Q_{rev}. \tag{18}$$

Since entropy is a state function, its variation depends only on the initial state 2 and the final state 3:

$$\Delta S_{rev} = S_3 - S_2 = \int_{2}^{3} dS = \int_{2}^{3} \frac{dQ_{rev}}{T} = \frac{Q_{rev}}{T} = nR \, ln\left(\frac{V_3}{V_2}\right) = nR \, ln\left(\frac{\frac{V_2}{2}}{V_2}\right) \tag{19}$$

$$= (1 \text{ mol}) \left(8.314 \text{ J.K}^{-1}.\text{mol}^{-1}\right) \, ln\left(\frac{1}{2}\right) = -5.76 \text{ J.K}^{-1}.$$

- *Variation of entropy of the thermostat*:

The received heat by the thermostat is:

$$\delta Q' = -\delta Q. \tag{20}$$

This heat is received at constant temperature $T_2 = 273$ K:

$$\Delta S_{th} = -\Delta S_{rev} = 5.76 \text{ J.K}^{-1}. \tag{21}$$

- *Entropy variation of the thermostat+gas set*:

The variation of the thermostat+gas set is the algebraic sum of the entropy variations of each constituent of this set:

$$\Delta S = \Delta S_{gas} + \Delta S_{th} = -5.76 \text{ J.K}^{-1} + 5.76 \text{ J.K}^{-1} = 0. \tag{22}$$

This result is predictable for a reversible transformation.

Exercise 42
Calculate the yield η of the Beau de Rochas cycle composed of two isochoric transformations and two adiabatic transformations with a compression ratio $\tau = 6$ and $\gamma \approx 1.4$. Describe this cycle. What do τ and γ represent? How does η vary when τ increases? What does the cycle area represent? We consider the air-fuel mixture as an ideal gas.

Solution
The Beau de Rochas cycle, named after the French engineer Alphonse Eugène Beau (1815–1893), is a four-stroke cycle. It is a theoretical thermodynamic cycle of the spark ignition internal combustion engines such as the gasoline engines of automobiles. This cycle is characterized by the linear movements of the piston: the admission or the first stroke, the compression or the second stroke, the combustion-relaxation or the third stroke, and finally the exhaust. The admission is modeled by an isobar $0 \rightarrow 1$ and the compression by a supposed adiabatic transformation $1 \rightarrow 2$. The explosion takes place at constant volume on the transformation $2 \rightarrow 3$ followed by an adiabatic relaxation in $3 \rightarrow 4$. The opening of the valve is modeled by the isochore $4 \rightarrow 5$ and the exhaust by the isobar $5 \rightarrow 0$.

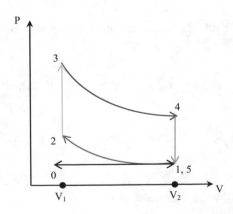

The theoretical yield calculated for an ideal cycle is maximal. It is the ratio between the work provided and the heat transfer. It is given by (see Eq. (4.80) of the course):

$$\eta = 1 - \frac{1}{\tau^{(\gamma-1)}} = 1 - \frac{1}{6^{(1.4-1)}} = 51\%. \tag{1}$$

In this expression, the gas is supposed ideal. The parameter τ represents the volumetric compression ratio. It is a key parameter of engines. It is given by:

$$\tau = \frac{V_2}{V_1} \tag{2}$$

where V_1 and V_2, respectively, represent the volumes of the isochoric transformations $2 \rightarrow 3$ and $4 \rightarrow 5$ and γ is the ratio of the thermal capacities at constant pressure and volume:

$$\gamma = \frac{C_P}{C_V} \approx 1.4. \tag{3}$$

Equation (1) shows that if the parameter τ increases, the efficiency η also increases, but a too strong compression favors explosions, and to remedy it, we must add an anti-knock. The area of the cycle corresponds to the work actually provided by the engine. The actual cycle departs significantly from the ideal cycle.

Exercise 43

A heat machine operates according to the Stirling cycle which comprises four reversible transformations, two isochores and two isotherms at temperatures T_1 and T_2 such that $T_1 < T_2$. The fluid that describes this cycle in the 12 341 direction is considered an ideal gas.
1. *What is the nature of each of the transformations $1 \rightarrow 2$, $2 \rightarrow 3$, $3 \rightarrow 4$, and $4 \rightarrow 1$?*
2. *For 1 mole of this fluid:*
   (a) *Express for each of the transformations described in 1) the work and the heat exchanged by the fluid with the external environment.*
   (b) *Calculate the values of the quantities expressed in a) for the transformations $1 \rightarrow 2$ and $2 \rightarrow 3$.*
   (c) *Express the total work exchanged per cycle between the fluid and the external environment. Is the machine a motor or a receiver? Justify your answer.*
   (d) *Calculate the entropies exchanged by the thermostats. Do they verify the Clausius inequality?*
   (e) *Diagrammatically represent the operating cycle of this machine.*
3. *What is the heat taken at the sink of low temperature and what is the useful heat? Quote a possible application of this machine.*
4. *What is the efficiency of this machine? We give:*
   • *$dU = C_V \, dT$ where C_V is the constant volume heat capacity.*
   • *Temperature of the cold reservoir $T_1 = 276$ K*
   • *Temperature of the hot reservoir $T_2 = 293$ K*
   • *Volumetric ratio $\tau = 3$*

- *Ideal gas constant R = 8.314 J/mol/K*
- *Constant volume capacity C_V = 21 J/mol/K*

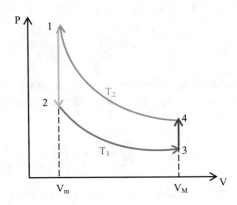

Solution
1. Nature of each transformation:

- $1 \to 2$: Isochoric relaxation
- $2 \to 3$: Isothermal expansion
- $3 \to 4$: Isochoric compression
- $4 \to 1$: Isothermal compression

2. (a) *Exchanged works and heat*:

Works

- *$1 \to 2$: Constant volume relaxation V_m:*

As the relaxation occurs at constant volume, the work is zero.

$$W_{1 \to 2} = 0. \tag{1}$$

- *$2 \to 3$: Expansion at constant temperature T_1:*

Since the gas is supposed ideal, it obeys the equation ($n = 1$ mol):

$$PV = nRT_1. \tag{2}$$

The work done during the isothermal expansion is:

$$W_{2\to3} = \int_{2}^{3} -PdV = -nRT_1 \int_{2}^{3} \frac{dV}{V} = -nRT_1 \ln\left(\frac{V_M}{V_m}\right). \tag{3}$$

- $3 \to 4$: *Constant volume compression V_M:*

  Since the compression occurs at constant volume, the work is zero.

$$W_{3\to4} = 0. \tag{4}$$

- $4 \to 1$: *Constant temperature compression T_2:*

  The work done during the isothermal expansion is:

$$W_{4\to1} = \int_{4}^{1} -PdV = -nRT_2 \int_{4}^{1} \frac{dV}{V} = -nRT_2 \ln\left(\frac{V_m}{V_M}\right). \tag{5}$$

Heats

- $1 \to 2$: *Constant volume relaxation V_m:*

  As the relaxation occurs at constant volume, the work is zero according to (1). On the basis of the hypotheses and the first principle of thermodynamics which gives us the variation of the internal energy of the fluid, it becomes:

$$\Delta U_{1\to2} = Q_{1\to2} + W_{1\to2} = Q_{1\to2} = nC_V\,\Delta T = nC_V\,(T_1 - T_2). \tag{6}$$

- $2 \to 3$: *Expansion at constant temperature T_1:*

  The internal energy of an ideal gas depends on temperature. For an isothermal process, its variation is zero:

$$\Delta U_{2\to3} = Q_{2\to3} + W_{2\to3} = 0 \quad \Rightarrow \quad Q_{2\to3} = -W_{2\to3} = nRT_1 \ln\left(\frac{V_M}{V_m}\right). \tag{7}$$

- $3 \to 4$: *Constant volume compression V_M:*

  As relaxation occurs at constant volume, the work is zero according to (4). On the basis of the hypotheses and the first principle of thermodynamics which gives us the variation of the internal energy of the fluid, we obtain:

$$\Delta U_{3\to4} = Q_{3\to4} + W_{3\to4} = Q_{3\to4} = nC_V\,\Delta T = nC_V\,(T_2 - T_1). \tag{8}$$

- $4 \to 1$: *Constant temperature compression T_2:*

The internal energy of an ideal gas depends on temperature. For an isothermal process, its variation is zero:

$$\Delta U_{4 \to 1} = Q_{4 \to 1} + W_{4 \to 1} = 0 \quad \Rightarrow \quad Q_{4 \to 1} = -W_{4 \to 1} = nRT_2 \; ln\left(\frac{V_m}{V_M}\right). \tag{9}$$

(b) *Calculation of values of the quantities expressed for the transformations 1 → 2 and 2 → 3:*

- *1 → 2: Constant volume relaxation V_m:*

$$W_{1 \to 2} = 0. \tag{10}$$

Therefore:

$$Q_{1 \to 2} = nC_V \; (T_1 - T_2) = (1 \; \text{mol}) \; (21 \; \text{J.mol}^{-1}.\text{K}^{-1}) \; (\; 276 - 293 \; \text{K}) \tag{11}$$
$$= -357 \; \text{J}.$$

These calculations were made for 1 mole which does not appear in (11), which explains the result in joules.

- *2 → 3: Expansion at constant temperature T_1:*

$$W_{2 \to 3} = -nRT_1 \; ln\left(\frac{V_M}{V_m}\right) = -(1 \; \text{mol}) \; (8.314 \; \text{J.K}^{-1}.\text{mol}^{-1}) \; (276 \; \text{K}) \; ln \, 3 \tag{12}$$
$$= -2520.9 \; \text{J}.$$

Therefore:

$$Q_{2 \to 3} = -W_{2 \to 3} = 2520.9 \; \text{J}. \tag{13}$$

(c) *Total work exchanged per cycle between the fluid and the external environment:*

The total work exchanged per cycle between the fluid and the external environment is the algebraic sum of the work done during each cycle:

$$W = W_{1\to2} + W_{2\to3} + W_{3\to4} + W_{4\to1} = 0 - nRT_1 \, \ln\left(\frac{V_M}{V_m}\right) + 0 - nRT_2 \, \ln\left(\frac{V_m}{V_M}\right)$$

$$= nR\ (T_2 - T_1)\ \ln\left(\frac{V_M}{V_m}\right)$$

$$= (1\ \text{mol})\left(8.314\ \text{J.K}^{-1}.\text{mol}^{-1}\right)(293\ \text{K} - 276\ \text{K})\ \ln 3$$

$$= 155.3\ \text{J}.$$

$$(14)$$

The total work is positive. The machine is a receiver. It receives work from the outside environment to produce heat.

(d) *Calculation of entropies exchanged by the thermostats*:

Since the four transformations are hypothetically reversible, the calculation of entropies exchanged by the thermostats can be done as follows:

$$\frac{Q_{1\to2}}{T_1} + \frac{Q_{2\to3}}{T_1} + \frac{Q_{3\to4}}{T_2} + \frac{Q_{4\to1}}{T_2}. \qquad (15)$$

In this equation, the expressions of heat $Q_{3\to4}$ and $Q_{4\to1}$ have been previously determined. Let us first determine their values. From (8), we obtain:

$$Q_{3\to4} = nC_V\ (T_2 - T_1) = (1\ \text{mol})\ \left(21\ \text{J.mol}^{-1}.\text{K}^{-1}\right)(293 - 276\ \text{K})$$
$$= 357\ \text{J}. \qquad (16)$$

From (9), we obtain:

$$Q_{4\to1} = nRT_2\ \ln\left(\frac{V_m}{V_M}\right) = (1\ \text{mol})\ \left(8.314\ \text{J.K}^{-1}.\text{mol}^{-1}\right)(293\ \text{K})\ \ln\left(\frac{1}{3}\right)$$
$$= -2676.2\ \text{J}. \qquad (17)$$

From (15), we deduce the entropies exchanged by the thermostats:

$$\frac{-357\ \text{J}}{276\ \text{K}} + \frac{+2520.9\ \text{J}}{276\ \text{K}} + \frac{+357\ \text{J}}{293\ \text{K}} + \frac{-2676.2\ \text{J}}{293\ \text{K}} = -0.075\ \text{J.K}^{-1} \neq 0 \qquad (18)$$

We obtain a non-zero result, whereas we expected after the Clausius statement (see the course) a zero result since the four transformations are hypothetically reversible. This calculation also reveals that the entropy of each cycle is an absolute value higher than the absolute value of the algebraic sum of the four entropies:

$$\left|\frac{-357 \text{ J}}{276 \text{ K}}\right| \gg \left|-0.075 \text{ J.K}^{-1}\right|$$

$$\left|\frac{+2520.9 \text{ J}}{276 \text{ K}}\right| \gg \left|-0.075 \text{ J.K}^{-1}\right|$$

$$\left|\frac{+357 \text{ J}}{293 \text{ K}}\right| \gg \left|-0.075 \text{ J.K}^{-1}\right| \tag{19}$$

$$\left|\frac{-2676.2 \text{ J}}{293 \text{ K}}\right| \gg \left|-0.075 \text{ J.K}^{-1}\right|.$$

Equation (18) shows that the algebraic sum of the entropies exchanged by the thermostats is very small (= 0.075 J/K) compared to the value of each term of this sum as shown in equation (19). Although this transformation is irreversible, we can consider as a first approximation that the assumptions (use of $dS = dQ/T$, see resolution of question (d) of this exercise) we made to make our calculations of entropies are acceptable. This exercise deals in fact with the case of almost reversible or quasi-reversible transformations.

(e) *Schematic representation of the operating cycle of this machine*:

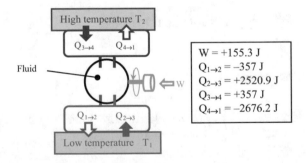

3. • *Heat taken at the sink of low temperature by 1 mole of the fluid*:

From the schematic representation of the machine, the heat taken from the low-temperature sink is $Q_{2\rightarrow3} = +2520.9$ J.

• *Useful heat per mole of fluid*:

The useful heat is that which is supplied by the fluid to the external environment, namely, $Q_{4\rightarrow1} = -2676.2$ J.

• *Possible application of this machine*:

This machine takes heat from a low-temperature sink ($T_1 = 276$ K ≈ 3 °C) by means of a contribution of work and provides heat to a high-temperature reservoir ($T_2 = 293$ K ≈ 20 °C). It thus functions as a heat pump to warm an apartment where the temperature T_1 corresponds to an outside temperature in the cold period and T_2 to the temperature of the inside of a habitable apartment.

4. *Yield of the machine*:

The efficiency of a thermal machine characterizes the efficiency with which the energy conversion is performed. It is defined by the ratio between the useful energy and therefore which interests us and the energy to be converted and thus to be paid.

$$\eta = \left|\frac{Q_H}{W}\right| = \left|\frac{Q_{4\to1} + Q_{3\to4}}{W}\right| = \left|\frac{-2676.2 \text{ J} + 357 \text{ J}}{+155.3 \text{ J}}\right| = 14.9 \qquad (20)$$

where Q_H is the heat actually supplied to the outside environment. It represents the heat supplied to the high-temperature reservoir minus the heat taken from it. The efficiency of such a machine is lower than a limit value which depends on the temperatures of reservoirs at low and high temperatures:

$$\eta_{\text{limit}} = \frac{T_2}{T_2 - T_1} = \frac{293 \text{ K}}{293 - 276 \text{ K}} = 17.2. \qquad (21)$$

The maximal efficiency is obtained for the limit case of a reversible machine. It would be obtained with an infinite number of sources or an oven that accompanies the evolution of the temperature of the system so as to ensure anytime the reversibility of the cycle.

Exercise 44

A thermal machine acts upon an ideal diatomic gas. This machine operates according to the Joule cycle consisting of two adiabatic transformations $1 \to 2$ and $3 \to 4$ and two isobaric transformations $2 \to 3$ and $4 \to 1$ during which the gas puts itself gradually in a thermal equilibrium with the hottest source at the temperature T_3 or with the cold sink at T_1. At state 1, the pressure is $P_1 = 10^5$ Pa and the temperature is $T_1 = 300$ K. At state 3, the pressure is $P_2 = 5 \times 10^5$ Pa and the temperature is $T_3 = 500$ K.

1. *The evolutions $1 \to 2$ and $3 \to 4$ being described in a reversible way, find a relationship between T_1, T_2, T_3, and T_4. Calculate T_2 and T_4.*
2. *Calculate for 1 mole of gas the quantity of heat $Q_{2\to3}$ exchanged as well as the variation of entropy during the evolution $2 \to 3$.*
3. *By applying the first principle of thermodynamics, calculate the work done by the motor during a cycle.*
4. *What is the efficiency η of this motor? We give:*
   - *$PV^\gamma = Cte$, the Laplace law*
   - *$PV = RT$, the ideal gas law for 1 mole*
   - *$\gamma = C_P/C_V$, the adiabatic index*
   - *$C_P - C_V = R$, the Mayer law with $C_V = 5/2\ R$*
   - *$R = 8.314$ J/K/mol, the ideal gas constant*
   - *$dH = C_P\ dT$*

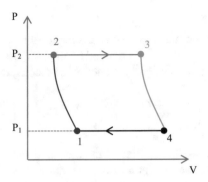

Solution

1. • *Relationship between T_1, T_2, T_3, and T_4:*

Starting from the Laplace law, we obtain:

$$P_1 V_1^\gamma = P_2 V_2^\gamma \quad \Rightarrow \quad \frac{P_1}{P_2} = \left(\frac{V_2}{V_1}\right)^\gamma. \tag{1}$$

Moreover, the ideal gas law applied to state 1 (P_1, T_1, V_1) and state 2 (P_2, T_2, V_2) gives:

$$\begin{aligned} P_2 V_2 &= nRT_2 \\ P_1 V_1 &= nRT_1 \end{aligned} \quad \Rightarrow \quad \frac{V_2}{V_1} = \frac{T_2}{T_1} \times \frac{P_1}{P_2}. \tag{2}$$

By injecting (2) into (1), we find:

$$\frac{P_1}{P_2} = \left(\frac{P_1}{P_2}\right)^\gamma \left(\frac{T_2}{T_1}\right)^\gamma \quad \Rightarrow \quad \left(\frac{P_1}{P_2}\right)^{1-\gamma} = \left(\frac{T_2}{T_1}\right)^\gamma \quad \Rightarrow \quad T_2 = T_1 \left(\frac{P_1}{P_2}\right)^{\frac{1-\gamma}{\gamma}}. \tag{3}$$

By proceeding in the same way for states 3 (P_2, T_3, V_3) and 4 (P_1, T_4, V_4), we arrive at the requested equation:

$$T_4 = T_3 \left(\frac{P_2}{P_1}\right)^{\frac{1-\gamma}{\gamma}}. \tag{4}$$

• *Calculation of T_2 and T_4:*

Let us first determine the constant pressure thermal capacity C_P based on the Mayer law:

$$C_P = C_V + R = \frac{5}{2}R + R = \frac{7}{2}R. \tag{5}$$

We deduce in a second time the adiabatic index γ:

$$\gamma = \frac{C_P}{C_V} = \frac{\frac{7}{2}R}{\frac{5}{2}R} = \frac{7}{5}. \tag{6}$$

In the end, the values of T_2 and T_4 are as follows:

$$T_2 = T_1 \left(\frac{P_1}{P_2}\right)^{\frac{1-\gamma}{\gamma}} = (300 \text{ K}) \left(\frac{10^5 \text{ Pa}}{5 \times 10^5 \text{ Pa}}\right)^{\frac{1-\frac{7}{5}}{\frac{7}{5}}} = 475.1 \text{ K.} \tag{7}$$

$$T_4 = T_3 \left(\frac{P_2}{P_1}\right)^{\frac{1-\gamma}{\gamma}} = (500 \text{ K}) \left(\frac{10^5 \text{ Pa}}{5 \times 10^5 \text{ Pa}}\right)^{\frac{1-\frac{7}{5}}{\frac{7}{5}}} = 315.7 \text{ K.} \tag{8}$$

2. • *Quantities of heat $Q_{2 \to 3}$ and $Q_{4 \to 1}$ exchanged for 1 mole of gas:*

Enthalpy, denoted H, comprises the internal energy of a system to which is added the work which this system must exert against the external pressure to occupy its own volume. Enthalpy is a chemical potential which predicts the evolution and the equilibrium of a thermodynamic system, and from which we can deduce all its properties. It is an extensive state function which plays a privileged role in the isobaric transformations which are very useful in chemistry. It is defined by:

$$H = U + PV. \tag{9}$$

Here the term PV corresponds to the energy of expansion or compression of the system. H is always greater than U. Enthalpy is in joules or in calories. For an infinitesimal transformation, we have, by starting from (4.24) and (4.27) of the course:

$$dU = \delta Q + \delta W = dQ - PdV. \tag{10}$$

By introducing the enthalpy function, we find:

$$dH = dU + d(PV) = dQ - PdV + PdV + VdP = dQ + VdP. \tag{11}$$

Since the $2 \to 3$ transformation is carried out at constant pressure, we have $dP = 0$ and taking into account the hypotheses ($dH = C_P \, dT$), we obtain:

$$Q_{2\to3} = nC_P(T_3 - T_2) = \frac{7}{2}nR\,(T_3 - T_2)$$
$$= \frac{7}{2}(1\ \text{mol})\ (8.314\ \text{J.K}^{-1}.\text{mol}^{-1})\ (500\ \text{K} - 475.1\ \text{K}) = 724.6\ \text{J}. \tag{12}$$

The value of C_P is given for 1 mole of gas which therefore does not appear in (12). This justifies the unity of the result in joules. By using the same reasoning, we arrive at the value of the quantity of heat $Q_{4\to1}$:

$$Q_{4\to1} = nC_P(T_1 - T_4) = \frac{7}{2}nR\,(T_1 - T_4)$$
$$= \frac{7}{2}\ (1\ \text{mol})\ (8.314\ \text{J.K}^{-1}.\text{mol}^{-1})\ (300\ \text{K} - 315.7\ \text{K}) = -456.9\ \text{J}. \tag{13}$$

- *Variation of the entropy during the isobar* $2 \to 3$:

  The infinitesimal entropy is given by:

$$dS = \frac{\delta Q}{T} = nC_P\frac{dT}{T}. \tag{14}$$

By integrating (13) between T_2 and T_3, we obtain the entropy variation $S_3 \to S_2$:

$$\Delta S = nC_P \int_2^3 \frac{dT}{T} = \frac{7}{2}nR\ \ln\!\left(\frac{T_3}{T_2}\right) = \frac{7}{2}(1\ \text{mol})\ (8.314\ \text{J.K}^{-1}.\text{mol}^{-1})\ \ln\!\left(\frac{500\ \text{K}}{475.1\ \text{K}}\right)$$
$$= 1.49\ \text{J.K}^{-1}. \tag{15}$$

3. *Calculation of the work done by the motor during a cycle*:

   By applying the first principle of thermodynamics, we obtain:

$$\Delta U_{\text{cycle}} = Q_{\text{cycle}} + W_{\text{cycle}} \quad \Rightarrow \quad W_{\text{cycle}} = -Q_{\text{cycle}}$$
$$= -(Q_{1\to2} + Q_{2\to3} + Q_{3\to4} + Q_{4\to1}). \tag{16}$$

The variation of the internal energy of a thermal machine during a cycle is always zero because whatever the mode of operation and the efficiency of this machine, it cannot either create or destroy energy according to the first principle of thermodynamics, and we shall always have:

$$Q_H + Q_B + W = 0 \tag{17}$$

where Q_H is the heat exchanged with the source at high temperature, Q_B the heat exchanged with the sink at low temperature, and W the work done during the cycle.

Moreover, $Q_{1\rightarrow2}$ and $Q_{3\rightarrow4}$ are null because the transformations $1 \rightarrow 2$ and $3 \rightarrow 4$ are adiabatic, and therefore, there is thus no exchange of heat with the outside environment. We finally get:

$$W_{\text{cycle}} = -(Q_{2\rightarrow3} + Q_{4\rightarrow1}) = -[724.6 \text{ J} - (-456.9 \text{ J})] = 1181.5 \text{ J}. \qquad (18)$$

This work is well negative because it is the useful work provided by the motor to the external environment.

4. *Efficiency η of the motor:*

The efficiency of a thermal machine characterizes the efficiency with which the energy conversion is performed. It is defined by the ratio between the useful energy and therefore which interests us and the energy to be converted and thus to be paid.

$$\eta = \left|\frac{W_{\text{cycle}}}{Q_{2\rightarrow3}}\right| = \left|\frac{-1181.5 \text{ J}}{724.6 \text{ J}}\right| = 1.6. \qquad (19)$$

$Q_{2\rightarrow3}$ is the heat taken from the hot source and thus to be paid.

Exercise 45
An internal combustion car engine operating according to the diesel cycle comprises four processes:
- *1 → 2: Reversible adiabatic compression of the air characterized by the volumetric ratio τ.*
- *2 → 3: Injection of the finely pulverized fuel into the hot compressed air causing its ignition. The combustion occurs at constant pressure.*
- *3 → 4: Reversible adiabatic relaxation of gases.*
- *4 → 1: Opening of the exhaust valve reducing the pressure to P_1, the gases undergo an isochoric cooling.*
   Since the quantity of fuel is small compared with the quantity of the inhaled air, we will consider that the total number of moles is not modified by the combustion and that the air-fuel mixture behaves like an ideal gas. The study will focus on 1 mole of this gas.
1. *This gas is admitted into the cylinders at pressure $P_1 = 1$ bar and at temperature $T_1 = 330$ K.*
   (a) *Calculate the volume V_1.*
   (b) *Calculate the pressure P_2 and the temperature T_2 at the end of the compression knowing that $\tau = 14$.*
2. *At the end of the combustion, the gas temperature is $T_3 = 2260$ K. Calculate the volume V_3 and the heat $Q_{2\rightarrow3}$ received by this gas during the transformation $2 \rightarrow 3$.*
3. *Calculate the pressure P_4 and the temperature T_4 at the end of the relaxation.*
4. (a) *Calculate the quantity of heat $Q_{4\rightarrow1}$ exchanged between the gas and the external environment during the isochoric transformation.*

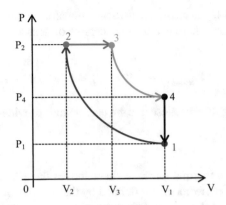

(b) *By applying the first principle of thermodynamics, calculate the work performed by the motor during a cycle.*

(c) *What is the efficiency η of this motor?*

We give:

- $PV^{\gamma} = Cte$
- $TV^{\gamma-1} = Cte$
- $\gamma = C_P/C_V$
- $R = 8.314\ J/K/mol$
- $C_P = 29\ J/K/mol$
- $\gamma = 1.4$

Solution

1. (a) *Calculation of the volume V_1:*

Since the gas is hypothesized to be ideal, let us apply the ideal gas law to 1 mole of this gas at state 1:

$$P_1V_1 = nRT_1 \quad \Rightarrow \quad V_1 = \frac{nRT_1}{P_1} = \frac{(1\ \text{mol})\ (8.314\ \text{J.K}^{-1}.\text{mol}^{-1})\ (330\ \text{K})}{10^5\ \text{Pa}} \tag{1}$$

$$= 27.4 \times 10^{-3}\ \text{m}^3 = 27.4\ \text{liters.}$$

(b) • *Calculation of the pressure P_2:*

Since the process $1 \rightarrow 2$ is adiabatic, let us apply the Laplace law to states 1 and 2:

$$P_1V_1^{\gamma} = P_2V_2^{\gamma} \quad \Rightarrow \quad P_2 = P_1\left(\frac{V_1}{V_2}\right)^{\gamma} = P_1\ \tau^{\gamma} = (10^5\ \text{Pa})\ (14^{1.4})$$

$$= 40 \times 10^5\ \text{Pa.} \tag{2}$$

- *Temperature T_2*:

  Let us apply the law of ideal gases to this gas in state 2:

  $$P_2V_2 = nRT_2 \quad \Rightarrow \quad T_2 = \frac{P_2V_2}{nR}. \tag{3}$$

  Knowing that the volumetric ratio τ has the following expression:

  $$\tau = \frac{V_1}{V_2} \quad \Rightarrow \quad V_2 = \frac{V_1}{\tau} \tag{4}$$

  and by injecting (4) into (3), we obtain the value of T_2:

  $$T_2 = \frac{P_2 \times \dfrac{V_1}{\tau}}{nR} = \frac{(40 \times 10^5 \ \text{Pa}) \times \dfrac{27.4 \times 10^{-3} \ \text{m}^3}{14}}{(1 \ \text{mol}) \left(8.314 \ \text{J.K}^{-1}.\text{mol}^{-1}\right)} = 941.6 \ \text{K}. \tag{5}$$

2. • *Calculation of the volume V_3*:

   Let us apply the law of ideal gases to this gas in state 3:

   $$P_3V_3 = nRT_3 \quad \Rightarrow \quad V_3 = \frac{nRT_3}{P_3}. \tag{6}$$

   Since the transformation $2 \to 3$ is isobaric, we have $P_3 = P_2$:

   $$V_3 = \frac{nRT_3}{P_2} = \frac{(1 \ \text{mol}) \left(8.314 \ \text{J.K}^{-1}.\text{mol}^{-1}\right) (2260 \ \text{K})}{40 \times 10^5 \ \text{Pa}} = 4.7 \times 10^{-3} \ \text{m}^3$$
   $$= 4.7 \ \text{liters}. \tag{7}$$

- *Heat $Q_{2\to3}$ received by this gas during the isobaric transformation $2 \to 3$*:

  $$Q_{2\to3} = nC_P (T_3 - T_2) = (1 \ \text{mol}) \left(29 \ \text{J.K}^{-1}.\text{mol}^{-1}\right) (2260 \ \text{K} - 941.6 \ \text{K})$$
  $$= 38.2 \ \text{kJ}. \tag{8}$$

3. • *Calculation of the pressure P_4*:

   Since the process $3 \to$ is adiabatic, let us apply the Laplace law to states 3 and 4:

   $$P_3V_3^{\gamma} = P_4V_4^{\gamma} \quad \Rightarrow \quad P_4 = P_3 \left(\frac{V_3}{V_4}\right)^{\gamma}. \tag{9}$$

The process $4 \to 1$ being isochoric $\Rightarrow V_4 = V_1$ and the transformation $2 \to 3$ isobaric $\Rightarrow P_3 = P_2$:

$$P_4 = P_2\left(\frac{V_3}{V_1}\right)^{\gamma} = \left(40 \times 10^5 \text{ Pa}\right)\left(\frac{4.7 \times 10^{-3} \text{ m}^3}{27.4 \times 10^{-3} \text{ m}^3}\right)^{1.4} = 3.4 \times 10^5 \text{ Pa}. \quad (10)$$

- *Temperature T_4*:

Let us apply the ideal gas law to this gas in state 4:

$$P_4 V_4 = nRT_4 \quad \Rightarrow \quad T_4 = \frac{P_4 V_4}{nR} = \frac{P_4 V_1}{nR} = \frac{\left(3.4 \times 10^5 \text{ Pa}\right)\left(27.4 \times 10^{-3} \text{ m}^3\right)}{(1 \text{ mol})\left(8.314 \text{ J.K}^{-1}.\text{mol}^{-1}\right)}$$

$$= 1120.5 \text{ K}.$$

$$(11)$$

4. (a) *Quantity of heat $Q_{4 \to 1}$ exchanged during the isochoric transformation*:

Taking into account the hypotheses of the exercise, the quantity of heat $Q_{4 \to 1}$ exchanged by the gas with the external environment during the isochoric transformation is equal for 1 mole of gas to:

$$Q_{4 \to 1} = nC_V(T_1 - T_4) = \frac{nC_P}{\gamma}(T_1 - T_4)$$

$$= \frac{(1 \text{ mol})\left(29 \text{ J.K}^{-1}.\text{mol}^{-1}\right)}{1.4}(330 - 1120.5 \text{ K}) \quad (12)$$

$$= -16.4 \text{ kJ}.$$

This quantity is negative, meaning that it is transferred by the gas to the outside environment during its cooling.

(b) *Work provided by the motor during a cycle*:

By applying the first principle of thermodynamics, we obtain:

$$\Delta U_{\text{cycle}} = Q_{\text{cycle}} + W_{\text{cycle}} \quad \Rightarrow \quad W_{\text{cycle}} = -Q_{\text{cycle}}$$

$$= -(Q_{1 \to 2} + Q_{2 \to 3} + Q_{3 \to 4} + Q_{4 \to 1}). \quad (13)$$

The variation of the internal energy of a heat machine during a cycle is always zero because whatever the mode of operation and the efficiency of this machine, it cannot either create or destroy energy according to the first principle of thermodynamics, and we will always have:

$$Q_H + Q_B + W = 0 \quad (14)$$

where Q_H is the heat exchanged with the source at high temperature, Q_B the heat exchanged with the sink at low temperature, and W the work done during the cycle. Moreover, $Q_{1\to2}$ and $Q_{3\to4}$ are null because the transformations $1 \to 2$ and $3 \to 4$ are adiabatic, so there is no heat exchange with the external environment. We finally get:

$$W_{cycle} = -(Q_{2\to3} + Q_{4\to1}) = -(38.2 - 16.4 \text{ kJ}) = -21.8 \text{ kJ}. \qquad (15)$$

This work is negative because it is the useful work which the motor provides to the car.

(c) *Efficiency η of this motor*:

The efficiency of a heat machine characterizes the efficiency with which the energy conversion is performed. It is defined by the ratio between the useful energy and therefore which interests us and the energy to be converted and thus to be paid.

$$\eta = \left| \frac{W_{cycle}}{Q_{2\to3}} \right| = \left| \frac{-21.8 \text{ kJ}}{38.2 \text{ kJ}} \right| = 57.1\%. \qquad (16)$$

$Q_{2\to3}$ is the heat taken from the hot source and thus to be paid.

4.9.4   Properties of Pure Bodies

Exercise 46
One liter of an alcoholic mixture contains some water and 447 ml of alcohol. Calculate the molar fractions of the constituents of this mixture. Verify that their sum is worth 1. We give:
- *Density of water: $\rho_{water} = 1 \text{ g/cm}^3$*
- *Density of alcohol: $\rho_{alcohol} = 0.794 \text{ g/cm}^3$*
- *Molar weight of water: $M_{water} = 18 \text{ g/mol}$*
- *Molar mass of alcohol: $M_{alcohol} = 46 \text{ g/mol}$*

Solution
This mixture contains by hypothesis 447 ml of alcohol on a total volume of 1 liter. The volume of water is:

$$V_{water} = 1000 \text{ ml} - 447 \text{ ml} = 553 \text{ ml} = 553 \text{ cm}^3. \qquad (1)$$

This water weighs:

$$\rho_{water} = \frac{m_{water}}{V_{water}} \quad \Rightarrow \quad m_{eau} = \rho_{eau} V_{eau} = (1 \text{ g.cm}^{-3}) (553 \text{ cm}^3) = 553 \text{ g}. \qquad (2)$$

The mass of the alcohol is:

$$\rho_{alcohol} = \frac{m_{alcohol}}{V_{alcohol}} \quad \Rightarrow \quad m_{alcohol} = \rho_{alcohol} V_{alcohol} = (0.794 \text{ g.cm}^{-3})(447 \text{ cm}^3)$$
$$= 355 \text{ g}.$$

(3)

The liquid molar fraction of a pure body represents the proportion of the quantity of the liquid of this body. It is given by:

$$x_L = \frac{\text{quantity of liquid}}{\text{total quantity}} = \frac{n_L}{n}.$$

(4)

• *Molar fraction of water*:

$$x_{water} = \frac{n_{water}}{n_{water} + n_{alcohol}}.$$

(5)

The number of moles of water n_{water} is:

$$n_{water} = \frac{m_{water}}{M_{water}} = \frac{553 \text{ g}}{18 \text{ g.mol}^{-1}} = 30.72 \text{ mol}.$$

(6)

The number of moles of alcohol $n_{alcohol}$ is:

$$n_{alcohol} = \frac{m_{alcohol}}{M_{alcohol}} = \frac{355 \text{ g}}{46 \text{ g.mol}^{-1}} = 7.72 \text{ mol}.$$

(7)

By injecting (6) and (7) into (5), we deduce the molar fraction of water:

$$x_{water} = \frac{n_{water}}{n_{water} + n_{alcohol}} = \frac{30.72 \text{ mol}}{30.72 \text{ mol} + 7.72 \text{ mol}} = 0.80.$$

(8)

• *Molar fraction of alcohol*:

$$x_{alcohol} = \frac{n_{alcohol}}{n_{water} + n_{alcohol}} = \frac{7.72 \text{ mol}}{30.72 \text{ mol} + 7.72 \text{ mol}} = 0.20.$$

(9)

The sum of the liquid fractions of the constituents is:

$$x_{water} + x_{alcohol} = 0.80 + 0.20 = 1.$$

(10)

Exercise 47

One mole of nitrogen assimilated to an ideal gas is raised from 20 to 100 °C. Calculate the quantity of heat received by this system in the following two cases:

1. *The transformation is isochoric.*
2. *The transformation is isobaric. We give:*
   - $C_P = 33$ *J/mol/K, the thermal capacity at constant pressure*
   - $R = 8.314$ *J/mol/K, the ideal gas constant*
   - $C_P - C_V = R$, *the Mayer relationship for 1 mole*
   - $dU = C_V dT$, *the variation of the internal energy at constant volume for 1 mole*
   - $dH = C_P dT$, *variation of enthalpy at constant pressure of 1 mole*

Solution

1. *Isochoric transformation*:

According to the first principle of thermodynamics:

$$dU = \delta Q + \delta W = \delta Q. \tag{1}$$

In Eq. (1), $\delta W = -PdV = 0$ because the transformation is carried out at constant volume. The quantity of heat received at constant volume is therefore equal to the variation of the internal energy of the supposed ideal gas:

$$
\begin{aligned}
Q_V = \Delta U = nC_V \int_{T_1}^{T_2} dT &= n(C_P - R)(T_2 - T_1) \\
&= (1\ \text{mol}) \left[(33\ \text{J.mol}^{-1}.\text{K}^{-1}) - (8.314\ \text{J.mol}^{-1}.\text{K}^{-1}) \right] (100 - 20\,^\circ\text{C}) \\
&= 1974.9\ \text{J}.
\end{aligned}
\tag{2}
$$

In Eq. (2), as it is about a temperature difference, T_2 and T_1 can be expressed indifferently in °C or in K. Since the number of moles is equal to 1, it does not appear in the equation which is well homogeneous to an energy in joules. As we could expect, this quantity of heat is positive because it is received by the system.

2. *Isobaric transformation*:

Enthalpy, denoted H, comprises the internal energy of a system to which is added the work which this system must exert against the external pressure to occupy its own volume. Enthalpy is a chemical potential that predicts the evolution and the equilibrium of a thermodynamic system, and from which we can deduce all its properties. It is an extensive state function that plays a privileged role in the isobaric transformations which are very useful in chemistry. It is defined by:

$$H = U + PV. \tag{3}$$

Here the term PV corresponds to the energy of expansion or compression of the system. H is always greater than U. Enthalpy is in joules or in calories. For an infinitesimal transformation, we have, by starting from (4.24) and (4.27) of the course:

$$dU = \delta Q + \delta W = dQ - PdV. \tag{4}$$

By introducing the enthalpy function, we obtain:

$$dH = dU + d(PV) = dQ - PdV + PdV + VdP = dQ + VdP. \tag{5}$$

Since the transformation is carried out at constant pressure, we have $dP = 0$ and taking into account the hypotheses $(dH = C_P dT)$, we obtain the quantity of heat received at constant pressure which is equal to the variation of the internal enthalpy of the supposed ideal gas:

$$Q_P = \Delta H = nC_P \int_{T_1}^{T_2} dT = nC_P (T_2 - T_1) = (1 \, \text{mol}) \left(33 \, \text{J.mol}^{-1}.\text{K}^{-1}\right) (100 - 20 \, ^{\circ}\text{C})$$

$$= 2640 \, \text{J}. \tag{6}$$

In Eq. (6), as it is about a temperature difference, T_2 and T_1 can be expressed indifferently in °C or in K. Since the number of moles is equal to 1, it does not appear in the equation which is well homogeneous to an energy in joules. As we could expect, this quantity of heat is positive because it is received by the system.

Exercise 48
Let us consider a balloon of 10 m^3 of water under 10 bars and of mass 169 kg.
1. *What is a vapor of saturation? Explain with a diagram (P,T).*
2. *Give the Clapeyron diagram which represents the state in which a body exists as a function of the pressure and the mass volume.*
3. *What do we call the liquid molar fraction and the vapor molar fraction?*
4. *What is an overheating?*
5. *What is the state of water in the balloon? Specify its pressure, its temperature, its fraction if it is a vapor of saturation, its overheating if it is a superheated vapor, and its total enthalpy. We give for a saturated water vapor under a pressure $P = 10$ bar, $T = 179.86 \,^{\circ}C$, $v_L = 1.1273 \times 10^{-3} \, m^3/kg$ (mass volume in the liquid state), $v_V = 0.1947 \, m^3/kg$ (mass volume of the vapor), $h_L = 182 \, kcal/kg$ (the mass enthalpy in the liquid state), and $h_V = 663.2 \, kcal/kg$ (the mass enthalpy in the vapor state).*

Solution
1. *Definition of a saturation vapor:*

At a given temperature T, the equilibrium of a liquid with its vapor is only possible under a pressure P called *saturation vapor pressure* (P_{sat}). The curve which represents the saturation vapor pressure as a function of temperature distinguishes the following cases:

- If $P < P_{sat}$: the fluid is in the gaseous state, called *dry vapor*.
- If $P > P_{sat}$: the fluid is in the liquid state.
- If $P = P_{sat}$: the fluid is in liquid-vapor equilibrium. In this case, the vapor is called *just saturation vapor* or *dry saturation vapor*. The set V–L is called *saturation vapor* or *wet vapor*. For the liquid, we are talking about *just saturation liquid*.

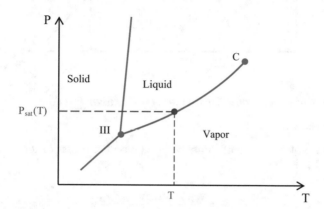

2. *Clapeyron's diagram*:

The Clapeyron diagram represents the state in which a body exists as a function of the pressure and the mass volume rather than the volume because the mass volume is an intensive quantity. It is composed of two curves: the boiling curve that separates the zone where the pure body is liquid from the zone where there is a vapor-liquid coexistence and the curve of dew that separates the zone where the pure body is a vapor of the zone where there is a liquid-vapor coexistence. These curves meet at the critical point C. By noting M the representative point of a pure body and E and R the points on the boiling and dew curves at the same pressure, we have

$$x_V = \frac{EM}{ER} \quad \text{and} \quad x_L = \frac{MR}{ER}. \tag{1}$$

The advantage of representing the liquid-vapor transition in the Clapeyron coordinates is that the reading of a point M on the change of the state plateau allows determining the composition of the pure two-phase body, its liquid and vaporing mass fraction.

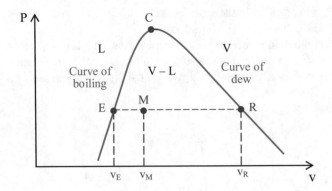

3. *Definitions of vapor molar fraction and liquid molar fraction*:

- *Vapor molar fraction*:

The vapor molar fraction of a pure body represents the proportion of the quantity of vapor of this body:

$$x_V = \frac{\text{vapor quantity}}{\text{total quantity}} = \frac{n_V}{n}. \tag{2}$$

- *Liquid molar fraction*:

The liquid molar fraction of a pure body represents the proportion of the quantity of liquid of this body:

$$x_L = \frac{\text{liquid quantity}}{\text{total quantity}} = \frac{n_L}{n}. \tag{3}$$

4. *Definition of an overheating*:

By an isobaric heating of an extremely pure body in the liquid phase, we reach a point on the vaporization curve and normally at this point the beginning of the vaporization of the liquid should appear. However, no vaporization takes place; the body remains in the liquid state beyond this point. The body is in a metastable state. It is said to be *overheated* because the formation of the first bubble is energetically disadvantageous. In this case, the infinitesimal modification of the pressure or the temperature by the intrusion of an external element destroys this equilibrium and the transition starts.

5. *State of water in the balloon*:

To find out if we have a vapor of saturation, let us determine the mass volume of water balloon:

$$v = \frac{10 \text{ m}^3}{169 \text{ kg}} = 59.2 \times 10^{-3} \text{ m}^3.\text{kg}^{-1}. \tag{4}$$

The volume v is intermediate between the mass volume in the liquid state $v_L = 1.1273 \times 10^{-3}$ m^3/kg and the mass volume of the vapor $v_V = 194.7 \times 10^{-3}$ m^3/kg for the same pressure $P = 10$ bar. We are necessarily on the plateau of the diagram (P,V). The fraction of the vapor is given by:

$$x_V = \frac{v - v_L}{v_V - v_L} = \frac{59.2 \times 10^{-3} \text{ m}^3.\text{kg}^{-1} - 1.1273 \times 10^{-3} \text{ m}^3.\text{kg}^{-1}}{194.7 \times 10^{-3} \text{ m}^3.\text{kg}^{-1} - 1.1273 \times 10^{-3} \text{ m}^3.\text{kg}^{-1}} = 0.3. \tag{5}$$

Since the sum of the liquid and the vapor contents is worth 1:

$$x_V + x_L = 1. \tag{6}$$

We deduce that there is 30% of vapor and 70% of liquid. The total enthalpy is:

$$h(x_V) = x_V h_V + (1 - x_V) h_L = (0.3) (663.2 \text{ kcal.kg}^{-1}) + (1 - 0.3) (182 \text{ kcal.kg}^{-1})$$
$$= 326.4 \text{ kcal.kg}^{-1}. \tag{7}$$

Exercise 49
Calculate the variation of enthalpy when 1 mole of iodine goes from 300 to 500 K under the pressure of 1 atmosphere. We give the molar thermal capacities of pure bodies and the latent heats:
- $C_P (I_2, \text{solid}) = 5.4 \text{ cal/mol/K}$
- $C_P (I_2, \text{liquid}) = 19.5 \text{ cal/mol/K}$
- $C_P (I_2, \text{gas}) = 9.0 \text{ cal/mol/K}$
- $L_{L \to G}$ *(vaporization, 475 K)* $= 6.10 \text{ kcal/mol}$
- $L_{S \to L}$ *(melting, 387 K)* $= 3.74 \text{ kcal/mol}$

Solution
The process of enthalpy variation when 1 mole of iodine goes from 300 to 500 K under 1 atmosphere can be schematized as follows:

- *Solid → Solid*: Iodine does not change the state. It remains in the solid state. Its temperature goes from $T_1 = 300$ K to $T_2 = 387$ K under a pressure of 1 atmosphere. The enthalpy variation of 1 mole of iodine during this isobaric heating is:

$$\Delta H_{S \to S} = \int_{T_1}^{T_2} n C_P(I_2, \text{solide}) \, dT = n C_P(I_2, \text{solide}) (T_2 - T_1)$$
$$= (1 \text{ mol}) (5.4 \text{ cal.mol}^{-1}.\text{K}^{-1}) (387 - 300 \text{ K}) = 469.8 \text{ cal} \tag{1}$$
$$= 0.4698 \text{ kcal}.$$

- *Solid $\rightarrow$ Liquid*: The iodine changes the state. It passes from the solid state to the liquid state at a fixed temperature $T_2 = 387$ K under a pressure of 1 atmosphere. The enthalpy variation of 1 mole of iodine $\Delta H_{S \rightarrow L}$ due to the change of state is, by hypothesis:

$$\Delta H_{S \rightarrow L} = L_{S \rightarrow L} = 3.74 \text{ kcal}. \tag{2}$$

- *Liquid $\rightarrow$ Liquid*: Iodine does not change the state. It remains in the liquid state. Its temperature goes from $T_1 = 387$ K to $T_2 = 457$ K under a pressure of 1 atmosphere. The enthalpy variation of 1 mole of iodine during this isobaric heating is:

$$
\begin{aligned}
\Delta H_{L \rightarrow L} &= \int_{T_2}^{T_3} nC_P(I_2, \text{liquid}) \, dT = nC_P(I_2, \text{liquid}) \, (T_3 - T_2) \\
&= (1 \text{ mol}) \left(19.5 \text{ cal.mol}^{-1}.\text{K}^{-1}\right) (457 - 387 \text{ K}) \\
&= 1.365 \text{ kcal}.
\end{aligned} \tag{3}
$$

- *Liquid $\rightarrow$ Gas*: Iodine changes state. It goes from the liquid state to the gaseous state at a fixed temperature $T_2 = 457$ K under a pressure of 1 atmosphere. The enthalpy variation of 1 mole of iodine $\Delta H_{S \rightarrow L}$ due to the change of state is, by hypothesis:

$$\Delta H_{L \rightarrow G} = L_{L \rightarrow G} = 6.10 \text{ kcal}. \tag{4}$$

- *Gas $\rightarrow$ Gas*: Iodine does not change state. It remains in the gaseous state. Its temperature goes from $T_3 = 457$ K to $T_4 = 500$ K under a pressure of 1 atmosphere. The enthalpy variation of 1 mole of iodine during this isobaric heating is:

$$
\begin{aligned}
\Delta H_{G \rightarrow G} &= \int_{T_3}^{T_4} nC_P(I_2, \text{gas}) \, dT = nC_P(I_2, \text{gas}) \, (T_4 - T_3) \\
&= (1 \text{ mol}) \left(9.0 \text{ cal.mol}^{-1}.\text{K}^{-1}\right) (500 - 457 \text{ K}) \\
&= 0.387 \text{ kcal}.
\end{aligned} \tag{5}
$$

The variation of enthalpy when 1 mole of iodine passes from 300 to 500 K under the pressure of 1 atmosphere is therefore:

$$
\begin{aligned}
\Delta H &= \Delta H_{S \rightarrow S} + \Delta H_{S \rightarrow L} + \Delta H_{L \rightarrow L} + \Delta H_{L \rightarrow G} + \Delta H_{G \rightarrow G} \\
&= 0.4698 \text{ kcal} + 3.74 \text{ kcal} + 1.365 \text{ kcal} + 6.10 \text{ kcal} + 0.387 \text{ kcal} \\
&= 12.062 \text{ kcal}.
\end{aligned} \tag{6}
$$

Exercise 50

Calculate the variation of enthalpy and the internal energy of 10 g of ice, of which the temperature varies from –20 to 100 °C under a pressure of 1 atmosphere. The water vapor is assimilated to an ideal gas. We give:

- C_P *(H_2O, solid) = 0.5 cal/g/K: the specific heat capacity of solid water*
- C_P *(H_2O, liquid) = 1 cal/g/K: the specific heat capacity of liquid water*
- *V (H_2O, solid) = 19.6 cm^3/mol: the molar volume of solid water*
- *V (H_2O, liquid) = 18 cm^3/mol: the molar volume of liquid water*
- $L_{S \to L}$ *(melting, 273 K) = 80 cal/g: the mass latent heat of melting of solid water*
- $L_{L \to G}$ *(vaporization, 273 K) = 539 cal/g: the mass latent heat of vaporization of liquid water*
- *R = 8.31 J/mol/K: the ideal gas constant*

Solution

The process of the variation of enthalpy and the internal energy when a mass of 10 g of ice passes from 353 to 373 K under 1 atmosphere can be schematized as follows:

- *Solid* → *Solid*: Ice does not change state. It remains in the solid state. Its temperature goes from $T_1 = 253$ K to $T_2 = 273$ K under a pressure of 1 atmosphere. The variation of enthalpy of this mass of ice during this isobaric heating is:

$$\Delta H_{S \to S} = \int_{T_1}^{T_2} mC_P(H_2O, \text{solid}) \, dT = mC_P(H_2O, \text{solid}) \, (T_2 - T_1)$$
$$= (10 \text{ g}) \left(0.5 \text{ cal.g}^{-1}.K^{-1}\right) (273 - 253 \text{ K}) = 100 \text{ cal.} \tag{1}$$

The variation of the internal energy of this mass of ice is given by:

$$\Delta H_{S \to S} = \Delta U_{S \to S} + \Delta(PV) = \Delta U_{S \to S} + P\Delta V + V\Delta P = \Delta U_{S \to S} = 100 \text{ cal.} \tag{2}$$

In Eq. (2), $P\Delta V = 0$ because the volume of the ice mass does not change during the transformation S → S (isochore transformation). In addition, $V\Delta P = 0$, because this transformation occurs at constant pressure of 1 atmosphere (isobaric transformation).

- *Solid* → *Liquid*: Water changes state. It passes from the solid state to the liquid state at a fixed temperature $T_2 = 273$ K under a pressure of 1 atmosphere. The enthalpy variation of 10 g of ice $\Delta H_{S \to L}$ due to the change of state is:

$$\Delta H_{S \to L} = mL_{S \to L} = (10 \text{ g}) \left(80 \text{ cal.g}^{-1}\right) = 800 \text{ cal.} \tag{3}$$

The variation of the internal energy of this mass of ice during its melting is given by:

$$\Delta H_{S\rightarrow L} = \Delta U_{S\rightarrow L} + \Delta(PV) = \Delta U_{S\rightarrow L} + P\Delta V + V\Delta P = \Delta U_{S\rightarrow L} + P\Delta V \quad \Rightarrow \quad \Delta U_{S\rightarrow L}$$
$$= \Delta H_{S\rightarrow L} - P\Delta V = \Delta H_{S\rightarrow L} - P(V_L - V_S). \tag{4}$$

In this equation $V\Delta P = 0$ because this transformation occurs at a constant pressure of 1 atmosphere (isobaric transformation). Moreover, since the volumes V_L and V_S are given in the statement in cm^3/mol, these quantities must be multiplied by the number of moles contained in the mass of 10 g of water. Given that 1 mole of H_2O weighs 18 g, the number of moles contained in $m = 10$ g is:

$$n = \frac{(1\ \text{mol})\,(10\ \text{g})}{18\ \text{g}} = 0.55\ \text{mol}. \tag{5}$$

By expressing, in the term $P\Delta V$, the pressure in Pa and ΔV in m^3, we obtain joules which must imperatively be converted into calories since $\Delta U_{S\rightarrow L}$ and $\Delta H_{S\rightarrow L}$ are in calories. Knowing that 1 cal amounts to 4.18 J, we get:

$$P\Delta V\ (\text{cal}) = \frac{P\Delta V\ (\text{J})\,(1\ \text{cal})}{4.18\ \text{J}}. \tag{6}$$

Taking into account (5) and (6), the variation of the internal energy of this mass of ice is:

$$\Delta U_{S\rightarrow L} = \Delta H_{S\rightarrow L} - P(V_L - V_S)$$
$$= 800\ \text{cal} - (1.013 \times 10^5\ \text{Pa})\,\left(18 \times 10^{-6}\ \text{m}^3.\text{mol}^{-1} - 19.6 \times 10^{-6}\ \text{m}^3.\text{mol}^{-1}\right)$$
$$\times \frac{(0.55\,\text{mol})\,(1\,\text{cal})}{4.18\,\text{J}}$$
$$= 800\ \text{cal} - 2.13 \times 10^{-2}\ \text{cal} \approx 800\ \text{cal} = \Delta H_{S\rightarrow L}. \tag{7}$$

- *Liquid* $\rightarrow$ *Liquid*: Water does not change state. It remains in the liquid state. Its temperature goes from $T_2 = 273$ K to $T_3 = 373$ K under a pressure of 1 atmosphere. The variation of enthalpy of this body of liquid water during this isobaric heating is:

$$\Delta H_{L\rightarrow L} = \int_{T_2}^{T_3} mC_P(H_2O, \text{liquid})\,dT = mC_P(H_2O, \text{liquid})\,(T_3 - T_2) \tag{8}$$
$$= (10\ \text{g})\,\left(1\ \text{cal.g}^{-1}.\text{K}^{-1}\right)(373 - 273\ \text{K}) = 1000\ \text{cal}.$$

The variation of the internal energy of this mass of ice is given by:

$$\Delta H_{L \to L} = \Delta U_{L \to L} + \Delta(PV) = \Delta U_{L \to L} = 1000 \text{ cal.} \tag{9}$$

In the Eq. (9), $\Delta(PV) \approx 0$ because for the condensed phases, this term is negligible compared to $\Delta U_{L \to L}$.

- *Liquid* → *Gas*: Water changes state. It passes from the liquid state to the gaseous state at a fixed temperature $T_3 = 373$ K under a pressure of 1 atmosphere. The enthalpy variation of 10 g of liquid water $\Delta H_{L \to G}$ due to the change of state is:

$$\Delta H_{L \to G} = m L_{L \to G} = (10 \text{ g}) \left(539 \text{ cal.g}^{-1}\right) = 5339 \text{ cal.} \tag{10}$$

The variation of the internal energy of this body of liquid water during its vaporization is given by:

$$\begin{aligned} \Delta H_{L \to G} &= \Delta U_{L \to G} + \Delta(PV) = \Delta U_{L \to G} + P\Delta V + V\Delta P = \Delta U_{L \to G} + P\Delta V \Rightarrow \Delta U_{L \to G} \\ &= \Delta H_{L \to G} - P\Delta V = \Delta H_{L \to G} - P\left(V_G - V_L\right). \end{aligned} \tag{11}$$

As by hypothesis, the water vapor is assimilated to an ideal gas, we obtain:

$$PV_G = nRT_3 \quad \Rightarrow \quad V_G = \frac{nRT_3}{P}. \tag{12}$$

In Eq. (12), the volume V_G is expressed in m³. As for the volume V_L, which is assumed to be expressed in m³/mol, it should be multiplied by the number of moles contained in this liquid, i.e., 0.55 moles. By expressing, in the term $P\Delta V$, the pressure in Pa and ΔV in m³, we obtain joules which must imperatively be converted into calories since $\Delta U_{L \to G}$ and $\Delta H_{L \to G}$ are in calories. Knowing that 1 cal amounts to 4.18 J, we get:

$$P\Delta V \text{ (cal)} = \frac{P\Delta V \text{ (J) (1 cal)}}{4.18 \text{ J}}. \tag{13}$$

Taking into account all these considerations, the variation of the internal energy during the transformation L → G is:

$$\begin{aligned} \Delta U_{L \to G} &= \Delta H_{L \to G} - P\left(V_G - V_L\right) = \Delta H_{L \to G} - P\left(\frac{nRT_3}{P} - nV_L\right) \times \frac{1}{4.18 \text{ J}} \\ &= 5339 \text{ cal} - \left(1.013 \times 10^5 \text{ Pa}\right) \left[\frac{(0.55 \text{ mol}) \left(8.31 \text{ J.K}^{-1}.\text{mol}^{-1}\right) (373 \text{ K})}{1.013 \times 10^5 \text{ Pa}} \right. \\ &\quad \left. - (0.55 \text{ mol}) \left(18 \times 10^{-6} \text{ m}^3.\text{mol}^{-1}\right) \right] \times \frac{1 \text{ cal}}{4.18 \text{ J}} = 4931 \text{ cal.} \end{aligned} \tag{14}$$

The variation of enthalpy and the internal energy when a mass of 10 g of ice passes from 353 to 373 K under 1 atmosphere are:

– *Enthalpy*:

$$\Delta H = \Delta H_{S \to S} + \Delta H_{S \to L} + \Delta H_{L \to L} + \Delta H_{L \to G}$$
$$= 100 \ \text{cal} + 800 \ \text{cal} + 1000 \ \text{cal} + 5339 \ \text{cal} = 7239 \ \text{cal}. \tag{15}$$

– *Internal energy*:

$$\Delta U = \Delta U_{S \to S} + \Delta U_{S \to L} + \Delta U_{L \to L} + \Delta U_{L \to G}$$
$$= 100 \ \text{cal} + 800 \ \text{cal} + 1000 \ \text{cal} + 4931 \ \text{cal} = 6831 \ \text{cal}. \tag{16}$$

Exercise 51
On the basis of the phase diagram of pure helium given in the statement, answer the following questions:
1. *What is the physical state of helium in a balloon inflated to 1.2 bars at room temperature?*
2. *Why might one fill a balloon with helium as opposed to air?*
3. *What happens if we gradually compress the previous balloon at constant temperature? Can we observe helium liquefaction or solidification?*
4. *Can the solid helium and the gaseous helium be in equilibrium?*
5. *What is the particularity of the line λ of the diagram?*
6. *Helium is cooled gradually under a constant pressure of 1 bar. Describe the observed successive phenomena.*
7. *Is helium dangerous to health?*
8. *What are the uses of liquid helium?*

Solution
1. *Physical state of helium*:

Under a pressure of 1.2 bar and at room temperature, helium is in the gaseous state (see diagram).

2. *Interest of having this balloon*:

It floats in the air because helium is seven times lighter than air.

3. *Progressive compression of the previous balloon at constant temperature*:

By compressing it at room temperature, the gaseous helium passes into its supercritical phase. It cannot be liquefied or solidified at room temperature. To make it liquid, it will be necessary to go down at very low pressure and at a temperature very close to the absolute zero; it becomes solid under a very higher pressure, greater than 25 bars!

4. *Equilibrium of the solid helium and the gaseous helium*:

The solid-gas equilibrium is totally impossible with helium. Note that helium has two liquid phases: He(I) and He(II).

5. *Line λ of the diagram*:

This line separates two liquid phases: He(I) and He(II).

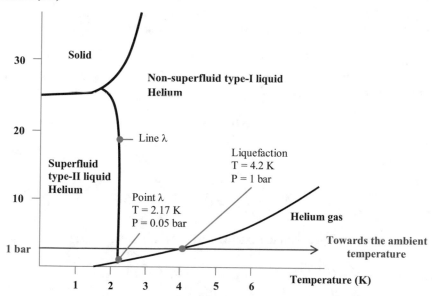

6. *Progressive cooling under a constant pressure of 1 bar*:

Under a pressure of 1 bar and from the ambient temperature, if we cool the gaseous helium, it will go to the liquid state at 4.2 K. At 2 K, we will arrive on line I and the liquid helium will pass through its superfluid phase with particular properties.

7. *Effects of helium on human health*:

Helium is a colorless, odorless, and nontoxic gas.

8. *Areas of the use of the liquid helium*:

Helium was the first gas used to fill balloons and airships and is still used for weather balloons. Its main application is providing an inert atmosphere. A large field of application is cooling at very low temperatures, below 15 K, especially in the superconductivity domain, because it remains in the gaseous or the liquid state. It is also used as a pressurizing gas for rocket fuels, in diving bottles mixed with oxygen, in the cooling of nuclear reactors, or in chromatography. The scientific and medical fields are also very dependent on this precious liquid, of which the cryogenic properties are highly exploited. Indeed, 15% of liquid helium produced in the world is used to cool MRI coils used daily in hospitals, and some 2% are used in basic research to study matter.

Exercise 52
Calculate the final temperature of liquid water when adiabatically mixing 1 mole of ice at − 15 °C with 4 moles of water at 25 °C under 1 atmosphere. We give:
- $L_{S \to L} = 6.056\ kJ/mol$: the melting enthalpy of ice
- $C_P\ (solid) = 37.62\ J/mol/K$: the molar heat capacity of ice
- $C_P\ (liquid) = 75.24\ J/mol/K$: the molar heat capacity of liquid water

Solution
The transformation is carried out at constant pressure; according to the first principle of thermodynamics, we have:

$$\Delta U = W + Q. \tag{1}$$

The work is worth:

$$W = -\int_i^f P_{ext}\, dV = -P_{ext}\, \Delta V. \tag{2}$$

The heat exchanged with the external environment is therefore:

$$\begin{aligned} Q = \Delta U - W = \Delta U + P_{ext}\, \Delta V = (U_f - U_i) + P_{ext}\, (V_f - V_i) \\ = (U_f + P_{ext}\, V_f) - (U_i + P_{ext}\, V_i). \end{aligned} \tag{3}$$

By introducing the enthalpy function, we obtain:

$$Q = H_f - H_i = \Delta H. \tag{4}$$

Since the total transformation is adiabatic, there is no heat exchange with the external environment:

$$Q = \Delta H = 0. \tag{5}$$

The transformation process of solid water in liquid water follows the following stages:

- *Solid → Solid*: The mole of ice does not change the state. It remains in the solid state. Its temperature goes from $T_1 = 258$ K to $T_2 = 273$ K under a pressure of 1 atmosphere. The variation of enthalpy of this mass of ice during this isobaric heating is:

$$Q_{S \to S}^1 = \Delta H_{S \to S}^1 = \int_{T_1}^{T_2} n_1 C_P(H_2O, \text{solid})\, dT = n_1 C_P(H_2O, \text{solid})\, (T_2 - T_1) \tag{6}$$
$$= (1 \text{ mol}) \left(37.62 \text{ J.mol}^{-1}.\text{K}^{-1}\right) (273 - 258 \text{ K}) = 564.3 \text{ J}.$$

- *Solid → Liquid*: Ice changes state. It passes from the solid state to the liquid state at a fixed temperature $T_2 = 273$ K under a pressure of 1 atmosphere. The enthalpy variation of 1 mole of ice $\Delta H_{S \to L}$, due to the change of state, is:

$$\Delta H_{S \to L}^1 = n_1 L_{S \to L} = (1 \text{ mol}) \left(6.056 \times 10^3 \text{ J.mol}^{-1}\right) = 6.056 \times 10^3 \text{ J}. \tag{7}$$

- *Liquid → Liquid*: The mole of liquid does not change state. It remains in the liquid state. Its temperature changes from $T_2 = 273$ K to the equilibrium temperature T_E of the mixture 1–4 moles under a pressure of 1 atmosphere. The variation of enthalpy of this mole during this isobaric heating is:

$$Q_{L \to L}^1 = \Delta H_{L \to L}^1 = \int_{T_2}^{T_E} n_1 C_P(H_2O, \text{liquid})\, dT = n_1 C_P(H_2O, \text{liquid})\, (T_E - T_2)$$
$$= (1 \text{ mol}) \left(75.24 \text{ J.mol}^{-1}.\text{K}^{-1}\right) (T_E - 273 \text{ K}). \tag{8}$$

- *Liquid → Liquid*: The four moles of liquid do not change state. They remain in the liquid state. Their temperature goes from $T_3 = 298$ K to the equilibrium temperature T_E of the mixture 1–4 moles under a pressure of 1 atmosphere. The variation of enthalpy of this mole during this isobaric cooling is:

$$Q_{L \to L}^4 = \Delta H_{L \to L}^1 = \int_{T_3}^{T_E} n_2 C_P(H_2O, \text{liquid})\, dT = n_2 C_P(H_2O, \text{liquid})\, (T_E - T_3)$$
$$= (4 \text{ mol}) \left(75.24 \text{ J.mol}^{-1}.\text{K}^{-1}\right) (T_E - 298 \text{ K}). \tag{9}$$

Since the transformation is adiabatic isobar, we obtain from (5):

$$Q = \sum_1^4 \Delta H = \Delta H_{S \to S}^1 + \Delta H_{S \to L}^1 + \Delta H_{L \to L}^1 + \Delta H_{L \to L}^1$$
$$= 564.3 \text{ J} + 6.056 \times 10^3 \text{ J} + (1 \text{ mol}) \left(75.24 \text{ J.mol}^{-1}.\text{K}^{-1}\right) (T_E - 273 \text{ K})$$
$$+ (4 \text{ mol}) \left(75.24 \text{ J.mol}^{-1}.\text{K}^{-1}\right) (T_E - 298 \text{ K}) = 0 \quad \Rightarrow \quad T_E = 275.4 \text{ K}. \tag{10}$$

Appendix A: The Michelson and Morley Experiment

Introduction

In the nineteenth century, light was considered as a vibration which propagated. This presupposed the existence of a vibrating medium, allowing its propagation just as air vibration which was at the origin of sound propagation. The problem was that light also propagated in the vacuum where, a priori, nothing vibrated; hence the hypothesis made at the time by scientists that the vacuum should in fact be a motionless substance, of which light was the manifestation of its vibrations. This medium received the name of *ether*. By analogy with the elements carried by a fluid, the speed of light had to change according to whether it descended or ascended the current created by the displacement of the Earth. Albert Michelson and Edward Morley attempted, between 1881 and 1887, to determine the speed of the Earth in the ether, the hypothetical medium in which light had to propagate. The principle of their experiment followed a remark of Maxwell that the time of return of light had to be different according to whether it propagated in the direction of the motion of the Earth or in the direction which was perpendicular to it if the law of the Galilean composition of speeds remained valid for light.

Interferometer

The accurate determination of light speed is a very laborious task because of its very high value, and to compare the difference in the speeds of two light rays requires even more precision which no measuring device can provide.

However, by mixing these same rays, we obtain some effects, detectable to the eye following their optical path difference provided that the rays are coherent.

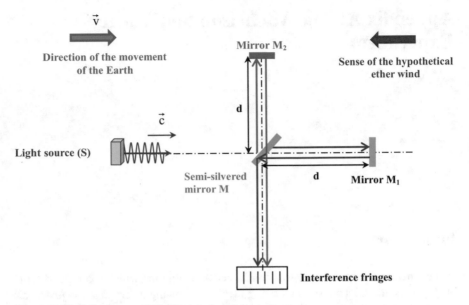

Fig. A.1 The Michelson and Morley interferometer

Michelson and Morley base themselves on the principle of interference fringes resulting from the interaction of the duplicated light rays. To do this, they design an apparatus so that the mirrors M_1 and M_2 are equidistant from a semi-silvered mirror oriented at 45°, of which the role is to separate the light ray, emitted by a source, into two rays, the first towards the mirror M_1 and the other in the perpendicular direction, that is to say, towards the mirror M_2. These two rays therefore travel the same return distance. The interference fringes appear at the recombination of the two rays of light at their arrival in Z (Fig. A.1). If the Earth is immobile with respect to the ether, the two paths in the two perpendicular directions must be equal (same distance traveled, same travel time). On the other hand, if the Earth is moving in relation to the ether, the two paths will not be done at the same speed and the distance traveled will not be the same in both directions.

Experiment

Let be a light ray coming from a very pure monochromatic source (S). It reaches the semi-silvered glass plate M through which it passes partly to the mirror M_1, whereas the reflected part reaches the mirror M_2 placed in a direction perpendicular to MM_1. After reaching the mirrors M_2 and M_1, the rays are reflected again in the semi-silvered mirror M where they must in principle interfere because they cannot have traveled the same path.

- Back and forth in the direction of the movement of the Earth (towards M_1):

During their emitting path, all the mirrors moved with the movement of the Earth: M_1 moved to M_1' and M_2 moved longitudinally but stayed at the same distance from M which also moved to M'.

On the emitted path, the ray MM_1 travels the distance $d + v\tau_1$ where $v\tau_1$ is the displacement of M_1 from the start of the ray from M to its impact on M_1 and τ_1 the time of this displacement:

$$c\tau_1 = d + v\tau_1 \quad \Rightarrow \quad \tau_1 = \frac{d}{c - v}. \tag{A.1}$$

On return, the ray $M_1'M'$ travels the distance $d - v\tau_2$ where $v\tau_2$ is the displacement of M_1' from the return of the ray from M_1' until its impact on M' and τ_2 the time of this displacement:

$$c\tau_2 = d - v\tau_2 \quad \Rightarrow \quad \tau_2 = \frac{d}{c + v}. \tag{A.2}$$

A back and forth in the direction of the movement of the Earth (towards M_1) therefore requires time:

$$t_1 = \frac{d}{c - v} + \frac{d}{c + v} = \frac{2dc}{c^2 - v^2} = \frac{2d}{c}\left(\frac{c^2}{c^2 - v^2}\right) = \frac{2d}{c}\left(\frac{1}{\frac{c^2 - v^2}{c^2}}\right)$$

$$= \frac{2d}{c}\left(1 - \frac{v^2}{c^2}\right)^{-1}. \tag{A.3}$$

Since the speed of the Earth is very low compared to the speed of light, the ratio v^2/c^2 is much less than 1. For ε is very small and using the following approximation:

$$(1 + \varepsilon)^n \approx 1 + n\varepsilon \tag{A.4}$$

we get:

$$t_1 = \frac{2d}{c}\left(1 + \frac{v^2}{c^2}\right). \tag{A.5}$$

- Path in the perpendicular direction:

For the path in perpendicular direction, it must be taken into account that the mirror initially in M_2 moved longitudinally before reflecting the light ray. If we decompose the isosceles triangle MM_2M' into two right triangles, we clearly have: $t_{MM2} = t_{M2M'} = t$ by symmetry, with $(ct)^2 = (vt)^2 + d^2$, i.e., $t^2 = d^2/(c^2 - v^2)$

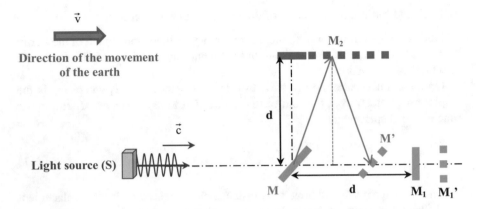

Fig. A.2 A back-and-forth perpendicular to the movement of the Earth (therefore to M_2)

(Fig. A.2). A back-and-forth perpendicular to the movement of the Earth (therefore to M_2) requires time:

$$t_2 = 2t = \frac{2d}{\sqrt{c^2 - v^2}}.$$ (A.6)

By using the same approximation as before, we obtain:

$$t_2 = \frac{2d}{\sqrt{c^2 - v^2}} = \frac{2d}{c}\left(1 - \frac{v^2}{c^2}\right)^{-\frac{1}{2}} \approx \frac{2d}{c}\left(1 + \frac{v^2}{2c^2}\right).$$ (A.7)

The difference of the travel time between the two paths is then:

$$t_1 - t_2 = d\frac{v^2}{c^3}.$$ (A.8)

An estimate of the order of magnitude of this difference can be obtained by considering the speed of light of the order of 3×10^8 m/s; the speed of the Earth with respect to the Sun, therefore with respect to the ether, of the order 3×10^4 m/s; and the dimension of the device of the order of 10 m. Let a measurement of $t_1 - t_2$ be of the order of $(10 \times 10^9)/10^{25}$, that is, 10^{-15} s. This is a far too small quantity to be measured with big precision. Michelson's genius was to predict that this effect could be highlighted in an experiment of optics. Indeed, the optical path (AB) between two points AB is given by:

$$(AB) = n \ AB = \frac{c}{v} \ AB = ct$$ (A.9)

where c is the speed of light in vacuum, v the speed of light in the medium joining the points A and B, n the refractive index of the medium (homogeneous), and t the time of the path of light between A and B at the speed v. In wave optics, the optical path difference δ between two light rays is the difference of the optical paths traveled by these two rays. It allows evaluating the phase shift between these two rays at any point M: $\delta(M) = MS_2 - MS_1$, where MS_2 is the distance traveled by a ray from a source S_2 to a point M and MS_1 corresponding to the distance traveled by a ray from a source S_1 to the same point M. The order of interference p at any point M of the figure of interference is an integer defined as the quotient of the optical path difference $\delta(M)$ by the wavelength λ. It allows designating the number of a fringe in a figure of interference. If this quotient takes an integer value, the interference is constructive with a local maximum of intensity. When the quotient takes a half-full value, the interference is destructive, resulting in a minimum of luminous intensity:

$$p = \frac{\delta(M)}{\lambda}. \qquad (A.10)$$

Given the orbital speed of the Earth in its orbit of 30 km/s and for a wavelength in the visible range $\lambda = 500$ nm and an arm of 10 m, we obtain an order of interference equal to 0.4, which is perfectly observable. However, the Michelson and Morley experiment was never able to highlight this interference as if there was no wind of the ether, by analogy with the wind felt when we move in the motionless air. The result was always negative, although the experiment was redone in several directions, and with several months apart to take advantage of the modification of the speed of the Earth compared to the ether and that the device was turned by a quarter of a turn to interchange the two paths and to throw off any unequal distances between the two mirrors with respect to the semi-silvered mirror M.

Interpretation

On the basis of (A.3) and to explain the fact that the two rays arrive without phase shift and that we always have $2\tau = \tau_{MM1} + \tau_{M1M}$, there are only two possibilities: either $v = 0$, meaning that the Earth is immobile (which is impossible), or $v \neq 0$. In the last case, it is necessary to intervene on the distance d or the time t^* by assuming that the objects shorten from L_1 to L_2 or that the clocks slow down in the direction of the movement from t_1^* to t_2^* such that:

$$L_2 = L_1\sqrt{1 - \frac{v^2}{c^2}} \quad \text{or} \quad t_2^* = \frac{t_1^*}{\sqrt{1 - \frac{v^2}{c^2}}}. \qquad (A.11)$$

However strange it can appear at first sight, it is these last two points of view which are correct; they reflect the fact that the notions of time and space are interrelated, and that we are immersed in a four-dimensional space (three coordinates of space and one temporal coordinate). This four-dimensional aspect of our universe cannot be naturally highlighted in our daily life, simply because we are incapable to move at the speed of light (remember that numerically $c = 3 \times 10^8$ m.s^{-1}). In our daily world, we always have $v \ll c$, and therefore, it is impossible for us to realize that the lengths get shorter and the watches slow down when we travel. We shall also notice that when $v = c$, the length becomes zero and time tends towards infinity translating the impossibility for humans to travel at the speed of light.

Appendix B: Useful Mathematical Reminders in Physics

Powers

In algebra, the power of a number is the result of the repeated multiplication of that number by itself. It is often noted by associating to the number an integer, typed in the exponent, which indicates the number of times which the number appears as a factor in this

$$a^4 = a \times a \times a \times a \quad \text{(four factors).} \tag{B.1}$$

It reads *a raised to the power of* 4 or *a to the power of* 4 or most briefly *the fourth power of a*. *a* is called *base* and 4 *exponent*. In particular, the square and the cube are powers of exponents 2 and 3, respectively:

$$
\begin{aligned}
a^0 &= 1 \\
a^1 &= a \\
a \times a &= a^2 \\
a \times a \times a &= a^3 \\
a^n \times a^m &= a^{(n+m)} \\
(a^n)^m &= a^{(n \times m)} \\
\frac{1}{a^n} &= a^{-n} \quad \text{if} \quad a \neq 0 \\
(a \times b \times c)^n &= a^n \times b^n \times c^n \\
\left(\frac{a}{b}\right)^n &= a^n \times b^{-n} \quad \text{if} \quad b \neq 0.
\end{aligned}
\tag{B.2}
$$

© The Author(s), under exclusive license to Springer Nature Switzerland AG 2021
S. Khene, *Topics and Solved Exercises at the Boundary of Classical and Modern Physics*, Undergraduate Lecture Notes in Physics,
https://doi.org/10.1007/978-3-030-87742-2

- A number raised to an even power gives a positive result. If n is even, then $(-a)^n = +a^n$. Example 1: $(-2)^4 = (-2) \times (-2) \times (-2) \times (-2) = +16$. Example 2: $(+2)^4 = (+2) \times (+2) \times (+2) \times (+2) = +16$.
- A number raised to an odd power gives a result of the same sign. If n is odd, then $(-a)^n = -a^n$. Let us consider the first example: $(-2)^3 = (-2) \times (-2) \times (-2) = -8$: (-8) has the same sign as (-2). In the second example: $(+2)^3 = (+2) \times (+2) \times (+2) = +8$. $(+8)$ has the same sign as $(+2)$.
- A positive number raised to a negative power gives a positive result. Example:

$$3^{-3} = \frac{1}{3^3} = \frac{1}{3 \times 3 \times 3} = +\frac{1}{27}.$$

- Do not confuse $(-a)^n$ and $-a^n$. Indeed:

$$(-2)^4 = (-2) \times (-2) \times (-2) \times (-2) = +16.$$
$$-2^4 = -(2 \times 2 \times 2 \times 2) = -16.$$

- Do not confuse a^2 and $2a$. Indeed:

$$a^2 = a \times a.$$
$$2a = a + a.$$

- Remember that:

$$3^{-1} \text{ is the multiplicative inverse of } 3.$$
$$-3 \text{ is the additive inverse of } 3.$$

- Also remember that:

$$2 + 3^2 = 2 + 9 = 11.$$
$$(2 + 3)^2 = 5^2 = 25.$$
$$a^2 + b^2 \neq (a + b)^2 \text{(unless one of } a \text{ or } b \text{ is zero)}.$$

- Standard form for numbers:
  In general, a number in standard form is written as $A \times 10^n$ where A is between 1 and 10. Below are specific examples:

$$0.000543 = 5.43 \times 10^{-4}.$$
$$0.001652 = 1.652 \times 10^{-3}.$$
$$564 \times 10^{34} = 5.64 \times 10^{36}.$$
$$-672 = -6.72 \times 10^2.$$

- We give below a summary table of powers of ten commonly used in physics exercises (Table B.1):

Table B.1 Table of powers of ten

Power of ten, negative or zero	Prefix	Power of ten, positive or zero	Prefix
$10^0 = 1$		$10^0 = 1$	
$10^{-1} = 0.1$	d (deci)	$10^1 = 10$	da (deca)
$10^{-2} = 0.01$ One hundredth	c (centi)	$10^2 = 10$ Hundred	h (hecto)
$10^{-3} = 0.001$ One thousandth	m (milli)	$10^3 = 1000$ Thousand	k (kilo)
$10^{-4} = 0.0001$		$10^4 = 10\ 000$	
$10^{-5} = 0.00001$		$10^5 = 100\ 000$	
$10^{-6} = 0.000001$ One millionth	μ (micro)	$10^6 = 1\ 000\ 000$ One million	M (mega)
$10^{-9} = 0.000000001$ One billionth	n (nano)	$10^9 = 1\ 000\ 000\ 000$ One billion	G (giga)
$10^{-12} = 0.000000000001$ One thousandth of a billionth	p (pico)	$10^{12} = 1\ 000\ 000\ 000\ 000$ One thousand billion	T (tera)

Roots

The square root of a positive number A is the positive number noted $\sqrt{A}$ of which the square is A, i.e., $\left(\sqrt{A}\right)^2 = A$. The number A is the *radicand* and the symbol "$\sqrt{}$" is called *radix*. $\sqrt{A}$ reads *square root of A* or more briefly *root of A*.

- The roots possess the properties of powers:

$$a^2 = A \quad \Rightarrow \quad a = \pm\sqrt{A} = A^{\frac{1}{2}}$$
$$\sqrt{A} \text{ exists if } A \geq 0$$
$$\left(\sqrt{A}\right)^2 = \sqrt{A^2} = A$$
$$\sqrt[m]{A} = A^{\frac{1}{m}}$$
$$\sqrt[m]{A^P} = A^{\frac{P}{m}} \tag{B.3}$$
$$\sqrt{A} \times \sqrt{B} = \sqrt{A \times B}$$
$$\sqrt{\frac{A}{B}} = \frac{\sqrt{A}}{\sqrt{B}} \quad \text{if} \quad B \neq 0.$$

Exponentials

The exponential function is a continuous and a strictly increasing function in **R**, the set of real numbers. We have:

Fig. B.1 Exponential
function variation as a
function of x

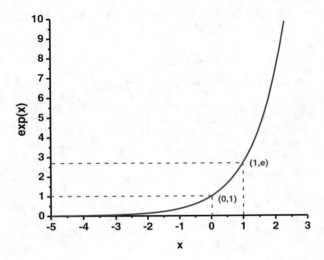

$$\lim_{x \to -\infty} exp(x) = 0 \quad \text{and} \quad \lim_{x \to +\infty} exp(x) = +\infty. \qquad (B.4)$$

For all $k \in]0; +\infty[$, the equation $exp(x) = k$ has a unique solution in **R**. This solution is denoted *ln k*. The exponential function is the function denoted *exp*. It is its own derivative:

$$\frac{d}{dx} exp(x) = exp(x). \qquad (B.5)$$

It takes the value 1 at 0 (Fig. B.1):

$$exp(0) = 1. \qquad (B.6)$$

We denote by e the value of this function at 1. This number is worth approximately 2.71828. It is called *basis of the exponential function*. It verifies: *ln e* = 1, and allows another notation of the exponential function:

$$exp(x) = e^x. \qquad (B.7)$$

The exponential function has the properties mentioned in Table B.2.

- $\forall$ the real numbers x and y, we have:

$$
\begin{aligned}
e^x \times e^y &= e^{x+y} \\
(e^x)^y &= e^{x \times y} \\
e^0 &= 1 \\
\frac{1}{e^x} &= e^{-x} \\
\frac{e^x}{e^y} &= e^{x-y}.
\end{aligned}
\qquad (B.8)
$$

Table B.2 Properties of the exponential function

Symbol	e^x
Reciprocal	$ln\ x$
Derivative	e^x
Primitive	e^x + Cte

Fig. B.2 Variations of the Napierian logarithm function and its reciprocal function (i.e., the exponential function) as a function of x

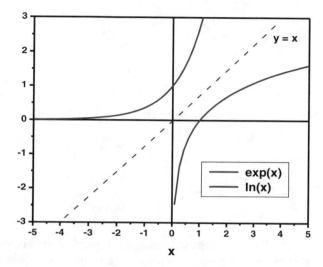

Logarithms

The Napierian logarithm function is a function which, to any strictly positive real, matches the unique y such that $e^y = x$. The natural logarithm function is noted ln. We have:

$$ln\ :\ ]0;\ +\infty[\ \ \rightarrow\ \ \mathbf{R}$$
$$x\ \ \rightarrow\ \ lnx. \tag{B.9}$$

The Napierian logarithm function is the reciprocal function of the exponential function (Fig. B.2).

- $\forall$ the positive real numbers x and y, we have:

$$ln(x \times y) = lnx + lny$$
$$lnx^y = y\,lnx$$
$$ln1 = 0$$
$$lne = 1$$
$$ln\left(\frac{1}{x}\right) = -lnx$$
$$ln\left(\frac{x}{y}\right) = lnx - lny$$
$$y = e^x \quad \Leftrightarrow \quad x = lny$$
$$e^{lnx} = x$$
$$ln\sqrt{x} = \frac{1}{2}lnx$$
$$(lnx)' = \frac{1}{x}$$
$$[lnu(x)]' = \frac{u'}{u}.$$

(B.10)

The Napierian logarithm function is particularly interesting because of its property of transforming a product in a sum. But, as we use, for writing numbers, the decimal system, we prefer in some cases to use another function, having the same property of transformation of a product in a sum but, taking the value 1 when $x = 10$ and thus the value 2 for $x = 100$, the value 3 for $x = 1000$, etc. This function is called *decimal logarithm function*. It is noted *log* and defined on $]0; +\infty[$ by (Fig. B.3):

Fig. B.3 Decimal logarithm function variation as a function of x

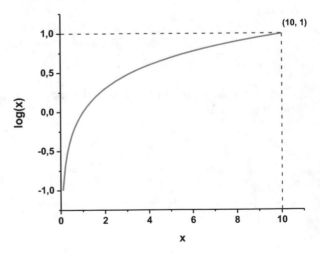

$$\begin{aligned} log \; : \;]0; \; +\infty[&\; \rightarrow \; \mathbf{R} \\ x \quad &\rightarrow \quad logx = \frac{lnx}{ln10}. \end{aligned} \qquad \text{(B.11)}$$

Quadratic Equation

A quadratic equation is a polynomial equation of the form:

$$ax^2 + bx + c = 0 \qquad \text{(B.12)}$$

where x is an unknown and a, b, and c are known numbers called *coefficients of the equation* such that a is not equal to 0. In $\mathbf{R}$, this equation has a maximum of two solutions. Let $\Delta = b^2 - 4ac$ be the discriminant of the equation.

I. If $a = 0$, we obtain an equation of the first degree, which admits one solution:

$$x = -\frac{c}{b}. \qquad \text{(B.13)}$$

II. If $a \neq 0$, we obtain a second-degree equation. Three cases can occur according to the sign of the discriminant.

(i) If $\Delta > 0$, the equation admits two real solutions in $\mathbf{R}$:

$$x_1 = \frac{-b - \sqrt{\Delta}}{2a} \quad \text{and} \quad x_2 = \frac{-b + \sqrt{\Delta}}{2a}. \qquad \text{(B.14)}$$

If the coefficient a is positive, the parabola opens upwards, and, if it is negative, it opens downwards (Fig. B.4).

Fig. B.4 If $\Delta > 0$, the quadratic equation admits two real solutions, x_1 and x_2

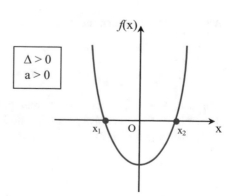

Fig. B.5 If $\Delta = 0$, the quadratic equation has a unique solution, x_0

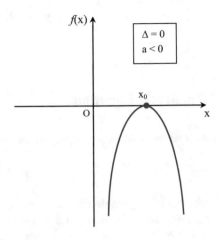

(ii) If $\Delta = 0$, the equation has a unique solution in $\mathbf{R}$ (Fig. B.5):

$$x_0 = \frac{-b}{2a}. \qquad (B.15)$$

(iii) If $\Delta < 0$, the equation has no solution in $\mathbf{R}$.

• Let us note that for $\Delta \geq 0$, we have:

$$S = x_1 + x_2 = -\frac{b}{a} \quad \text{and} \quad P = x_1 \times x_2 = \frac{c}{a}. \qquad (B.16)$$

Systems of Two Equations with Two Unknowns

A system of two equations with two unknowns is of the form:

$$\begin{cases} ax + by = c \\ a'x + b'y = c' \end{cases} \qquad (B.17)$$

where a, b, c, a', b', and c' are real numbers. x and y are the unknowns to be determined. To solve this system is to find all the pairs (x,y), which are both solutions of the two equations. There are several methods to solve this equation system.

Resolution by Combination

Let us consider the following system:

$$\begin{cases} 4x + 3y = 2 \\ 6x + 7y = 13. \end{cases} \qquad \text{(B.18)}$$

- We identify the coefficients in front of one of the two unknowns, for example, those of x, and we determine one of their non-zero common multiples. In our case, 12 is a common multiple of 4 and 6.
- We make the coefficients equal in front of the chosen unknown. In our case, we make the coefficients equal in front of x. So, to have 12, we multiply the first equation by 3 and the second by 2:

$$\begin{cases} 12x + 9y = 6 \\ 12x + 14y = 26. \end{cases} \qquad \text{(B.19)}$$

- We subtract the equalities member to member, and we solve the obtained equation (we can add the two equations as appropriate):

$$(12x - 12x) + (9y - 14y) = 6 - 26$$
$$0 - 5y = -20$$
$$y = \frac{-20}{-5} \qquad \text{(B.20)}$$
$$y = 4.$$

- The value of the other unknown is determined by replacing the found value in one of the two equations. For example, we replace $y = 4$ in the first equation:

$$4x + 3 \times 4 = 2$$
$$4x + 12 = 2$$
$$4x = -10$$
$$x = \frac{-10}{4} \qquad \text{(B.21)}$$
$$x = -2.5.$$

- We verify if the found couple is a solution of the starting system. In our case, we verify if the couple $(-2.5, 4)$ is a solution of the system:

$$4 \times (-2.5) + 3 \times (4) = -10 + 12 = 2$$
$$6 \times (-2.5) + 7 \times (4) = -15 + 28 = 13. \tag{B.22}$$

Resolution by Substitution

Let us consider the following system:

$$\begin{cases} 3x + y = 1 \\ 2x + 3y = -4. \end{cases} \tag{B.23}$$

- We express one of the unknowns as a function of the other. In our case, the simplest option is to express y in terms of x in the first equation:

$$y = 1 - 3x. \tag{B.24}$$

- We replace this unknown by its new expression in the other equation. In our case, we replace y by the expression $(1 - 3x)$ in the second equation:

$$2x + 3(1 - 3x) = -4$$
$$2x + 3 - 9x = -4$$
$$-7x + 3 = -4$$
$$-7x = -4 - 3 = -7$$
$$-7x = -7$$
$$x = \frac{-7}{-7}$$
$$x = 1. \tag{B.25}$$

- Let us determine the value of the other unknown. We know that $y = 1 - 3x$. Just replace x by 1 in this expression:

$$y = 1 - 3 \times (1)$$
$$y = -2. \tag{B.26}$$

- We verify if the found couple $(1, -2)$ is the solution of the system:

$$\begin{cases} 3 \times (1) + (-2) = 3 - 2 = 1 \\ 2 \times (1) + 3 \times (-2) = 2 - 6 = -4. \end{cases} \tag{B.27}$$

Resolution by the Determinant Method

Let us consider the following system:

$$\begin{cases} ax + by = c \\ a'x + b'y = c', \end{cases} \tag{B.28}$$

for which the determinant D is non-zero:

$$D = \begin{vmatrix} a & b \\ a' & b' \end{vmatrix} = ab' - a'b \neq 0. \tag{B.29}$$

The solutions of this system are given by:

$$x = \frac{\begin{vmatrix} c & b \\ c' & b' \end{vmatrix}}{\begin{vmatrix} a & b \\ a' & b' \end{vmatrix}} = \frac{cb' - c'b}{ab' - a'b} \quad \text{and} \quad y = \frac{\begin{vmatrix} a & c \\ a' & c' \end{vmatrix}}{\begin{vmatrix} a & b \\ a' & b' \end{vmatrix}} = \frac{ac' - a'c}{ab' - a'b}. \tag{B.30}$$

As an example, let us consider the following system:

$$\begin{cases} 3x + y = 1 \\ 2x + 3y = -4. \end{cases} \tag{B.31}$$

The solutions of this system are given by:

$$x = \frac{\begin{vmatrix} 1 & 1 \\ -4 & 3 \end{vmatrix}}{\begin{vmatrix} 3 & 1 \\ 2 & 3 \end{vmatrix}} = \frac{1 \times 3 - 1 \times (-4)}{3 \times 3 - 1 \times 2} = \frac{7}{7} = 1 \tag{B.32}$$

and

$$x = \frac{\begin{vmatrix} 3 & 1 \\ 2 & -4 \end{vmatrix}}{\begin{vmatrix} 3 & 1 \\ 2 & 3 \end{vmatrix}} = \frac{3 \times (-4) - 1 \times (2)}{3 \times 3 - 1 \times 2} = \frac{-14}{7} = -2. \tag{B.33}$$

The solution $(1, -2)$ is well the one that we found by substitution method.

Graphical Method

Let us consider the following system:

$$\begin{cases} 2x - y = 1 \\ -x + 2y = 4. \end{cases} \qquad (B.34)$$

To graphically solve this system to two equations, let us rewrite it as follows:

$$y = 2x - 1 \quad \text{and} \quad y = \frac{1}{2}x + 2 \qquad (B.35)$$

Equation (B.35) clearly shows that these two equations are those of two straight lines, which are intersecting because their directing coefficients are not equal $2 \neq 1/2$. Graphically, these two lines intersect at point (2,3), which constitutes the solution of this system (Fig. B.6). This determination, although approximate, allows verifying the existence or not of solutions:

- Intersecting straight lines $\Rightarrow$ 1 solution.
- Parallel straight lines $\Rightarrow$ 0 solution.
- Superimposed straight lines $\Rightarrow$ infinity of solutions.

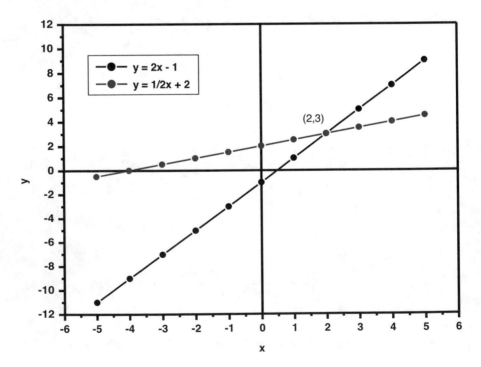

Fig. B.6 The two straight lines intersect at point (2,3)

Trigonometry

Trigonometry is a part of geometry. Geometry deals with the measurement of the earth, while trigonometry concerns the measurement of the three-angle bodies.

Radian

An angle of one radian, denoted *rad*, intercepts on the circle circumference an arc of length equal to the radius. The angle of a complete revolution is 2π (rad), of a half turn π (rad), of a quarter turn $\pi/2$ (rad), etc. The angles are also measured in degrees and in grades. If α is the measure of the angle in radians, β the measure of the same angle in degrees, and γ that of the sasme angle in grades, we have:

$$\frac{\alpha}{2\pi} = \frac{\beta}{360} = \frac{\gamma}{400}. \tag{B.36}$$

Astronomers and astrophysicists measure the angles in minutes or in seconds of arc such as:

$$1° = 60'(\text{minutes of arc}) \qquad 1' = 60''(\text{seconds of arc}). \tag{B.37}$$

Trigonometric Functions of the Circle

By application of the Pythagoras theorem, we obtain:

$$\left|\overrightarrow{OM}\right| = R = \sqrt{x^2 + y^2} \tag{B.38}$$

where R is the radius of the circle. From this representation, we can define the trigonometric functions of the circle (Fig. B.7).

Fig. B.7 Trigonometric
functions of the circle

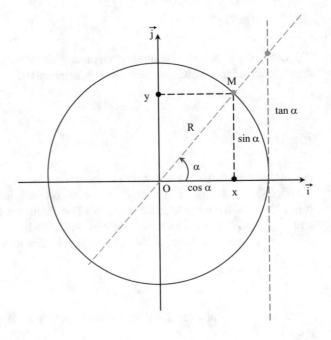

$$cos\,\alpha = \frac{x}{R} \quad \Rightarrow \quad \alpha = arccos\frac{x}{R}$$

$$sin\,\alpha = \frac{y}{R} \quad \Rightarrow \quad \alpha = arcsin\frac{y}{R}$$

$$tan\,\alpha = \frac{y}{x} \quad \Rightarrow \quad \alpha = arctan\frac{y}{x}$$

$$cotan\,\alpha = \frac{x}{y} \quad \Rightarrow \quad \alpha = arccos\frac{x}{y}$$

$$sin^2\alpha + cos^2\alpha = 1$$

$$cos(\alpha \pm 2k\pi) = cos\,\alpha \quad \forall \alpha \in \mathbf{R}, \forall k \in \mathbf{N}$$

$$sin(\alpha \pm 2k\pi) = sin\,\alpha \quad \forall \alpha \in \mathbf{R}, \forall k \in \mathbf{N}$$

$$cos(-\alpha) = cos\,\alpha \qquad\qquad\qquad\qquad\text{(B.39)}$$

$$sin(-\alpha) = -sin\,\alpha$$

$$sin(n\pi) = 0 \quad \forall n \in \mathbf{Z}$$

$$cos(n\pi) = (-1)^n \quad \forall n \in \mathbf{Z}$$

$$sin\left[(2n+1)\frac{\pi}{2}\right] = (-1)^n \quad \forall n \in \mathbf{Z}$$

$$cos\left[(2n+1)\frac{\pi}{2}\right] = 0 \quad \forall n \in \mathbf{Z}$$

$$tan\,\alpha = \frac{sin\,\alpha}{cos\,\alpha}$$

$$cotan\,\alpha = \frac{1}{tan\,\alpha}.$$

Here, **N** is the set of natural numbers and **Z** that of integers.

Algebraic Expressions (Table B.3)

Table B.3 Algebraic expressions

α	0	$\dfrac{\pi}{6}$	$\dfrac{\pi}{4}$	$\dfrac{\pi}{3}$	$\dfrac{\pi}{2}$
$sin\ \alpha$	0	$\dfrac{1}{2}$	$\dfrac{\sqrt{2}}{2}$	$\dfrac{\sqrt{3}}{2}$	1
$cos\ \alpha$	1	$\dfrac{\sqrt{3}}{2}$	$\dfrac{\sqrt{2}}{2}$	$\dfrac{1}{2}$	0
$tan\ \alpha$	0	$\dfrac{\sqrt{3}}{3}$	1	$\sqrt{3}$	∞
$cotan\ \alpha$	∞	$\sqrt{3}$	1	$\dfrac{\sqrt{3}}{3}$	0

Notable Formulas

$$sin(\alpha + \beta) = sin\,\alpha\,cos\,\beta + sin\,\beta\,cos\,\alpha$$
$$sin(\alpha - \beta) = sin\,\alpha\,cos\,\beta - sin\,\beta\,cos\,\alpha$$
$$cos(\alpha + \beta) = cos\,\alpha\,cos\,\beta - sin\,\alpha\,sin\,\beta$$
$$cos(\alpha - \beta) = cos\,\alpha\,cos\,\beta + sin\,\alpha\,sin\,\beta$$
$$tan(\alpha + \beta) = \frac{tan\,\alpha + tan\,\beta}{1 - tan\,\alpha\,tan\,\beta}$$
$$tan(\alpha - \beta) = \frac{tan\,\alpha - tan\,\beta}{1 + tan\,\alpha\,tan\,\beta}$$
$$sin\,2\alpha = 2\,sin\,\alpha\,cos\,\alpha$$
$$cos\,2\alpha = cos^2\alpha - sin^2\alpha$$
$$tan\,2\alpha = \frac{2tan\,\alpha}{1 - tan^2\alpha}.$$

(B.40)

Metric Relationships in a Triangle

Let us consider a triangle ABC. S denotes its area and a, b, and c are the opposite sides of the angles α, β, and γ, respectively. We have the following three basic equations:

- The Al-Kashi formula:

Fig. B.8 Triangle ABC: S denotes its area and a, b, and c are the opposite sides of the angles α, β, and γ, respectively

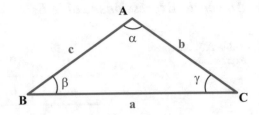

$$a^2 = b^2 + c^2 - 2bc \cos \alpha. \tag{B.41}$$

- Formula of the triangle area

$$S = \frac{1}{2} bc \sin \alpha. \tag{B.42}$$

- Sinus formula (Fig. B.8):

$$\frac{a}{\sin \alpha} = \frac{b}{\sin \beta} = \frac{c}{\sin \gamma}. \tag{B.43}$$

Approximate Calculation

Let ε be an infinitely small quantity. We have:

$$
\begin{aligned}
(1 + \varepsilon)^n &\approx 1 + n\varepsilon \\
\sin \varepsilon &\approx \varepsilon \\
\cos \varepsilon &\approx 1 - \frac{\varepsilon^2}{2} \\
\tan \varepsilon &\approx \varepsilon \\
\ln(1 + \varepsilon) &\approx \varepsilon \\
e^\varepsilon &\approx 1 + \varepsilon.
\end{aligned}
\tag{B.44}
$$

Derivatives

Definition

A function $f(x)$ is differentiable at point a if and only if the limit exists, and is finite:

$$\lim_{h \to 0} \frac{f(a+h) - f(a)}{h} = A. \qquad \text{(B.45)}$$

Its value A is the derivative of $f(x)$ at point a.

Geometric Interpretation of the Derivative

Let (C) be the curve representative of a function $f(x)$ differentiable at a. Let us consider point A $[a, f(a)]$ and point B $[a + h, f(a + h)]$. The gradient of the straight line AB is given by:

$$\frac{f(a+h) - f(a)}{a+h-a} = \frac{f(a+h) - f(a)}{h}. \qquad \text{(B.46)}$$

When point B approaches point A (i.e., when h tends towards 0), the gradient of AB tends to the derivative $f'(a)$ and the straight line becomes tangent to (C) (one point of contact). The derivative $f'(a)$ is the gradient of the tangent to the curve (C) at point A of the abscissa a (Fig. B.9).

Equation of the Tangent

We have for every point $M(x,y)$ of the tangent to (C) in A:

Fig. B.9 Tangent to (C) in A

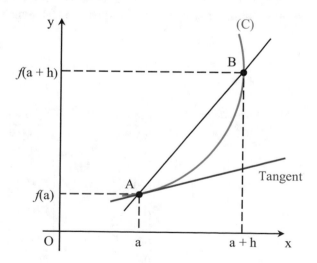

$$f'(a) = \frac{y - f(a)}{x - a} \tag{B.47}$$

which gives the equation:

$$y = f(a) + f'(a)(x - a). \tag{B.48}$$

By extension, we denote $y' = f'(x)$, the derivative of the function $y = f(x)$ at any point M of abscissa x of the interval I. Geometrically, $f'(x_o)$ represents, as we have just seen, the slope of the tangent to the curve (C) at point A of abscissa a. The sign of $f'(x)$ gives the direction of the variation of the function:

- If $f'(x) > 0$, the function is increasing.
- If $f'(x) < 0$, the function is decreasing.
- If $f'(x) = 0$, the function is constant, the graph admits an extremum, which can be a maximum if the second derivative is negative or a minimum if the second derivative is positive.

In physics, the notion of derivative is very important in the study of the variation of the potential energy E_p as a function of the distance x. So, for $dE_p/dx = 0$, we have an equilibrium. This equilibrium is stable (E_p minimum) for $d^2E_p/dx^2 > 0$ and unstable for $d^2E_p/dx^2 < 0$.

Derivatives of Usual Functions

In Table B.4, we give the derivatives of the functions, frequently encountered in the exercises of physics, $\mathbf{R}^*$ is the set of positive real numbers.

Table B.4 Derivatives of usual functions

Function f	Derived function f'	Definition domain
$f(x) = k$ (constant)	$f'(x) = 0$	$\mathbf{R}$
$f(x) = ax + b$	$f'(x) = a$	$\mathbf{R}$
$f(x) = x^n$ $(n \geq 1)$	$f'(x) = n\,x^{n-1}$	$\mathbf{R}$
$f(x) = \frac{1}{x^n}$ $(n \geq 1)$	$f'(x) = \frac{-n}{x^{n+1}}$	$\mathbf{R}^*$
$f(x) = \sqrt{x}$	$f'(x) = \frac{-1}{2\sqrt{x}}$	$]0; +\infty[$
$f(x) = \ln x$	$f'(x) = \frac{1}{x}$	$]0; +\infty[$
$f(x) = e^x$	$f'(x) = e^x$	$\mathbf{R}$

Table B.5 Derivatives of composite functions

Function	Derived function	Example
$u + v$	$u' + v'$	$f(x) = x + \dfrac{1}{x} \;\Rightarrow\; f'(x) = 1 - \dfrac{1}{x^2}$
uv	$u'v + v'u$	$f(x) = x\sqrt{x} \;\Rightarrow\; f'(x) = (1)(\sqrt{x}) + \left(\dfrac{1}{2\sqrt{x}}\right)x = \dfrac{3}{2}\sqrt{x}$
u^n	$n\,u'\,u^{n-1}$	$f(x) = (3x^2 - 2)^5 \;\Rightarrow\; f'(x) = (5)(6x)(3x^2 - 2)^4 = 30x(3x^2 - 2)^4$
$\dfrac{1}{v}$	$-\dfrac{v'}{v^2}$	$f(x) = \dfrac{1}{x^2 - 1} \;\Rightarrow\; f'(x) = -\dfrac{2x}{(x^2 - 1)^2}$
$\dfrac{u}{v}$	$\dfrac{u'v - v'u}{v^2}$	$f(x) = \dfrac{2x + 3}{4x - 4} \;\Rightarrow\; f'(x) = \dfrac{2(4x - 4) - 4(2x + 3)}{(4x - 4)^2} = -\dfrac{20}{(4x - 4)^2}$
$\ln u$	$\dfrac{u'}{u}$	$f(x) = \ln(x^2 - 1) \;\Rightarrow\; f'(x) = \dfrac{2x}{x^2 - 1}$

Derivatives of Composite Functions

In Table B.5, we give the derivatives of composite functions, frequently encountered in the exercises of physics.

Differentials

Differential of a Function of One Variable

$y = f(x)$ being a function of one variable defined, continuous, and differentiable on an interval I, we call differential of the function $y = f(x)$ the function:

$$df = \frac{df}{dx}dx \tag{B.49}$$

where df and dx are the infinitesimal increments.

Differential of a Function of Several Variables

Let be the three-variable function $U = f(x,y,z)$. We call differential of the function U the equation:

$$dU = \frac{\partial U}{\partial x} dx + \frac{\partial U}{\partial y} dy + \frac{\partial U}{\partial z} dz \qquad \text{(B.50)}$$

where $\partial U/\partial x$, $\partial U/\partial y$, and $\partial U/\partial z$ are the partial derivatives of the function U, i.e., the derivative with respect to the variable indicated in the denominator while keeping the two other variables constant.

Total Differential

In the case of a two-variable function $U = f(x,y)$, the differential form:

$$dU = \frac{\partial U}{\partial x} dx + \frac{\partial U}{\partial y} dy \qquad \text{(B.51)}$$

is called *total* if and only if the crossed second derivatives of the function U with respect to x and y are equal:

$$\frac{\partial^2 U}{\partial x \partial y} = \frac{\partial^2 U}{\partial y \partial x}. \qquad \text{(B.52)}$$

The function U is called *state function* (see Chap. 4). Its variation does not depend on the followed path to go from point A to point B.

Integrals

Definition

$F(x)$ and $f(x)$ being two functions defined over an interval I, we say that $F(x)$ is a primitive or an integral function of $f(x)$ if and only if $F(x)$ admits for derivative $f(x)$ on this interval (Fig. B.10):

Fig. B.10 The primitive of $f(x)$ represents the area limited by the curve (C) and the straight lines of equations $x = a$ and $x = b$

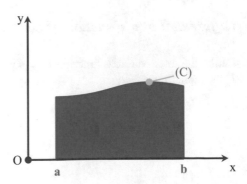

$$\int f(x)dx = F(x) + \text{Cte.} \tag{B.53}$$

If a and b represent the limits of this interval, we write:

$$\int_a^b f(x)dx = F(b) - F(a). \tag{B.54}$$

If $f(x)$ is a continuous function, its primitive represents the area limited by the curve (C), representative of the function $f(x)$, and the straight lines of equations $x = a$ and $x = b$.

Usual Primitives

In Table B.6, we give the primitives of the functions, frequently encountered in the exercises of physics.

Table B.6 Usual primitives

Function f	Primitive F (+ Cte)	Domain of definition
$f(x) = k$ (real)	$F(x) = k\,x$	$\mathbf{R}$
$f(x) = x$	$F(x) = \dfrac{x^2}{2}$	$\mathbf{R}$
$f(x) = x^n \ (n \in \mathbf{N}^*)$	$F(x) = \dfrac{x^{n+1}}{n+1}$	$\mathbf{R}$
$f(x) = \dfrac{1}{x^2} \quad (x \neq 1)$	$F(x) = -\dfrac{1}{x}$	$\mathbf{R}^*$
$f(x) = \dfrac{1}{x^n} \quad (n \in \mathbf{N}, n \geq 2)$	$F(x) = -\dfrac{1}{n+1}\dfrac{1}{x^{n-1}}$	$\mathbf{R}^*$
$f(x) = \dfrac{1}{\sqrt{x}}$	$F(x) = 2\sqrt{x}$	$]0; +\infty[$
$f(x) = \dfrac{1}{x}$	$F(x) = \ln x$	$]0; +\infty[$
$f(x) = e^x$	$F(x) = e^x$	$\mathbf{R}$
$f(x) = e^{ax} \ (a \in \mathbf{R}^*)$	$F(x) = \dfrac{e^{ax}}{a}$	$\mathbf{R}$
$f(x) = \sin x$	$-\cos x$	$\mathbf{R}$
$f(x) = \cos x$	$\sin x$	$\mathbf{R}$
$f(x) = \dfrac{1}{\cos^2 x}$	$F(x) = \tan x$	$]-\pi/2 + k\pi; +\pi/2 + k\pi[$

Table B.7 Primitives of composite functions

Function	Derived function	Domain of definition
$u'\, u^n\ (n \in \mathbf{N}^*)$	$\dfrac{u^{n+1}}{n+1}$	$u(x) > 0$
$u'\, u^a\ (a \in \mathbf{R}^*)$	$\dfrac{u^{a+1}}{a+1}$	$u(x) > 0$
$\dfrac{u'}{u^2}$	$-\dfrac{1}{u}$	$u(x) \neq 0$
$\dfrac{u'}{u^2}\quad (n \in \mathbf{N},\ \ n \geq 2)$	$-\dfrac{1}{n-1}\dfrac{1}{u^{n-1}}$	$u(x) \neq 0$
$\dfrac{u'}{\sqrt{u}}$	$2\,\sqrt{u}$	$u(x) > 0$
$\dfrac{u'}{u}$	$ln\ u$	$u(x) > 0$
$u'\, e^u$	e^u	

Table B.8 Universal constants

Speed of light in vacuum	$c \approx 3 \times 10^8$ m/s
Electrical permittivity of vacuum	$\varepsilon_0 \approx [1/(36\pi)]\ 10^{-9}$ F/m
Vacuum magnetic permeability	$\mu_0 \approx 4\,\mu\ 10^{-7}$ H/m
Constant of the universal gravitation	$G \approx 6.67 \times 10^{-11}$ Nm^2/kg^2
Boltzmann's constant	$k_B \approx 1.38 \times 10^{-23}$ J/K
Number of Avogadro	$N_A \approx 6.02 \times 10^{23}$ particles/mole
Planck's constant	$h \approx 6.63 \times 10^{-34}$ J.s
Constant of ideal gases	$R = 8.32$ J.K^{-1}.mol^{-1}

Primitives of Some Composite Functions

In Table B.7, we give the primitives of the composite functions, frequently encountered in the exercises of physics.

Universal Constants

In Table B.8, we give some universal constants, frequently encountered in the exercises of physics.

Atomic Constants

In Table B.9, we give some atomic constants, frequently encountered in the exercises of physics.

Table B.9 Atomic constants

Elementary charge	$e \approx 1.6 \times 10^{-19}$ C
Electron mass (charge $-e$)	$m_e \approx 9.1 \times 10^{-31}$ kg
Proton mass (charge $+e$)	$m_p \approx 1836\, m_e = 1.67 \times 10^{-27}$ kg
Neutron mass (charge 0)	$M_n \approx 1839\, m_e = 1.68 \times 10^{-27}$ kg
Classical electron radius	$r_e \approx 2.8 \times 10^{-5}$ Å
The Bohr radius of hydrogen	$r_0 \approx 0.53$ Å

Table B.10 Astronomical constants

Earth mass	$M_E \approx 6 \times 10^{24}$ kg
Earth radius	$R_E \approx 6.37 \times 10^6$ m
Acceleration due to gravity	$g_0 \approx 9.81$ m/s^2
Sun-Earth distance	$D_{S-E} \approx 1.5 \times 10^{11}$ m $\approx 10^5\, R_E$
Sun mass	$M_S \approx 1.99 \times 10^{30}$ kg $\approx 10^6\, M_E$
Moon radius	$R_M \approx 1.74 \times 10^6$ m $= R_E/3.7$
Sun-our galaxy distance	$D_{S-G} \approx 27\,000$ light years

Table B.11 Some usual units

1 micron (1 μ) $= 10^{-6}$ m
1 angstrom (1 Å) $= 10^{-8}$ cm $= 10^{-10}$ m
1 light year $\approx 9.46 \times 10^{15}$ m
1 unified atomic mass unit (1 u) $\approx 1.65 \times 10^{-27}$ kg
1 liter $= 1$ dm^3 $= 10^3$ cm^3 $= 10^{-3}$ m^3
1 electronvolt (1 eV) $\approx 1.6 \times 10^{-19}$ J
1 calorie ≈ 4.18 J
1 horsepower (1 hp) ≈ 736 W
1 atm $\approx 1.013 \times 10^5$ Pa ≈ 760 mm Hg $= 760$ Torr
Absolute zero $= 0$ K ≈ -273 °C
1 degree $= (\pi/180)$ radian $= 17.45 \times 10^{-3}$ radian

Astronomical Constants

In Table B.10, we give some atomic constants, frequently encountered in the exercises of physics.

Some Usual Units

In Table B.11, we give some usual units, frequently encountered in the exercises of physics.

Table B.12 Greek alphabet

A	α	Alpha	N	ν	Nu
B	β	Beta	Ξ	ξ	Xi
Γ	γ	Gamma	O	o	Omicron
Δ	δ	Delta	Π	π	Pi
E	ε	Epsilon	P	ρ	Rho
Z	ζ	Zeta	Σ	σ	Sigma
H	η	Eta	T	τ	Tau
Θ	θ	Theta	Υ	υ	Upsilon
I	ι	Iota	Φ	φ	Phi
K	κ	Kappa	X	χ	Chi
Λ	λ	Lambda	Ψ	ψ	Psi
M	μ	Mu	Ω	ω	Omega

Greek Alphabet

In Table B.12, we give the Greek alphabet which is frequently used in the exercises of physics.

Bibliography

1. Achuthan, M. (2009). *Engineering thermodynamics* (2nd ed.). Prentice Hall India Learning Private Limited.
2. Baeyens, D., & Warzee, N. (2015). *Chimie générale*. Dunod.
3. Bailey, A. A. (1998). *La conscience de l'atome*. Dervy.
4. Bardez, E. (2009). *Chimie générale*. Dunod.
5. Bergman, T. L., Lavine, A. S., Incropera, F. P., & DeWitt, D. P. (2011). *Introduction to heat transfer* (6th ed.). John Wiley & Sons.
6. Bey, P., & Gérard, J. P. (2013). *Faut-il avoir peur de la radioactivité ?* Odile Jacob.
7. Blanc, D. (1973). *Physique nucléaire*. Masson.
8. Callen, H. B. (1985). *Thermodynamics and an introduction to thermostatistics* (2nd ed.). Wiley.
9. Cengel, Y. A., & Boles, M. A. (2017). *Thermodynamics: An engineering approach* (8th ed.). McGraw Hill Education.
10. Cohen-Tannoudji, C., Dupont-Roc, J., & Grynberg, G. (1996). *Processus d'interaction entre photons et atomes 2*. EDP Sciences.
11. Condon, E. U., & Odabasi, O. (1980). *Atomic structure*. Cambridge University Press.
12. Debu, P. (2017). *Noyaux et radioactivité: une introduction à la physique des particules et à la physique nucléaire*. Transvalor–Presses des mines.
13. DeHoff, R. (2006). *Thermodynamics in materials science* (2nd ed.). CRC Press.
14. De la Souchère, M. C. (2005). *La radioactivité: mécanismes et applications*. Ellipse.
15. Desit-Ricard, I. (2010). *L'atome en clair*. Ellipse.
16. Friedli, C., & Sahil-Migirdicyan, A. (2010). *Exercices de chimie générale, 400 exercices avec solutions – 140 QCM corrigés*. Presses Polytechniques et Universitaires Romandes.
17. Fröbrich, P., & Reinhard, L. (1996). *Theory of nuclear reactions*. Oxford University Press.
18. Holman, J. P. (2010). *Heat transfer*. Mcgraw Campagnies.
19. Hughes, A. L., & DuBridge, L. A. (1932). *Photoelectric phenomena*. McGraw Hill Book Company.
20. José Ouin, J. (1998). *Transferts thermiques: Rappels de cours et applications*. Casteilla.
21. Khene, S. (2002). *Électricité: Rappels de cours et exercices corrigés*. Publication de l'université Badji Mokhtar d'Annaba (Algérie).
22. Khene, S. (2003). *Mécanique des fluides: Cours et exercices avec solutions*. Publication de l'université Badji Mokhtar d'Annaba (Algérie).
23. Khene, S. (2015). *Mécanique du point matériel: Cours et 201 exercices corrigés*. Éditions Connaissances et Savoirs (France).

24. Khene, S. (2016). *Mécanique du solide: Cours, exercices et problèmes corrigés*. Presses Internationales Polytechnique (Canada).
25. Kondepudi, D., & Prigogine, I. (2014). *Modern thermodynamics: From heat engines to dissipative structures* (2nd ed.). Wiley.
26. Kuhn, S. T. (1987). *Black-body theory and quantum discontinuity*. University of Chicago Press.
27. Lambert, G. (2004). *Une radioactivité de tous les diables: Bienfaits et menaces d'un phénomène naturel . . . dénaturé*. EDP Sciences.
28. Lumbroso, H. (1973). *Thermodynamique: 100 exercices et problèmes résolus*. McGraw Hill.
29. Lovérini, M. J. (1996). *L'atome, de la recherche à l'industrie – Le commissariat à l'énergie Atomique*. Gallimard.
30. Moran, M. J., Shapiro, H. N., Boettner, D. D., & Bailey, M. B. (2014). *Fundamentals of engineering thermodynamics* (8th ed.). Wiley.
31. Nag, P. K. (2013). *Engineering thermodynamics* (5th ed.). McGraw Hill Education.
32. Nellis, G., & Klein, S. (2012). *Heat transfer*. Cambridge University Press.
33. Nikjoo, H., Uehara, S., & Emfietzoglou, D. (2012). *Interaction of radiation with matter*. CRC Press.
34. Pitts, P., & Sissom, L. E. (1998). *Heat transfer* (2nd ed.). McGraw Hill.
35. Queyrel, J. L., & Mesplède, J. (1996). *Précis de physique, thermodynamique, cours et exercices résolus*. Edition Réal.
36. Reuss, P. (2016). *Fission nucléaire, réaction en chaine et criticité*. EDP Sciences.
37. Sacadura, J. P. (2015). *Transferts thermiques: Initiation et approfondissement*. Tec et Doc – Lavoisier.
38. Sandler, S. I. (2006). *Chemical, biochemical, and engineering thermodynamics* (4th ed.). Wiley.
39. Schieck, H. P. (2014). *Nuclear reactions: An introduction*. Springer.
40. Servin, A. (2016). *Chimie générale – Tout le cours en fiches*. Dunod.
41. Stuewer, R. H. (1975). *The Compton effect: Turning point in physics* (1st ed.). Science History Publications.
42. Thompson, I. J., & Nunes, F. M. (2009). *Nuclear reactions for astrophysics: Principles, calculation and applications of low-energy reactions*. Cambridge University Press.
43. Van Ness, H. C. (2012). *Understanding thermodynamics*. Dover Publication.
44. Willett, E. (2005). *The basics of quantum physics: Understanding the photoelectric effect and line spectra*. The Rosen Publishing Group.

Index

Printed in the United States
by Baker & Taylor Publisher Services